Early American Gardens

"For Meate or Medicine"

EARLY AMERICAN GARDENS

"For Meate or Medicine"

By

Ann Leighton

WITH 84 ILLUSTRATIONS OF THE PERIOD

The University of Massachusetts Press

Amherst, 1986

Library of Congress Cataloging-in-Publication Data

Leighton, Ann, 1902?–1985.
Early American gardens.

Reprint. Originally published: Boston : Houghton Mifflin, 1970.
Bibliography: p.
Includes index.
1. Gardening—New England—History.
2. Gardens, American—New England—History.
3. Herb gardening—New England—History.
4. Medicinal plants—New England—History.
I. Title.
SB451.34.N38L45 1986   712'.6'0974   86–6980
ISBN 0–87023–530–3 (pbk.)

Portions of this book have previously appeared as articles in various publications:
American Heritage, October 1966, "Take a Handful of Buglosse . . ." Copyright ©
1966 by American Heritage Publishing Co., Inc.: *Antiques* Magazine, August
1966, "Look Down at the Plants." Copyright 1966 by Straight Enterprises, Inc.:
History Today, June 1968, "Meate and Medicine in Early New England": *Horticulture* Magazine, August 1968, "Whipple House—Ipswich: A Seventeenth
Century Garden": *The New-England Galaxy,* Summer 1964 and Winter 1969,
respectively, "Of Puritans and Pinks" and "The Prospect and the Rarities." Copyright © 1964, 1969 Old Sturbridge Inc.

TO MY GRANDCHILDREN

Julia Herrick Macaulay Smith
Ann Leighton Macaulay Smith
Sarah Stuart Macaulay Smith
Patrick Eliot Macaulay Cain
and those to come

Acknowledgments

Many names deserve grateful mention for help with this book, as I should like to include all those who pointed out even the simplest of facts or flowers, outdoors or in. Especially have I valued threads of reference from the past, still holding strong in phrases and jokes and even deep aversions, which furnished a connection by human voice and practice with material taken from the documented page.

My greatest thanks are due to my husband, A. William Smith, a gifted gardener and a "curious" one in the seventeenth-century sense. From him I learned all I know of practical gardening and the pleasures of growing any plant one has heard of, just to see what it looks like. It was his splendid 1633 copy of Gerard's *Herball*, as edited by Thomas Johnson, and his gifts of the reprint of Parkinson's 1629 *Paradisi in Sole Paradisus Terrestris* and Geoffrey Grigson's *Englishman's Flora* that opened up for me the possibility of repeating the plants of the past, just as they grew.

My family has been all-enduring and all-obliging. My son, Thomas W. M. Smith, arranged for me to work in the Bibliothèque Nationale. My son, James M. Smith, was equally helpful on New England ground with New England references and constant encouragement. He found the quotation from *Tom Jones* which explains my labors to grow all the plants as well as record them. His wife, Katharine Archer Smith, has helped me with the final proofs. My daughter, Emily Cain, took me to the countryside of "Lady Fenwick" and helped me to site her garden. My daughter-in-law, Jane McDill Smith, explored with me for old roses still growing around Paris, and my granddaughter Julia ran happily about the Jardin des Plantes while I searched there for New World plants. My son-in-law, Thomas Cain, an expert horticulturist, has helped with plants escaped into New England

and Canada. And all my English in-laws and friends have been untiring in seeing me into and about old gardens, the Chelsea Physic Garden and the British Museum. I owe also a special debt to friends here who said, on hearing I was working on a book about Puritan gardens in New England, "But I did not know those people *had* any gardens. . . ." Nothing is more stimulating than a defense of one's ancestors.

The Ipswich Historical Society allowed me to grow each plant in the garden of the 1640 Whipple House, designed by Arthur A. Shurcliff who instructed me, finally, to plant it "symmetrically with authentic material." It was this challenging directive which, due to the lack of any existing lists and plans, led me into many happy years of research and practical gardening. I am grateful to the curators of the Whipple House, Elizabeth and James Newton, for never-flagging interest, and Elizabeth Newton especially for several good clues and for helping decipher Gothic handwriting.

I thank Miss Ruth Bradstreet for introducing me to the works of Dr. Andrew Boorde and allowing me to own his two little books from her aunt's house in Newbury. I thank Miss May Underwood of Fayette, Maine, for so easily and amusingly remembering her mother's home medicines and giving me a living link with the past. I am grateful to Mrs. Joseph Ross of Ipswich for her constant contributions of both plants and references and for expeditions to unlikely spots, old railway lines and rocky town hilltops and Rowley meadows, to collect live materials. To Katherine Thompson, incomparable searcher of records, I owe Captain Hammond and the story of the young man and the juniper berries. To Lovell Thompson, patient and understanding first editor of this book, I owe many years of quiet confidence which I can only hope to justify, now that he has left me with Houghton Mifflin Company and started his own ultimately equally distinguished publishing firm of Gambit, Incorporated.

The rare book sellers to whom I owe most for providing me with materials unobtainable in libraries are my friends Everett D. Jewett of Ipswich and George T. Goodspeed and his partner, Michael J. Walsh, of Boston.

The Bibliography will show how grateful I am, and should be, to all the kind librarians in London and Paris, Ipswich, Boston, Philadelphia, Cambridge, New Haven and New York, but most of all to those of the Massachusetts Horticultural Society, the Boston Athenaeum, the Massachusetts Historical Society, the Houghton Library of Harvard and the Countway Library of the Harvard Medical School.

I am grateful, too, to the various publications which have entertained portions of this book and enabled me to see in advance how it would look in print and how various its appeal might be.

I thank my photographers of illustrations in old books, especially George M. Cushing who has been patient and skillful.

And, especially, I must thank Mr. Palmer Perley of the Bay Road Gardens for furnishing plants year after year to keep the Ipswich Historical Society garden beds "in fashion." For, as Fielding says, "the true practical system can be learnt only in the world. . . . Neither physic nor law are to be practically known from books. Nay, the farmer, the planter, the gardener, must perfect by experience what he hath acquired the rudiments of by reading. How accurately soever the ingenious Mr. Miller may have described the plant, he himself would advise his disciple to see it in the garden."

And, finally, may I thank my typists, chiefly Constance Nelson who, having typed my husband's *Gardener's Book of Plant Names*, was happy to help me with my own book about plants.

Foreword

The intent of this book is to make the gardens of the early settlers of New England, founders of our country, grow again. To find out what they grew and why is to rediscover the people and their times and consequently the origins of much of the American character today. Among their flowers and within the confines of their gardens we may come to know not only the people but in some small ways ourselves.

The first chapter sketches the background of gardening, that "primitive" first employment of the human race and earliest concern of the settlers after landing. The second chapter shows what they found already growing in "this wildernesse" with its "signals of fertility." The third chapter briefs the findings of the two chief chroniclers of New England as concerned with growing things. The fourth tries to show the stature of the early gardeners. The fifth deals with the sum total of background knowledge applied by the settlers to their task of growing and distilling and preserving all they would need for both "meate and medicine." The sixth deals with the "meate" and the seventh with the "medicine" for which they felt sure so many plants were intended. The eighth chapter deals with the three leading authorities to whom they had recourse, whose works they owned and from whom they learned all they knew about gardening and medicine. The ninth deals with the forms and shapes of their gardens, howbeit too honestly to be very rewarding to those who begin to read here. And the tenth deals with the lists of plants, even with the names of single plants embedded in seemingly unexpected surroundings which have been collected from books and letters and accounts and recipes written by the people of and in these times. To these I have added a few further names, the "escaped" wild-flowers never specifically mentioned but obviously brought here with a purpose. As, for instance, Bouncing Bet. The Appendix gives each plant its place alphabetically, with its botanical names as decreed by L. H. Bailey and Asa Gray, and the names of its sponsors or growers in seventeenth-century New England, together with a brief description of each plant and its uses drawn from the three contemporary seventeenth-century authorities, Johnson-upon-Gerard, Parkinson and Culpeper, with a few recommendations by that recognized specialist in salads, John Evelyn.

And now to our cabbages . . . .

Contents

Illustrations

PART II

The illustrations of flowers in the Appendix, which are arranged alphabetically, are all taken from the 1633 edition of John Gerard's *Herball.*

Part I

1

"That Primitive Imployment of Dressing a Garden"

The gardens of any period in history are its most intimate spirit, as immediate as its breath, and as transient. Yet, unlike all else about a particular time, they are capable of being recaptured and recreated today, in essence and in fact. Styles of speech have disappeared; modes of pronunciation must be guessed at; tones and voices can be only surmised; sounds and timing of laughter are gone forever. But in living materials, identical with those which delighted and sustained people in the past, we can replant their gardens. Their plants bear direct testimony to the tastes and needs, the whims and joys, even to the most secret hopes and fears, of the people themselves. In recreating gardens of other times we come as close as is possible to those who worked and walked in them.

Bringing back the physical surroundings of the early gardeners and garden owners, the fragrances and colors and arrangements of the plants upon which these people so depended, clearly reveals what they personally thought, and knew, and felt. Each plant has its own witness to bear. As William Cole said of gardening in his seventeenth-century *Art of Simpling*, "The pleasure that is received from it no man knows but he that is acquainted with it. What a pleasant thing it is for a man (whom the ignorant think to be alone) to have plants speaking Greek and Latin to him, and putting him in mind of stories which otherwise he would never think of." What pleasure for us today, to learn from their plants about those who founded our nation.

Men and their gardens have forever been interdependent and inseparable, each other's true revelation. To know one we must know the other. Our country's origins are most deeply entrenched in seventeenth-century New England. Later centuries and other sections of the country have produced their special contributions to American gardens and the American character, but to search out our roots we cannot do better than cultivate New England's wild soil. In gardening, as in all else, it is wise to select the most rewarding area and to concentrate within it, to define boundaries and work inside them, and not to distract ourselves by gazing too widely at the harvests of others. Flowers are easier to study in the hand than in distant fields.

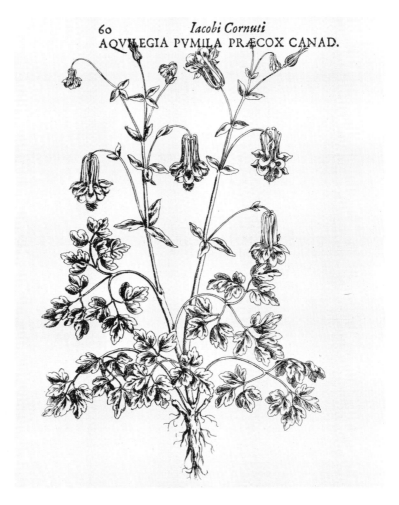

Columbine was described by Gerard as looking like doves drinking. This resembles the European variety more closely than our less compact wild species. Jacques Philippe Cornut, 1635.

To get our particular segment of time and its gardens into focus, we may well begin by "taking a prospect" of gardening generally, to view from our vantage point in time "that primitive imployment of dressing a garden" as Captain Fenwick, founder of the new Connecticut colony of Saybrook, described his and his wife's especial enjoyment.

The history of gardens is a well-worn path. Unhappily for those who would work with living plants, histories of gardens and gardening are usually presented like histories of art, with influences traced, inspirations credited, achievements marked, and the subject left as inanimate as paint and stone. Which is a pity because, with the exception of the Garden of Eden and the Hanging Gardens of Babylon, many of the great gardens of all time are still growing or within our powers to restore. Where walls have crumbled we can at least put back the plants, as fresh and gay as ever. Where some of the plants might be expected to have changed with time and the effects of natural forces, it is still possible to find the old progenitors growing exactly as they are depicted in ancient stone and frescoes and manuscripts. Few plants have totally disappeared. Even trees once glimpsed and collected, though never again seen growing wild — like the Franklinia discovered in Georgia by William Bartram — are growing today in our gardens. We may even observe ancient and primitive gardens, if we go where they are still growing, caught in time.

Conventional accounts of the development of gardens begin with a description of the garden flanking a prehistoric hut. We can see such a garden today, part of an uninterrupted sequence from earliest times, in, for instance, South India. The earth is stamped smooth at the entrance of the palm-thatched mud hut. On either side of the door, a bright pattern is picked out in little stones and colored sand. To one side of the hut is the garden, fenced with palm fronds. In its center there is a tall stick topped with an overturned pot spotted white to ward off evil spirits — possibly the origin of all garden ornaments. And who cannot see the future parterre in the design of sand and stones?

To understand garden design we must be willing to try to understand the gardeners. We cannot afford to ignore even those who loved the seemingly incongruous and irrelevant adoptions and adaptations of foreign and bizarre philosophies of gardening. "Curious" travelers have always been taken with odd conceits and fancies and have wished to transport them in toto and intact to their homes and gardens. Some of these "new" ideas seem to have appealed chiefly for their shock value, like the surprise tree-fountains for dousing unsuspecting guests which so amused European nobles, and post-tops made of animal skulls noted by an early traveler in upstate New York, or shells stuck like an upside-down seashore in the roof and walls of an inland garden grotto in Rhode Island. Designs for a painted backdrop of a ruin, to supply a vista for a French town garden, occupied famous eighteenth-century architects and landscape gardeners. Whatever we think of them all today, they offer us valuable clues to the relationship of men to their gardens.

But no excesses of eccentricity can distract us from detecting a strong consistency in garden design, a principle present in all the gardens men have ever made. To a tireless observer of others equally taken by "that primitive imployment," it seems that, all over the world, every man's garden is what that man feels most lacking in his immediate environment. Reduced to its absurd but nonetheless true extreme: a nearby flower is a weed — a distant weed is a flower. A garden, to be a garden, must represent a different world, however small, from the real world, a source of comfort in turmoil, of excitement in dullness, security in wildness, companionship in loneliness. Gardening offers a chance for man to regulate at least one aspect of his life, to control his environment and show himself as he wishes to be.

Looking back to prehistory and the usual next step in the sequence of garden development, what royal Egyptian gardeners, with unlimited gold and manpower, took the greatest pains to achieve in their gardens was just what they did not naturally possess: water in unfailing quantities, bright flowers in constant supply, and dark tall accents. Desert dwellers have always craved relief from sand and sun, as they sang of the "shadow of a great rock" in the Psalms. Not for them are expanses of artistically raked pebbles.

The Hanging Gardens of Babylon, which achieved fame second only to the Garden of Eden, must have offered sumptuous relief from stony desert living, but their massive receding pile of tree-planted terraces was too much of a once-in-a-lifetime idea to succeed as a popular garden plan.

On the other hand, the so-called Persian garden, another carefully considered contrast to desert living, but capable of modifications, did catch on, almost everywhere and forever. An enclosed oblong, geometrically divided by strips of water and stabbed at regular intervals with shafts of green-black cypresses, decorated with constantly changed tubs and pots of the brightest and most non-desert flowers, the Persian garden plan pleased everyone. But it belongs in a landscape where there is nothing for miles to detract from its exquisite design and its startling assurance that, "if there be a Paradise, it is this, it is this, it is this." The Persian garden loses its beauty and its point draped over European hills. A prayer rug is made for flat, bare ground. And those who protest that deserts blossomed as the rose before they were rendered into deserts by men and goats are referring to a state before recorded time and recorded gardens.

Our theory of what men want in gardens holds true when we look farther East. Chinese gardens are, again and also, the exact opposite to anything provided by the Chinese countryside: peace and seclusion in the midst of turbulence and overcrowding; a small and complete world in surroundings on such a vast and rugged scale that Chinese artists present man as a least little figure content merely to contemplate immensity. Within high walls are quiet water, arching bridges, trees trailing flowered limbs at the pool's edge, teahouses appearing to overhang valleys with such beguiling views and vistas that all normal scale is forgotten. To look over the walls of a Chinese garden at the teeming streets or tortuously cultivated countryside is enough to make one realize the extent of this supreme achievement in garden design.

Alchemy, and its philosophy of the achievement of the universal panacea in one small scientifically concocted "stone," had roots in China. The microcosm-macrocosm definition of man, reflecting in detail the vast elemental powers of the universe, is not unlike the concept of the Chinese private garden. The enormous landscaped approaches to the Ming Tombs, with their gigantic staircases and alleys of different animals in stone, larger than life and neatly paired, are no nearer to true Chinese private gardens than Japanese temple gardens are to the private Japanese gardens derived from the Chinese idea of a small world self-contained.

Curiously, for a thousand years no travelers seem to have fancied the design of the Chinese and Japanese garden for export. When the Chinese idea did finally arrive, it was through a French monk's startling display, to his symmetrically oriented European world, of the Chinese flair for asymmetrical naturalness. Ironically, this exciting revelation resulted first in a giddy vogue for ill-assorted and out-of-place ornaments, like the Pagoda in Kew Gardens. Later this initial release from rigid formality fostered the rise of the "English landscape school." But for a long time the problems and pleasures of Eastern gardens remained strange to European visitors. Marco

Polo had been, rather provincially, most impressed by similarities to Venice in the Chinese canal cities, and his compliments to this effect are still quoted in China. Later, British explorers and traders contented themselves with transporting cuttings, roots and seeds, leaving the garden designs behind. They found their "discoveries" cultivated in pots for hundreds of years or growing wild on the roughest of countrysides. Only today do we have direct Oriental influence in garden design as represented by the raked gravel and protruding boulders of the Japanese temple gardens, introduced, sadly enough, as needing "minimum upkeep."

Gardens conceived for relatively barren landscapes, however, were early taken to the European heart. The Persian garden carpet was joined to the Greek court within the walls of the house and became the plan which so pleased the Romans; open-sided rooms surrounding a Persian garden under the sky. The Romans developed this idea and painted the inner wall of the court with all the extra trees and birds and shrubs in pots for which there was no room in the courtyard. Just as the Greeks had felt the ideal was to be able to move about and converse — Aristotle's had been called the "Peripatetic school" with good reason — the younger Pliny's description of his Roman gardens leaves no doubt that his prime consideration also was to be able to walk and talk. Pliny's box cut into amusing shapes and his live roses growing on dining-room walls were intended to divert his guests. One wonders about the effect on high thinking and serious discussions of dining in the open air around a marble pool, floating marble plates back and forth across the surface.

Pliny's letter about his country villa and its garden ranks, with Bacon's essay on his ideal for all gardens, as a supreme delight for reading gardeners. Pliny's description is, practically, the more rewarding, as it is founded upon fact and not upon a horticulturally impossible dream. Pliny endears himself to us also by being the first among recorded gardeners to welcome the view of wild country beyond his garden's confines and to make the most of it with vistas showing his meadow from many angles. When he happily describes rooms as "gloomy," he gives us an early inkling of the Italian Renaissance obsession for dripping grottoes. In a country where light is brilliant, dark itself has charm. Similarly — still pursuing the theory of opposites and contrasts — the Italian passion for stone staircases and terraces, sculptured cypresses, clipped box borders, and jars planted with small trees, lent architectural order to a landscape of sudden rocky hills and patchwork areas of cultivation. The Italian fancy for geometrical designs in miles of clipped box was the best of the garden conceits and architectural innovations which so captivated François Premier and sadly changed French taste.

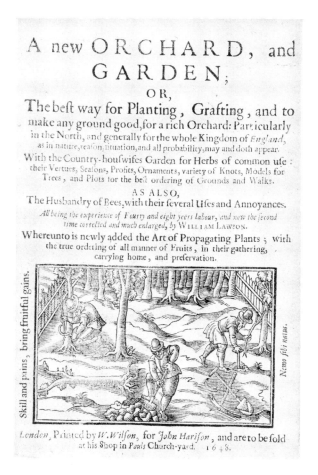

Title page from William Lawson's early seventeenth-century book of instructions in good husbandry and housewifery, with an interesting indication of who took care of the herb gardens, and an illustration of the planting and grafting of fruit trees, an art which our early settlers practiced with great hopes.

As for Bacon, whose essay *On Gardens* is a literary masterpiece and almost the only work of his read today, people have been trying hopelessly to tread upon variously scented plants of all sizes and seasons ever since.

While the monks of the Dark and Middle Ages were determinedly saving even the most modest plants, as well as all else, from barbarian hordes, they evolved a formal layout of a useful and much-used garden with room also for walking and talking. In the meantime, the barbarians, who, after all, also had to eat and "physick," were busily developing the same knowledge — especially in the sciences — which the monks were so assiduously preserving. The barbarians, however, could afford to grow their herbs outside their walls, and retain their formally beautiful and purely decorative gardens intact. And intact they still are in Spain today. By the time the rather self-consciously civilized European community emerged from hiding, and the Moors were able to turn over to it a considerable contribution in chemistry, medicine, philosophy and horticulture, the knowledge was gratefully received. By mid-sixteenth century in England, Dr. Andrew Boorde's *Breviary of Health* was giving each recorded disease first its Greek, then its "barbarous," and then its English name. And when European scholars began to go down into Spain to find out what they had missed, there were the gardens of true desert dwellers with simple contrasts of light and dark, water and stone, pierced walls and pointed cypresses.

While we have the monasteries to thank for the perfection of bright and aromatic gardens of constant usefulness where one could walk and talk and work, every fortress and manor house had exactly the same problem and coped with it in much the same way. Design was made subject to convenience. No one has ever wanted a garden which he must keep up, to be, or even appear to be, difficult. Religious or secular, those who have needed gardens to live with, or by, have always adjusted plans to practical considerations, proportions to efficiency. And the plants have made them beautiful.

While we owe to careful monks the conception of a useful garden being also attractive, it is to the extreme restlessness and extravagance of the French kings, their mistresses and nobles, that we owe the ultimate perfection of the French formal garden. Nothing was ever too much for any of them — not even two identical gardens, an extravagant subterfuge suggested for one French king by his landscape gardener to keep his fruits safe from his nobles. To be able to call upon the best horticultural and artistic

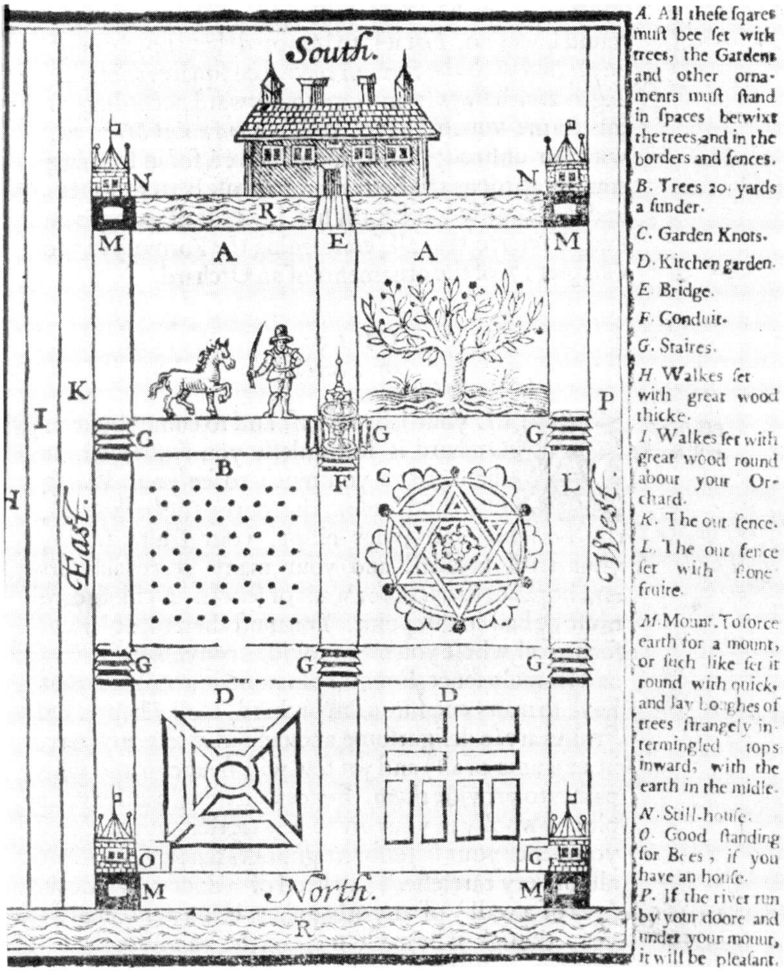

South.
North.
East.
West.
A
B
C
D
E
F
G
H
I
K
L
M
N
O
P
R
A. All theſe ſqares muſt bee ſet with trees, the Gardens and other ornaments muſt ſtand in ſpaces betwixt the trees, and in the borders and fences.
B. Trees 20. yards aſunder.
C. Garden Knots.
D. Kitchen garden.
E. Bridge.
F. Conduit.
G. Staires.
H. Walkes ſet with great wood thicke.
I. Walkes ſet with great wood round about your Orchard.
K. The out fence.
L. The out fence ſet with ſtone fruite.
M. Mount. To force earth for a mount, or ſuch like ſet it round with quick, and lay boughes of trees ſtrangely intermingled tops inward, with the earth in the midle.
N. Still-houſe.
O. Good ſtanding for Bees, if you have an houſe.
P. If the river run by your doore and under your mount, it will be pleaſant.

In William Lawson's *New Orchard and Garden*, the 1648 edition, "falling gardens" are shown below two large squares devoted to fruit trees, a "conduit" in the center, two more large squares, one for trees and one for an elaborate design of garden knots, and, after another drop, two squares for kitchen gardens laid out in completely utilitarian designs which we would do well to emulate today. Two still-houses occupy the corners nearest the house. One of the matching houses at the foot of the garden is devoted to beekeeping and the other to stairs leading down to the canal. Ideally, says Lawson, a canal flows also between the house and garden. Governor Spotswood must have been happy to have achieved at least a canal, if not a fishpond.

This is a well-known plan for such an establishment as Governor Spotswood may have had in mind for Williamsburg.

talents of the nation, to support the top geniuses of all time in formal garden design, to have each strive to exceed his predecessor for a century of the
most extravagant expenditure upon the art of garden-making which the
world has ever seen, could not but result in something as memorable and
monumental as the pyramids. As, indeed, it did. For, whatever one may
think of the pyramids as an art form, or even as a solution to an age-old
human problem, they are the ultimate in what they are. Similarly, after the
Versailles of Le Nôtre, there was nowhere left to go in the design of a magnificent symmetrical formal garden.

Even when massive attempts were made, elsewhere and later, merely to
imitate Versailles, they resulted in looking only exactly that. Somehow, the
same again never amounts to as much as the same in the first place. Reduc-
tions, arrangements after, and *souvenirs de* never turn out well. To choose
any detail of a grand design to be a feature all alone is to end up only with
something quite literally out of place. A *boulingrin*, a closely mowed,
sunken oblong with sloping sides designed by the English for the game of
bowls and adapted by the French, without credit to its origin or purpose, to
contain a pool or a fountain, is at best an oddity by itself. A *pâte d'oie*, or
the goose-foot design of the forest which ended the great French gardens
and afforded a place where duchesses in gigs could follow the hunt by dash-
ing across the splays — and what a nuisance they must have been — is only
effective in forest trees and green rides. The French garden is a success
chiefly in its resolution of problem after problem, each inevitably a part of
the great overall solution.

The pity of it is that, with all the information available, the parterres of
Versailles upon which Le Nôtre intended the nobles to look from the up-
stairs windows of the palaces, and which he said only nursemaids would
ever observe closely, are today solidly planted with annuals he never knew
existed, shimmering like the palettes of the *surrealist* school. The bright and
clashing blooms of nineteenth-century Brazil and Mexico, all crowded together, flourish for their season and are replaced by equally closely packed
chrysanthemums from Japan, also unknown to Le Nôtre.

At Vaux-le-Vicomte, which was Le Nôtre's first triumph and the cause of
his being snatched away to design Versailles, the parterres are left in their
embroidery of clipped grays and greens against a background of colored
stones and gravel and sand. But it is admittedly less stimulating to bus-
loads of tourists than the flashing brilliance of the parterres of Versailles.

At La Malmaison, it would be easy, and of the greatest possible interest and delight to gardeners, to bring in all the roses of the Empress Josephine as painted by Redouté. Here again, however, the modern planting is the hideously varicolored confusion of late nineteenth-century public parks, which is to say English mid-Victorian and of no pertinent interest.

In contrast to most of the European world, the English people as a whole, and as a whole they are gardeners, have always been more interested in plants than in design. Mrs. Tuggy, whoever she was, is immortalized in "her" carnations on an equal basis of respect and fame with other more elegantly named sixteenth-century gardeners. The English have also been more willing than others to bring wild plants into their gardens. With "everything to do with," as cooks used to say of good situations, the English were free to develop in any way they pleased in garden design, but their first interest has always been in plants and shrubs and trees. English kings and queens made royal progress from one great house to another; but they felt no urge to outdo their predecessors. Their living quarters were confined to traditional sites where the gardens, with only a few modifications, had been handed on from monarch to monarch, like the kingdom. Henry VIII fancied one style, and Elizabeth another, and Queen Anne had all the box removed because she disliked the scent — or perhaps only the Dutch who introduced it — but it was not within their conception of their rights to create ever new extravagances in garden design, as was being done across the Channel. Still, as always, their gardens had to be places to walk in, although the English climate excused them from the obligation of entertaining thousands out of doors.

From the time of the Druids, influences had been brought to bear upon English horticulture and had been absorbed. The Celts wrote an herbal which started a succession of books telling the English what each of their plants was good for. Magic properties were ascribed to the most common English flowers, and their medicinal virtues were carefully recorded. Aristotle's theory of the four humors was incorporated in the planning and planting of every Englishman's garden. Astrological overtones were heeded in harvesting the plants. Primitive beliefs in "signatures" were respected. And foreign plants were welcomed. Romans introduced the cabbage and it became the most British of vegetables. Norsemen and Danes brought their indispensable plants and they became English wildflowers. Normans brought carnations and lilies and pears and apples for cider. Monks brought their simples. The Crusaders brought back roses heavily scented, some striped, and all with more petals than the briery English and Scottish sorts. Russian steppes and Persian foothills yielded roots and bulbs to English gardens through the marts of Constantinople and Vienna. And everything grew in England better than it had grown anywhere else. Or was grown, for English climate and English soil cannot account for all *that* success with everyone else's plants.

But there are pitfalls in all gardens. Being able to grow anything well can lead to as regrettable excesses as a propensity for innovation in garden ornaments. After quiet periods in plant importation, when only the most

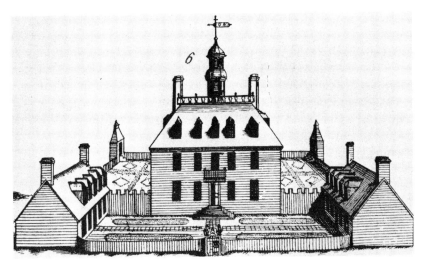

The picture of the Governor's Palace and Gardens in Williamsburg is taken from the January 1937 copy of *Landscape Architecture* which was devoted to an article by Arthur A. Shurcliff on "The Gardens of the Governor's Palace, Williamsburg, Virginia." The sketch, which was made about 1740, is from a copperplate found in the Bodleian Library at Oxford University. The building was started in 1706 and illustrates the time lag in garden design, as it is a conventional seventeenth-century layout on what was actually a rather simple scale, in spite of the complaints that the Governor was "lavishing away the country's money." Governor Spotswood, whose fancy was obviously taken by the idea of a "Fish-Pond and Falling Gardens," was willing to pay for them himself. That he achieved at least a canal, we know from an account in 1724, describing "fine Gardens, Offices, Walks, a fine Canal, Orchards, etc."

earnest explorers and settlers of the New World could be relied upon for anything new, the introduction of Chinese and Indian plants began to stir the English garden world. And the American. It was actually an early nineteenth-century gentleman in South Carolina, Mr. Champney, who first succeeded in crossing an ever-blooming Chinese rose with a once-blooming European rose. (Called "Champney's Pink Cluster," this was taken over by the Noisette brothers, nurserymen in Paris, to start the long line of Noisette Roses.) When the brilliant and exotic contributions to gardens from South Africa and Brazil and Mexico flooded in to aid Victorian gardeners in their passion for space-filling at all costs, cultivation under glass had become a recourse for high and low alike, and "bedding-out" could be indulged in to an endless and hideous degree. Even in the only country where Pliny's idea of seeing his meadow as part of his garden design had finally caught on, and all England had been announced to be a garden, the beautiful rolling vistas of naturally redesigned countryside perfected by those giants of landscaping, Kent and Brown and Repton, were cut up with plantings which, as at Versailles today, excelled only in their intensity. It was not until the late Miss Gertrude Jekyll and William Robinson protested that painting and gardening were not unallied that subtle and beautiful effects were achieved or desired.

A unique characteristic of the English in relation to gardening is worth noting, if we are to replant the gardens of our early settlers. Due to their being the most loyally home-loving of people, no matter how far or how long the English live away from England, they, alone, except perhaps for the Romans, appear to have taken their gardens with them wherever they went. Surely, nothing could look stranger or more homesick than the English suburban gardens in Karachi, the Victorian London gardens in Bombay, the English park planting on the great and refreshingly gloomy compounds of South Indian planters, and the sunken herbaceous garden in the Hope Gardens in Jamaica. English governors' wives, ardent gardeners to a lady, carrying pretty plants from one post to another, are responsible for many of the horticultural invasions for which today they are not so fondly remembered. And yet all these gardeners, the noble and ignoble, the creators and the destroyers, are part of the history of gardening, which, unlike most of them, is happily, still alive. We have only, as it were, to choose our field, however small, to recapture the past within it.

Reconstructing a Norse barge from the fragments of wood in a peat bog,

or a Bronze Age galley from bolts and bars on the bottom of the sea, or a
Greek trading ship from sunken amphorae still containing grape seeds, are
adventures in history closed to all but a skillful and near-amphibious few.
How much easier to reconstruct a garden.

Information on the gardens of our country's earliest settlers, from whom
so much of our national character still derives, may seem scant and difficult
to glean from the angular Gothic script of plant lists, seed bills and books
worn with use. But we have another more engaging recourse for filling the
gardens of our ancestors again with their own flowers. Every day in early
summer, in New England country roads and meadows, gaily spread before
the eyes of passersby, are quantities of living clues to the daily life of this
country's founders. Called "adventives" or "escapes," or described as
"introduced" by botanists and native wildflower collectors, these flowers
were brought to a new world by those who knew they would have need of
them. Some of them stretch back in time to antedate bronze or stone arti-
facts. They traveled on Norse barges and on Greek and Roman ships. They
lived in medieval monasteries and were traded from caravans in Turkey.
They grew and were grown throughout the Old World, and when the
whole great continent of the New was opened up, they came to these shores
on little high-pooped ships like the *Arbella* and the *Lion* and the *Mayflower*
and the *John and Mary*, and were grown in the gardens of the newcomers
"for meate and medicine."

The story of our country's early gardens comes alive as gardening and as
history when we flash past our roadsides and meadows in the New England
summers. There is yarrow, originally recommended by Chiron the centaur
and collected by Samuel Sewall; Bouncing Bet, brought by the fullers of
cloth as their especial aid; teasel, unexcelled even today for fluffing wool;
St. Johnswort, oldest of all charms against the witches and said to be splendid against sciatica; and the little scarlet pimpernel or poor-man's-hourglass
considered so good for whooping cough. They and many more are all
there, some dignified even today by modern medical use, though the use
may not be that for which they were imported. How sad to pass such history by and furbish up historic houses with plants their owners never knew.
How sad, in passing, to miss the opportunity of understanding our country's founders through their gardens.

2

"Forerunning Signals of Fertilitie"

Gardens were literally of the first importance to the early settlers of New England. Too sensible to expect to be fed by either the natives or the never-failing lushness of tropical lands, they knew from early explorers' tales that they would have to grow their own food as soon as their ships' stores ran out. Information about the new land had been avidly awaited and read in the old. Firsthand accounts seem to have been fairly accurate. Invariably they declared a determination to be frank and to stick to "the naked truth." Even in the light of today's knowledge, there was very little dishonest elaboration or willful deceit. "Travelers' tales" are few in accounts of early New England, and there is an astonishing lack of the supernatural or monstrous, compared with the reports of other lands. New England maps lack rich elaborations of land or sea monsters. Even the Indians are kept down to size, although one might suppose that a little exaggeration of the difficulties to be encountered with some of them would have been justified. Reading the accounts of the early explorers and of those anxious to encourage "planters" or brethren left behind to come over, one is struck by the earnest self-discipline of the writers. So the religious leaders of the Puritan sect, anxious to set their "city upon a hill" far beyond the persecutions of Bishop Laud and his followers, were not deceived about the difficulties of sustaining themselves in "this wildernesse." While historians have pointed out that, what with all the wars on the continent of Europe, there was very little otherwhere to flee, the migration to New England was still a forethoughtful move, planned with attention to all the information possible to obtain.

And it is natural that "garden" is a word at the tip of every seventeenth-century tongue or pen seeking to convey high hopes or good reports of the

New World. It is the first word that occurs to those prudent enough to arrive at a favorable season on the coasts of what was beginning to become New England. The poor Pilgrims, who had arrived ten years before the great Puritan migration began, betrayed by both circumstances and their conductors, could have seen little to remind them of a garden in the frosty and desolate shores they encountered. But the letters and accounts of early explorers and the more provident and faithfully served Puritan settlers are redolent and bright with carefully documented names of the fruits and plants they found upon landing or later grew. "Sallets" especially are mentioned, that regular diet of green stuffs which, usually cooked and with a

Champlain's own sketch of his settlement at St. Croix, showing the disposition of the houses and gardens. From the first edition of *Les Voyages*, 1619. The original sketch has been lost.

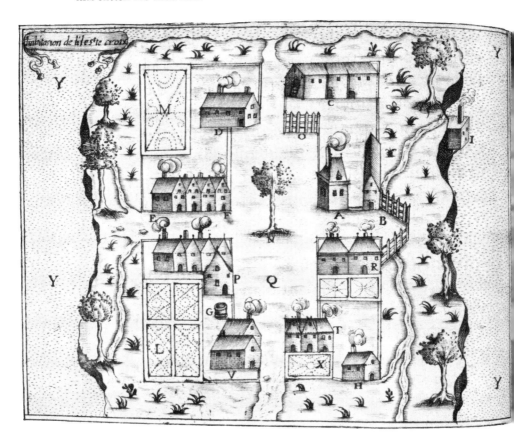

dressing of oil and vinegar, had been brought to England by the ancient ⌐
Romans and thereafter figured prominently in the fare of even the simplest
Englishmen.

The first garden planted in seventeenth-century New England was actually in what was then New France. This garden was set up by Champlain
and his men at the beginning of the century on an island in the St. Croix
River, which island now, happily for us, belongs to the State of Maine.
Since Maine itself is a French word, we cannot feel too guilty to snatch an
early French garden for New England. The whole of this part of the New
World was such a palimpsest of national claims and names and counter-
claims and renamings that credits are due wherever clues survive. Even the
Vikings are finally being awarded recognition. But no suggestions of Vi-
king primacy could have had such interest to or impact upon the early settlers as the appearance of a welcoming Indian dressed in Breton fisherman's
smallclothes. Incongruous to us, this apparition must, to the contemporary
Englishmen, have sharply pointed up the warnings of Samuel Purchas in
Purchas his Pilgrimes that, if the English did not hurry, the rest of the world
would be claimed by other nations before the English got there.

But to Champlain's garden. Incredibly, and fortunately for this book,
we have not only an account but a picture in which the garden beds are
plotted out in a design worthy of any parterre before a French or English
country house. Though a diligent European engraver may possibly have
added the arabesques, we can at least be sure of Champlain's siting and
the proportions of the gardens to the houses. We are told of an additional
garden on the mainland, but it is not sketched and we may presume that it
was for field crops, as was the case with later layouts in more southern New
England. Champlain's sketches of Indian villages and gardens where
Gloucester and Plymouth now stand have an air of such careful exactness
that we are inclined to believe the elaborations of his garden plots on the
island settlement may indeed be his and not the engraver's. And we feel almost his own pleasure and sympathy to read that, a year after the settlement
was abandoned, a party went back to see how the buildings had fared (very
well, they said, the Indians had touched nothing) and dined off a salad
picked from the overgrown tangle that had been the garden. They cooked
and enjoyed lettuce, sorrel and cabbage.

A few years later Captain John Smith prudently planted a garden upon a
rocky island near the coast of Maine, which grew while he was sailing down

the coast claiming the land for England. Proudly he wrote: "I made a garden upon the top of a Rockie Ile in 43.1/2, 4 leagues from the Main in
May, that grew so well, as it served us for sallets in June and July." ("Main"
here presumably refers to the ocean main and not to the province named by
a French sailor for his native province in France.) During his trip down the
coast Smith recounts naming Strawberry Bank where Portsmouth now
stands, noticing Indian gardens on the hilltops in what is now Ipswich,
calling a great cape Tragabigzanda for his Turkish patroness, and naming
three little islands off the shore for three savage Turks' heads he had severed
from their bodies under greater difficulties than any he encountered in New
England. Rallying his readers to think well of this part of the world, he
says the soil is "so fertil, that questionless it is capable of producing any
Grain, Fruits, or seeds you will sow or plant. . . . And surely by reason of
those sandy cliffes and cliffes of rocks, both of which we saw so planted
with Gardens and Corne fields . . . who can but approve this a most excellent place, both for health and fertilitie?" Approaching the future Cape
Tragabigzanda, regrettably renamed Cape Anna by King Charles (to honor
his mother), Captain John Smith passed "Agoam." Later this place would
be named Ipswich for the busy market town of the English Winthrops, perhaps in order to hearten the few men who helped John Winthrop, Junior,
hold the post against the Indians and the French.

Captain John Smith found the future Ipswich's river mouth barred with
sand and the harbor "inbayed too far from the deep sea," but he noted
"many rising hills, and on their tops and descents many corne fields and
delightful groves." The captain is exact in most things, but occasionally his
wishful thinking and lack of botanical knowledge play him false. He sees
on Plum Island "many faire high groves of mulberrie trees," for which at
that season he may have mistaken the beach plum; and, in a list of advantages to be reaped, between salt production and muskrat furs, he places
"certain red berries called Alkermes." Cochineal insects were what these
"berries called Alkermes" are supposed to replace, though for what purpose, medicinal or dyeing, the good captain does not specify. One suspects
it is for dyeing and wonders if the berries may have been cranberries. The
Norsemen are said to have so approved of these berries in their abundance
as to call the land Vineland, a name which does not actually have anything
to do with grapes.

Trees and berries always caught the eyes of the early visitors. In Rosier's

True Relation of Waymouth's Voyage and Exploration, published in 1615 but written about the year 1605, Monhegan Island is described as "woody, growen with Firre, Birch and Oke and Beach, as far as we saw along the shore . . . on the verge grow gooseberries, strawberries, wild pease and wild rose bushes."

And they also gardened: "Wednesday the 22 of May, we . . . sowed pease and barley, which in sixteen days grew eight inches above ground. . . ." And further, "All along the shore and some space within where the wood hindreth not, grow plentifully raspberries, gooseberries, strawberries, roses, currants, wild vines . . . angelica. . . . Within the ilands grows wood of sundry sorts, some very great and all tall. Birch, Beech, Ash, maple, spruce, cherry-tree, yew, oke very great and good. Firre tree out of which issueth turpentine in so marvelous plenty and so sweet as our chirurgeon and others affirm they never saw so good in England." While sailors regarded the forest chiefly as a source of shipbuilding materials, the ship's surgeon "and others" were obviously impressed by the quality of the turpentine, then highly approved of as the treatment for open wounds, poured on hot.

Rosier also relates "Fruites, Plants and Herbs. . . . Tobacco, excellent sweet and strong . . . wild-vines, strawberries, gooseberries, hurtleberries, currant trees, rosebushes, pears, ground-nuts, angelica, a most soveraigne herbe. An herbe that spreadeth the ground and smelleth like sweet Marjoram, gret plenty." They also find the Indians had "Very good Dies, which appear by their painting, which they carry with them in bladders." And, still in Maine, "by both the syd of the river the grapes grow in abundance and also very good Hoppes and also Chebolls and Garleck."

The gay and rather suspect Thomas Morton of Merry Mount, in his *New English Canaan*, published in 1637, says: "The country there naturally affordeth very good pot herbs and sallet herbes and those of a more masculine vertue than any of the same species in England, as pot marjoram, time, alexander, angellica, purslane, violets and anniseeds in very great abundance. For the pot I gathered in summer, dried and crumbled into a bagg to preserve for winter store." He mentions also: "Hunnisuckle balm and divers other good herbes are there that grow without the industry of man that are used when occasion serveth very commodiously." And he says there are: "divers arematicall herbes and plants, as sassafras, muske roses, violets, balme, laurell, hunnisuckles and the like that with their vapours perfume the air." His mention of the "more masculine vertue" antedates

John Josselyn's later remarks on the herbs of New England as being: "generally of a more masculine virtue than any of the same species in England
but not in so terrible degree as to be mischievous or ineffectual to our English bodies."

We know how fortunate we are to have these earnestly factual accounts
when we consider the Reverend Mr. Morrell's "Poem in New England":

> In brief survey, here water, earth and ayre
> A people proud, and what their orders are.
> The fragrant flowers, and the vernaunt groves,
> The merry shores, and storm astranting coves,
> In brief, a briefe of what may make men blest.

ending with the last lines:

> If heaven graunt these, to see here built I trust
> An English Kingdome from this Indian dust.

Morrell, an Episcopal clergyman, arrived with Sir Ferdinando Gorges who,
having what his English compatriots today would call "got the wind up,"
came to lay claim to his grants in the New World as soon as he realized
others were succeeding in colonizing them. Morrell was left behind by
Gorges' party to spend a year in Plymouth to look over the prospects,
though he never divulged his mission until he was about to leave.

In Morrell's poem he mentions also groundnuts and lists berries, filberts,
cherries, grapes, and—unfortunately for our credulity—palms. One can
only believe that he leaned heavily for inspiration upon some other poem
about some other place. On the other hand, as his poem originally appeared
in Latin, we may be doing him a disservice to criticize the English version.

A far more exact soul and honest clergyman, the Reverend Francis Higginson, arrived in Salem in 1629 and was elected "reader" for the Reverend
Mr. Skelton who was the minister chosen for Endecott's little colony. Higginson's secondary position is probably what gave him time to become our
most lyric describer of the garden possibilities in the new land.

Higginson's voyage, begun in April 1629, was undertaken at the behest
of the Company of New England "ayming at the glory of God, the propagation of the Gospell of Christ, the conversion of the Indians, and the en-
largement of the King's . . . dominions in America. . . ." After a "swift"
voyage of two months, during which he saw many great sights vouchsafing

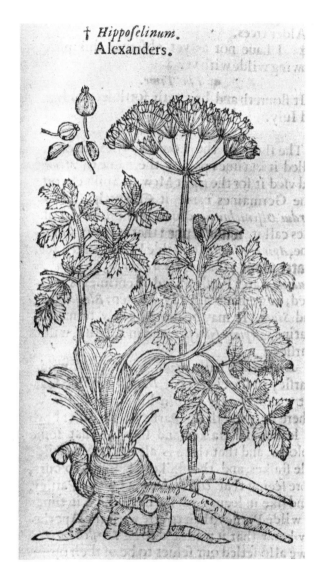

† *Hippoſelinum.*
Alexanders.

the wonders of God and buried one of his children at sea after she died of smallpox, he sighted land, "a cleare and comfortable sight of America." "Now we saw abundance of makrill, a great store of whales puffing up water as they goe," and then, "The sea was abundantly stored with rockweed and yellow flowers like gilly-flowers. By noon we were within three leagues of Capan, and as we sayled along the coasts we saw every hill and dale and every island full of gay woods and high trees. The nearer we came

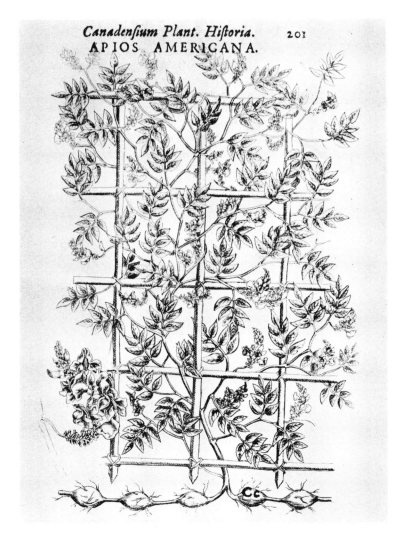

Canadenſium Plant. Hiſtoria.   201
APIOS AMERICANA.

Quite possibly the "groundnut" on which the Pilgrims sub-sisted their first years in the New World. By Jacques Philippe Cornut, 1635.

to the shore the more flowers in abundance, sometimes scattered abroad, sometimes joyned in sheets nine or ten yards long, which we supposed to be brought from the low meadows by the tide. Now what with fine woods and greene trees by land and these yellow flowers painting the sea, made us all desirous to see our new paradise of New England whence we saw such forerunning signals of fertilitie afarre off."

Higginson's short dissertation, published in England in 1630, is entitled *New England's Plantation, or a Short and True Description of the Commodities and Discommodities of that Countrye, Written by a reverend Divine now there resident.* He begins by "Letting passe our voyage by sea . . ." — which, under the circumstances, would seem wise, although he found it "swift" in two months. He then proceeds to establish the universe upon a strictly Aristotelian system which he does not appear to feel conflicts at all with his religious views. As this philosophy, transmitted through Galen, is the basis for much of the gardening to be done, we may let the Reverend Francis Higginson describe it for us, in the shortest possible terms.

"Because," he says, "the life and welfare of everie Creature here below, and the commodiousness of the Countrey where the Creatures live, doth by the most wise ordering of God's providence, depend next unto himselfe, upon the temperature and disposition of the foure Elements — Earth, Water, Aire and Fire (For as of the mixture of all these, all sublunarie things are composed; so by the more or lesse comfortable measure in all Countreys under the Heavens) Therefore . . ." he proposes in his turn to "report nothing but the naked truth. . . ."

Good Galenist that he is, he takes up the naked truth about the new country element by element. Earth is first. "The earth," he says, "is fat black Earth or clay," from which bricks are even then being made "and Tyles and Earthen-Pots. . . . There is also sand or gravel and the land is situated neither too high nor too low. . . ." With earth he includes what "aboundeth naturally" as well as what the settlers have already planted. "Our Turnips, Parsnips and Carrots are here both bigger and sweeter than is ordinarily to be found in England. Here are also store of Pompions, Cowcumbers, and other things of that nature which I know not of. Also divers excellent pot-herbs grow abundantly among the Grasse, as Strawberrie leaves in all places of the Countrey, and plenty of Strawberries in their time, and Pennyroyall, Winter Savorie, Sorrell, Brooklime, liverwort, Carwell and Watercresse, also Leakes and onions are ordinarie, and divers Physicall

Herbes. Here also are abundance of other sweet Herbes delightful to the smell, whose names we know not, &c. and Plentie of single Damaske Roses verie sweete; and two kinds of Herbes that bear two kinds of Flowers very sweet, which they say, are as good to make Cordage or Cloath as any Hempe or Flax we have." He lists fruits, "Excellent Vines — the Governour has already planted a Vinyard with great hopes of encrease — and Mulberries, Plums, Raspberries, Currance, Chestnuts, Filberts, Walnuts, Smalnuts, Hurtleberries and Hawes of Whitethorne neere as good as our Cherries in England. . . ." He describes the trees, especially those yielding "Turpentine, Pitch, Tarre, Masts and other materials for building both of Ships and Houses . . . Sumacke Trees . . . for dying and tanning of leather . . . a precious gum called White Benjamin that they say is excellent for perfumes . . . divers Roots and Berries wherewith the Indians dye excellent holyday colours that no rain or washing can alter. . . ." And, still with the first element, he notes that "English corn" has already been tried at "new Plimouth Plantation" and concludes that "all our Graines will grow here very well." "Our Governour," he says "hath store of greene Pease growing in his Garden as good as ever I eat in England," and with a description of the animals, all of whose skins he has seen except the reputed lion's, he finishes with the first element and comes to the second element, or water.

Water, of course, includes fish, which Higginson found plentiful "beyond imagining." He includes here also the "excellent temper" of the country for making salt; and mentions, too, the fresh water in "daintie Springs" and the many easily-dug wells.

Of air, the third element, he says, "the Aire of New England . . . is one specially thing that commends this place . . . there is hardly a more healthful place to be found in the World that agreeth better with our English Bodyes." In fact, "Many that have been weake and sickly in old England, by coming hither have been thoroughly healed and growne healthfull and strong." Among these Higginson hopefully sees himself who can now, contrary to all the wrapping up he had to do in Old England, go capless in the daytime. Where he used to be unable to eat almost anything, he can now digest coarse foods, and even "oftentimes drink New England water verie well." This points up a fact it is well to remember here, that everyone drank beer because of the dangers in drinking water, and it explains the joyful surprise of the early settlers at the springs of pure water by their doorstones. But beer remained the standard by which to judge water, as witness William

Wood's highest praise of New England water — that, while it cannot be said to be better than the best beer, it is certainly better than the worst and rates easily above whey. Higginson goes so far as to say, in fact, "a sup of New England's Aire is better than a whole draft of old England's Ale." And, before finishing his description of the third element by taking up the fowls of the air, he surmises that this "cleere and dry Aire . . . is of a most

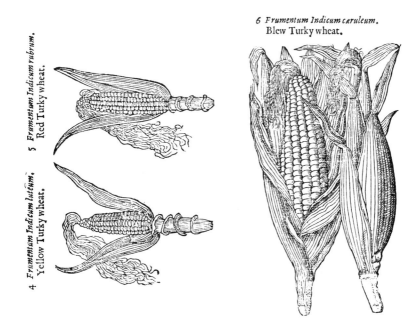

6 *Frumentum Indicum cæruleum.*
Blew Turky wheat.

5 *Frumentum Indicum rubrum.* Red Turky wheat.

4 *Frumentum Indicum luteum.* Yellow Turky wheat.

healing nature to all such as are of a Cold, Melancholy, Flegmatick, Reumaticke temper of Body." As who should hope to know better than poor Mr. Higginson, who was certainly of this temper himself and died after only a year in this country, in spite of all the fine air and water.

But his account has still to deal with the fourth element, or fire, of which he must say something "proportionable to the rest," and so he does indeed, extolling the quantity and quality of the firewood, "so that a poor servant here that is to possess but fifty acres of land . . . can give more wood, for fire . . . than many Noble Men in England can afford to doe. Here is good living for those that love good Fires." And so for light, there are so many fish all New England can afford oil for lamps and the pine trees furnish "such candles as the Indians commonly use . . . the wood of the Pine Tree cloven in two little slices something thin, which are so full of the moysture of Turpentine and Pitch that they burne as clear as a Torch."

So much for the elements, but the naked truth demands he mention "musketoes," snow and rattlesnakes. Indians, however, he places apart in a brief essay on the "Inhabitants." This *True Description*, published in London in 1630, appears to have had great influence, but it is his other little book, called a *True Relation* of his voyage, which rings the truest to us today and contains his lyric outburst upon "our new Paradise of New England." Like Samuel Sewall's beautiful apostrophe to Plum Island, it is as immediate and graphic today as it was then. Higginson's lyric words are an introduction to all lovers of this country. Samuel Sewall's description belongs later in time and this chapter.

The year after Francis Higginson had landed and found things good, John Winthrop sailed into the harbor of Salem, also in early summer. After a voyage which would have filled any man of less heroic mold with serious misgivings, he sighted land for the first time near the island Champlain had called Mont Desert but which Winthrop loyally identified as Mount Monsell, since it had been renamed for one of the members of the Council for New England. He was much refreshed by the "pleasant sweet air" and "a smell off the shore like the smell of a garden." Sailing on down the coast past three low hills on the mainland, he identified them as the Three Turks Heads, which shows that Captain John Smith had succeeded in his efforts to leave his escutcheon immortalized somewhere. Rounding what Winthrop knew as Cape Anne, he was greeted in the harbor by that other great gardener in the New World, "Mr. Endicutt," who took the newcomer ashore to sup on "a good venyson pastye and good beere," while "most of our people went on shore upon the land of Cape Anne . . . and gathered store of fine strawberries." It is charming to think of the first meeting of these two great, kind and industrious men as marked by the hospitality of the new land itself. They were to meet again many times and exchange what they grew or hoped to grow. Endecott who spelled his own name with the second *e* which it retained for a century, although it was variously spelled throughout it, had a large grant of land called "Birches" for many years until it was suddenly named "Orchards." Winthrop took over an island in Boston Harbor, where he planted such a garden, orchard and vineyard that the place was soon known as Governor's Garden. Ships entering the harbor paying a courtesy call on the Governor were there given a first taste of New England apples. Some may say these were fruits of previous plantings by the few "Old Planters" found already settled, but Winthrop would have had a garden sooner and better than most wherever he found himself. He was obviously one of those who gardened for love as well as necessity.

Early proof of his passion for growing things is apparent in a love letter he wrote to his third wife when he was courting her in England. She later shared his new life in the New World. This letter has been quoted as derived from the Song of Solomon. Actually it is the love letter of a really dedicated gardener.

Written from "Groton where I wishe thee, April 4, 1618," it begins, "Grace, mercie and peace etc." And then, "My Onely beloved spouse, my most sweet friend, and faithful companion of my pilgrimage. . . . Being filled with the joy of thy love . . . to ease the burthen of my mind my scribblings penne," and then he really begins to write:

And now, my sweet love, lett me a whyle solace my selfe in the remembrance of our love, of which this springe tyme of our acquaintance can putt forthe as yet no more but the leaves and blossoms whilst the fruit lyes wrapped up on the tender budds of hope, a little more patiense will disclose this good fruit, and bring it to some maturytye; let it be our care and labour to preserve these hopefull buddes from the beasts of the fields, and from frosts; and other injuryes of the ayre, least our fruit fall of ere it be ripe, or lose ought in the beautye and pleasantness of it; lett us pluck up such nettles and thornes as would defraud our plants of their due nourishment; let us not sticke at some labour in watering and manuring them, the plentye and goodness of our fruit shall recompense us abundantly; our trees are planted in fruitful soyle.

At least this will prepare her for such missives as these, nearly twenty years later, when her husband is detained by business for the General Court and writes to "Mrs. Winthrop in Boston," dated sometime in November 1637. He first announces his own good health, wishes that peace be on his family, says he kisses her and hopes shortly to see her, and suggests that she quickly send away Scarlett and gather the turnips. The second letter is pure affection, in which he sends her a token of his love and kisses her "a second tyme farewell." The third follows swiftly as he will not see her as soon as he expected and he needs some fresh linen — "a fresh band or 2 and cuffes." He asks her to send also "6 or 7 leaves of Tobacco dried and powdred." He begs her to take care of herself in this cold weather and asks her to "speak to the folks to keep the goats well out of the garden." In closing he says, if the "sheepe ramme" is still with them and has not been called for, "let them looke him up and give him meate, the green pease in the Garden etc. are good for him." And more blessings.

Governor William Bradford of the Plymouth Plantation may not have been much of a poet, even in an age when it was considered both elegant and educated to take to versifying, and when many Puritan divines like Wigglesworth perceived, or fancied for their own ends to perceive, that what was said in verse often reached its target quicker than what was said in prose. However, Bradford's *Descriptive and Historical Account of New England in Verse*, found in manuscript in his letter book, contains our first naming of any of that "variety of flowers" later mentioned by Edward Johnson and others. Like most of the early chroniclers, he concentrates upon the vegetables and grains, but we are truly indebted to him for mention of "the fair white lily and sweet fragrant rose" in an ornamental role. The poem begins: "Famine once we had — " and the end of that line is missing. In my copy a neat hand with a quill pen has written in "most sore." Elsewhere a bolder pen has written in deferential parentheses "(wanting corn and bread)." Bradford continues:

> But other things God gave us in full store,
> As fish and ground nuts, to supply our strait
> That we might learn on providence to wait . . .

and having waited:

> And all did flourish like the pleasant green
> Which in the joyful spring is to be seen.

The date of this writing can be roughly estimated by the next line:

> Almost ten years we lived here alone

when he describes the settlement of Salem, after which:

> Multitudes began to flow,
> More than well knew themselves where to bestow

and Boston takes the lead.

> And truly it was admirable here to know
> How greatly all things here began to grow,
> New plantations were in each place begun
> And with inhabitants were filled soon.
> All sorts of grain which in our land doth yield
> Was higher brought, and sown in every field;
> As wheat and rye, barley, oats, beans and pease.

And then he gets to what grows in the gardens:

> All sorts of roots and herbes in gardens grow,
> Parsnips, carrots, turnips, or what you'll sow,
> Onions, melons, cucumbers, radishes,
> Skirrets, beets, coleworts, and fair cabbages.
> Here grow fine flowers many and mongst those,
> The fair white lily and sweet fragrant rose.

So we have had the fields and the gardens, and now he comes to the fruits.

> Many good wholesome berries here you'll find,
> Fit for men's use, almost of every kind,
> Pears, apples, cherries, plumbs, quinces and peach,
> Are now no dainties; you may have of each,
> Nuts and grapes of several sorts are here,
> If you will take the pains them to seek for.

Perhaps the most one can say for this list as a poem is that he got it all in. As far as gardens are concerned, we are indebted to him for placing what they grew in order of first importance. After this the poem goes on to treat of the high price of cattle, horses, swine and goats, the number of churches, the worthy leaders, the "prudent Magistracy," the deplorable gain of "whimsy errors" among the "mixt multitude," and the mischief of selling firearms to the Indians, about which he cares so passionately that it would seem to be his chief reason for writing the entire poem. And how right he was.

For the general layout of gardens in a town we have Samuel Maverick's description of Newbury, in Massachusetts, in 1660. "The houses stand at a good distance each from the other, a field and a garden between each house and so on both sides the street for four miles or thereabouts."

The last of what we may call lay witnesses to the promise of the new land and its potential gardens, is Edward Johnson. His *Wonderworking Providence* _of Sions Saviour in New England_, published in London in 1654, contains among all its historical and religious ponderosities several quick and refreshing references to gardens and orchards in those places where the churches — set forth in order and by number — were "gathered." The ninth, the Church of Christ at Ipswich, where "Rocks hinder not the course of the Plow" consisted of about 160 souls, among a settlement of 140 families in houses "many of them very faire built with pleasant Gardens and Orchards." Many of the other settlements were equally blessed. Summing up, Johnson says that in Massachusetts in 1642 ". . . the Lord hath been pleased to turn all the wigwams, huts, and hovels the English dwelt in at their first coming, into orderly, fair and well-built homes, well-furnished many of them, together with Orchards filled with goodly fruit trees, and gardens with variety of flowers." He estimates "near a thousand acres of land planted for Orchards and Gardens, besides their fields are filled with garden fruit, there being, as is supposed in this Colony, about fifteen thousand acres in tillage. . . . Thus has the Lord incouraged his people. . . ."

Encouraged, indeed, and the promise held good, as witness the testimony of that loyal citizen of the seventeenth century, Samuel Sewall, anxious for New England to be recognized in all its power, promise, purity and beauty as a possible seat for the New Jerusalem. A typical Puritan, he can combine truth and beauty and good husbandry, fact and emotion, with ringing exposition from the Old Testament, into one fused and fusing whole.

As long as Plum Island shall faithfully keep the commanded Post; notwithstanding all the hectering Words, and hard Blows of the proud and boisterous Ocean; as long as any Salmon, or Sturgeon shall swim in the streams of the Merrimack; or any Perch or Pickeril, in Crane Pond; as long as the Sea-fowl shall know the Time of their coming, and not neglect seasonably to visit the Places of their Acquaintance; as long as any Cattel shall be fed with the Grass growing in the Medows, which do humbly down themselves before Turkey-Hill; as long as any Sheep shall walk upon Old Town Hills, and shall from thence pleasantly look down upon the River Parker and the fruitful Marishes lying beneath; as long as any free and harmless Doves shall find a White Oak, or other Tree within the Township, to perch, or feed, or build a careless Nest upon; and shall voluntarily present themselves to perform the office of Gleaners after Barley-Harvest; as long as Nature shall not grow Old and date, but shall constantly remember to give the rows of Indian Corn their education by Pairs; so long shall Christians be born there; and being first made meet shall from thense be Translated, to be made Partakers of the Inheritance of the Saints in Light. Now seeing the inhabitants of Newbury, and of New England upon the observance of their Tenure, may expect

that their Rich and Gracious Lord will continue and confirm them in the Possession of these valuable Privileges. Let us have Grace whereby we may serve God
acceptably with Reverence and Godly Fear, for our God is a consuming Fire.
Heb. 12, 28, 29.

3

The Prospect *and the* Rarities

John Milton, even more hypersensitive to episcopacy than many Puritans who suffered personally from it, felt that, without it and with presbyterianism established, England itself could hope to become "the City of God." He had, however, the greatest sympathy for "free-born Englishmen" who "had been constrained to forsake their dearest home . . . whom nothing but the wide ocean, and the savage deserts of America could hide and shelter from the fury of the Bishops."

Early settlers, planters and hopeful founders of the ideal state in seventeenth-century New England may all have been dedicated to presenting the truth about these "savage deserts" as a refuge from the rest of the troubled and troubling world, but they were not scientists and their accounts are in every case those of amateurs. There are, however, two men of the seventeenth century who set themselves the task of describing the part of the New World known as New England in the most methodical ways at their command. Where Captain John Smith could say airily that he was no alchemist but believed the resources in metal and minerals to be very great, these two men, William Wood and John Josselyn, endeavored to report only what they knew to be facts. With very few obviously outlandish references excepted, their books, though quite dissimilar, still stand as what they were intended to be, reference books for intelligent and "curious" readers.

And the curious readers were many. Although the early Puritans were a dedicated group, intent upon proving themselves, their underwriters and overseers were not in the least above hoping to make a profit. The expectation of gold, so stimulated by earlier Spanish conquests and later Spanish wrecks, had soon faded under the practical reports of men like Captain John Smith. It early became clear that they must be content with fish and turpentine and ships' masts as forms of fortune. But there was also the chance that they might emulate another example of Spanish luck by discovering new medicinal plants, or plants of such usefulness for food or fibers as would amply repay what was, at best, an extremely risky venture of capital.

There was a good deal of confusion about the climate in different parts of the New World. Even the Pilgrims had toyed with the idea of landing in real tropics as advertised to all by Sir Walter Raleigh. Virginia had seemed to many a likely compromise between jungles and Arctic wastelands. But Virginia had not proved self-supporting until the discovery that only hard labor and tobacco could make it so. New England remained an enigma behind its rocky coastlines. Misconceptions bloomed and bore fruit. Pomegranate seeds were sent as well as other semitropical plants which could not be brought to fruition in England. Grapes and mulberries for silkworms were considered sound exports to the settlers. In return, rare and rewarding plants were expected to be sent back to England, some in very large quantities. "A ton of silk grass seed" was one of the earliest requests from the English authorities to their colonizers in New England, a hard order to fill, as the English venturers quite possibly meant yucca, unheard of in New England. But hopes ran high for profits, both monetary and physical, from plants. As the Winthrop party left England, loads of tobacco from Virginia had sailed in past them. Ships loaded with nothing but sassafras were not unwelcome as return loads from the New World. Who knew what a wealth of new plant material might not be opened up. In 1637, Sir Drew Deane, in a polite note to "John Winthrop Esqr. governor of new Inglande," in one long sentence wishes everyone well and beseeches "if you find any curious flowers to favour me with some."

And they did send seeds "home." Robert Child, one of whose many facets seems to have been alchemical, wrote from Gravesend in 1644-45 to thank John Winthrop, Junior, for seeds which he had delivered to "the Gardiner of Yorke garden and to Mr. Tredescham, who are very thankefull to you for them and have returned diverse sorts which you shall receive by the hands of Mr. Willoughby." Mr. Child sent also five or six sorts of grapevines, some "prun grafts . . . some pyrocanthus trees, and very many of our common plants and seeds," desiring that they be planted with "all Expetition." He is confident that good wine will result in three years' time — indeed he intends to come over and "undertake a vineyard with all care and industry." Of the pyrocanthus trees sent to John Winthrop, Junior, we do not hear again. Perhaps because, although Parkinson greatly admired them and gave them space at the very end of his *Paradisus* with other new and unproved plants, he had to admit that they were of no use he had heard

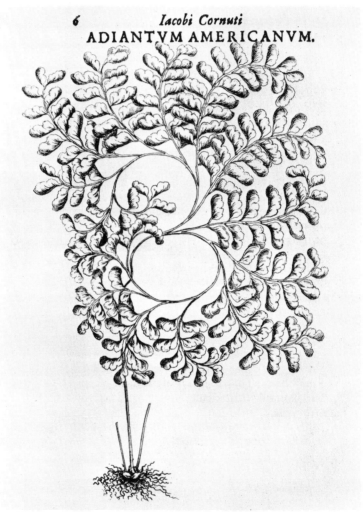

The American maidenhair noticed by Josselyn as growing in such quantities that apothecaries need never adulterate it with wall rue. By Jacques Philippe Cornut, 1635.

of. From the date, "Mr. Tredescham" would appear to be John Tradescant the Younger who had himself voyaged to the New World searching for plants. His father, John Tradescant the Elder, friend of Captain John Smith, had been willed several books from the captain's sea chest.

The Pilgrims, seeking desperately to defray some of the expenses of their passage, had attempted to send home a load of marketable herbs, in which, as in so much else, they were disappointed. The French in Canada had set an early example by sending home collections of the plants they found there which explains why so many New England wildflowers are botanically labeled *canadense*. Many of the plants our settlers "found" were already growing in the royal botanical garden in Paris, later to be known as the Jardin des Plantes. Due to the friendship between the Tradescants, father and son, and the overseers of the royal French botanical garden, many of these American plants had reached England even before the younger Tradescant's voyages to, and friends in, the New World could prove productive. In London, a large garden by the river (later known as the Chelsea Physic Garden) became the repository for whatever plants of value or possible "vertues" were discovered in the New England Wilderness. Scientific interest and acquisitive instincts were both strong. Everyone was in a hurry to find and to find out. All that was lacking was informed authority on the nature of this extraordinary part of the New World wedged between the precarious acquisitions of the French to the north and, to the south, the uneasy holdings of the Dutch, with other, more commercial, English interests and the ever-greedy Spaniards beyond them. It was high time for someone to undertake to explain the true situation of New England, its climate and its flora and fauna. Fortunately for us, those two able and dissimilar writers, William Wood and John Josselyn, undertook to do so.

The first of these, in point of publication, was William Wood, whose *New Englands Prospect* was printed in London in 1634. Among all the explorers' and planters' and promoters' accounts of the New World, this book has an evenness of style which others lack, and an impetuosity in rendering information which gives the reader the feeling Wood knows even more than he hastens to tell. He serves up the features of the new land and its future possibilities in a clear and rapid style without affectations. In a punning age, he scorns to pun. His recourse to Latin quotations is for our convenience and not his own aggrandizement. He calls neither upon the Creator, nor the "ancients," nor accounts of others for his authority. His

The South part of Nevv-England, as it is Planted this yeare, 1635.

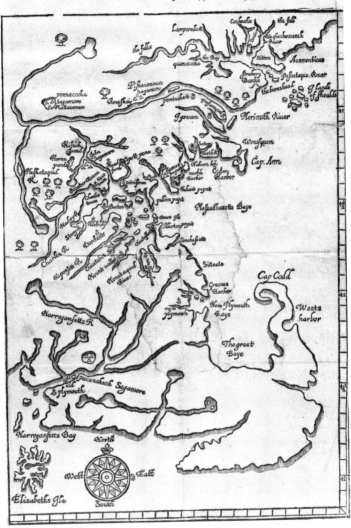

A map used to illustrate the reprint of William Wood's *New Englands Prospect*, showing the country he studied so painstakingly. The reprint has the map dated 1634. This illustration is taken from a copy of the map in the archives of the Massachusetts Historical Society and is dated a year after Wood's book was published.

four poems, modestly interspersed in the text, neatly and wittily sum up the kinds and characteristics of New England trees, beasts, birds and fowls, and fish. While he declares he has rendered the second part of his book, an account of the Indians, in a "more light and facetious stile" than the first part, since the Indians' "carriage and behaviour hath afforded more matter of mirth and laughter than gravity and wisdom," he still seems to give us a complete and understanding account of the truly simple savages with whom the early settlers had to deal. And he ends with a little glossary of the Indian language as he knows it.

After the inclusion of a crude map of "The South part of New-England as it is Planted this Yeare, 1634," William Wood describes his book upon the title page as "A True, lively and experimentall description of that part of America commonly called New England; discovering the state of that Countrie, both as it stands to our new-come English Planters; and to the old Native Inhabitants. Laying downe that which may both enrich the knowledge of the mind-traveling Reader or benefit the future Voyager."

In his "frontispeece" Wood mentions the "well deserving name" of "the Right Worshipfull, my much honored friend, Sir William Armyne, Knight and Baronet," acknowledging his "bounteous favor and love towards my selfe in particular" and wishing him "a confluence of all the blessings both of the throne and the foot-stoole." And he turns next to "the Courteous Reader," to whom, though he promises "no such voluminous discourse as many have made upon a scanter subject though they have travailed no further than the smoake of their own native chimneys," he yet presumes to present "the true and faithfull relation of some few yeares travels and experiences." He does this, he says, for the sake of his countrymen because, "there hath some relations heretofore past the Presse which have been very imperfect; as also there have been many scandalous and false reports past upon the country," wherefore, he says, "I have laid down the nature of the Country without any partial respect unto it. . . . What I speake," he declares, "is the very truth, and this will informe thee almost as fully concerning it, as if thou wentest over to see it. Thus, thou mayest in two or three houres travaile over a few leaves, see and know that, which cost him that writ it, yeares and travaile over Sea and Land before he knew it. . . ."

There follows a short poem by "S. W." "To the Author, his singular good friend, Mr. William Wood" which in twelve lines thanks the author for packing so much knowledge so completely "in so small roome." And here, for those who are searching for clues to gardens, it is enticing to come across a possible reference to a common feature in formal designs for the sixteenth- and seventeenth-century English and French gardens—the mount. In England every nobleman and even some gentlemen could hope to have a mount. Situated as far as possible from the house and made, perhaps, from soil removed to create a fish pond, canal or pool, the mount was a symmetrical mound with graded sides hopefully planted to look natural. Crowned by a little summer house, pavilion, pagoda or even just a seat protected by a hedge, the mount was a vantage point from which the relation of house to garden, as well as the whole garden itself, might be studied and enjoyed. "S. W." very charmingly asserts that the combination of William Wood's experience, travel and knowledge "thus a Mount doth make, From whence we may New Englands Prospect take."

And, finally, Wood is able to begin his book. His "very truth" he divides into two well-organized sections. The first section consists of twelve short chapters concerned with: the situation, the seasons, the climate, the soil, "Hearbes," fruits, woods, waters, minerals, beasts, birds, fowl and fish. He lists the several plantations already made and their individual advantages. He candidly admits also, in their own chapter, all the evils, among which he includes rattlesnakes, mosquitoes, gnats, green-heads and lazy men. He describes the provisions to be taken for the sea journey and the quantity necessary to ensure enough food for a year and a half after landing, when the settlers' own crops should be enough to sustain them.

In the second section of his book, he follows the same system of short chapters on each subject. While showing the Indian men as coming a long second best after the women in industry, he asserts his is a very fair summary of "their persons, cloathings, diet, natures, customs, lawes, marriages, worships, conjurations, warres, games, huntings, fishings, sports, language, death, and burials." It is here that we gratefully gain, in oblique fashion, a glimpse of the position of the early English women settlers whose lot, by comparison, so upset the Indian women. William Wood gladly includes this vignette of social customs to help explode yet one more aspersion reputedly cast upon the English in New England—that they had learned from the Indians how to treat their wives. On the contrary, he declares, the squaws "made miserable by seeing the kind usage of the English to their wives . . . doe as much condemne their husbands for unkindness and commend the English for their love." The Indian men, never particularly attractive creatures domestically, "doe condemn the English for their folly in spoyling good working creatures." As one of the aspersions quoted included English women's having to carry water, Wood declares in one splendid last word, "For what need they carry water, seeing everyone hath a

Bee-balm or horsemint, which the Indians were reputed to use against stings. By Jacques Philippe Cornut, 1635.

Spring at his door, or the Sea by his House?" and he concludes, "Thus much for the satisfaction of women, touching this entrenchment upon their prerogative . . . ," giving us a touchstone for the spirit of English women in those times.

Much of what Wood says will turn up in other accounts later, verbatim or added to, but still recognizably his. While "the smoake of their own chimneys" rings through early accounts as a measure of inclination to adventure, Wood has fewer phrases in common with others than most. He, indeed, seems to originate later clichés. It is his claim that the country is so healthy that, "in public assemblies it is strange to hear a man sneeze or cough as ordinarily they doe in old England." And to him we seem to owe the first indignant claim that "whereas many died at the beginning of the plantations, it was not because the Country was unhealthful, but because their bodies were corrupted with sea-diet, which was naught, their Beefe and Porke being tainted, their Butter and Cheese corrupted, their Fish rotten, and voyage long, by reason of Crosse Windes. . . ."

After lush descriptions of the natural fertility of the soil and the splendid grasses which flourish, he comes in his fifth chapter to what most concerns us: "Of the Hearbes, Fruites, Woods, Waters. . . ."

The ground, affords very good kitchin Gardens, for Turneps, Parsnips, Carrots, Radishes and Pumpions, Muskmillions, Isquoutersquashes, Coucumbers, Onyons, and whatever grows well in England grows well there, many things being better and larger; there is likewise growing all manner of Hearbes for meate, and berries, Treackleberries, Hurtleberries, Currants, which being dryed in the Sunne medicine, and that not only in planted Gardens, but in the Woods, without eyther the art or the help of man, as sweet Marjoram, Purselane, Sorrel, Peneriall, Yarrow, Mirtle, Saxifarilla, Bayes, etc. There is likewise Strawberries in abundance, very large ones, some being two inches about; one may gather halfe a bushell in a forenoone. In other seasons there bee Gooseberries, Bilberries, Res- are little inferiour to those that our Grocers sell in England. This land likewise affords Hempe and Flax, some naturally, and some planted by the English, with Rapes if they may be well managed.

He then goes on to praise the water, though in a tempered manner.

For the Countrey it is as well watered as any land under the Sunne, every family, or every two families having a spring of sweet waters betwixt them, which is farre different from the waters of England, being not so sharpe, but of a fatter sub-

stance, and of a more jetty color; yet dare I not prefer it before good Beere, as
some have done, but any man will choose it before bad Beere, Wheay, or Buttermilk.

It is worth quoting Wood's verses listing the trees, if only to pin down
the line about sumach which comes toward the end. He says he recites this
for "mechanicall artificers to know what Timber and wood of use is in the
Countrey":

> Trees both in hills and plains in plenty be,
> The long liv'd Oake, and mourneful Cypris tree,
> Skie towering pines, and Chestnuts coated rough,
> The lasting Cedar, with the Walnut tough;
> The rozin dropping Firre for masts in use,
> The boatmen seeke for Oares light, neat-grown Sprewse,
> The brittle Ash, the ever trembling Aspes,
> The borad-spread Elme, whose concave harbours waspes,
> The water spungie Alder good for nought,
> Small Elderne by the Indian Fletchers sought,
> The knottie Maple, pallied Birtch, Hawthornes,
> The Horne bound tree that to be cloven scorns;
> Which from the tender Vine oft takes his spouse
> Who twines embracing arms about his boughs.
> Within this Indian Orchard fruites be some,
> The ruddie Cherrie, and the jettie Plumbe,
> Snake murthering Hazell, with sweet Saxaphrage,
> Whose spurnes in beere allayes hot fevers rage.
> The Diars Sumach, with more trees there be,
> That are both good to use and rare to see.

This is the origin of the most frequently quoted words from William
Wood's entire book. "Dear sumach," he is supposed to have said, and
gentle lady writers have loved him the more for it. Nineteenth-century
botanists picked it up and say he must have intended "deer sumach" by
which, of course, they say, is meant the stag-horn sumach, which is indeed
what Wood intended, botanically, although it was the virtues of the shrub
as a source for dyes that he referred to and not its velvety horns. So do we
pursue our own tales.

All Wood's countrymen had dreams of gold and silk and wine. Wood
disposes of the first by saying that "the Spaniards blisse" may lie hid in the

barren mountains, but he dares not confidently so conclude. The possibilities of growing mulberry trees for silkworms have not reached him. But he thinks well of wine-making, the grapes being "so delectable that there is no knowne reason why as good wine may not be made in those parts as well as in Burdeaux in France." It is when he describes our chokecherries that we realize what an honest and graphic man he is. "The Cherrie trees," he says, "yeeld great store of Cherries, which grow on clusters like grapes; they be much smaller than our English cherries, nothing near so good if they be not very ripe; they so furre the mouth that the tongue will cleave to the roofe and the throat wax horse with swallowing those red Bullies as I may call them being little better in taste. English ordering may bring them to be an English Cherrie, but yet they are as wild as the Indians." Faith in "English ordering" was not misplaced. While the English in the end had to accept the fruits of the country for what they were, they swiftly introduced their own sorts. The idea of improving the wild cherry proved as unprofitable as another expectation Wood mentions — that of putting a yoke on the moose and teaching it to be a beast of burden.

"The Plummes," says Wood, "be better for Plummes than the Cherries be for Cherries. . . . The white Thorn affords hawes as bigge as an English Cherrie which is esteemed above a Cherrie for his goodness and pleasantness to the taste."

It is from Wood that we learn the English manure their land with fish "not because the land could not bring corn without, but because it brings more with it." He tells us the Indians hoed with clamshells before the English came. And he sees wheat and beans growing well in gardens, although they have not yet been tried in newly plowed land prepared by oxen and horses, "where English Corne, especially Rie and Oates and Barley," have done well.

When he lists the "severall plantations in particular," to be often told over and added to by later visitors, there are only a very few upon whose gardens he can remark. Dorchester has "faire corne-fields and pleasant Gardens, with Kitchin-gardens. . . . Roxberry . . . the inhabitants have faire houses, store of Cattle, impaled Corne-fields, and fruitful Gardens." Boston has "fruitful Gardens." Governors Island "an Orchard and a Vine-yard." "In Mysticke" is "Maister Craddock's plantation . . . where he hath impaled a parke where he keeps his Cattle till he can store it with Deere."

So we leave William Wood, honest enthusiast.

The second self-appointed and eagerly earnest authority is a man about whom we know more personally than we do about William Wood, because he was, happily for us, unable to leave himself out of anything he wrote. John Josselyn carefully styled himself "Gent." upon his title page, pre-sumably to distinguish himself from the mass of Puritan settlers who strove to erase any pretensions to social position beyond what they felt their elective offices demanded. He was the younger son of Sir Thomas Josselyn, whose name headed the list of supporters for the charter of Sir Ferdinando Gorges for the Province of Maine. Although Sir Thomas Josselyn never visited New England, his elder son, Henry, spent most of his life there. In 1634 Henry appears in records as assistant to Captain John Mason at Piscat-aqua, and in 1643 as owner of a patent to a tract of land at Black Point, by the will of Captain Thomas Cammock, whose widow he married. Magis-trate, deputy governor and soldier, he died in the early 1680's deserving, in the words of Governor Andros, "fitting respect" after a lifetime of defend-ing his province from both the Indians and the acquisitive Puritans of Massachusetts Bay. It was to visit this brother that John Josselyn made his two voyages to New England.

His first visit began in July of 1638, when he arrived in Boston, paid his respects to Mr. Cotton and Governor Winthrop, to whom he delivered some translations of the Psalms by the poet Quarles, and sailed up the coast past "a mere wildernesses" to Scarborough where he remained for a year and a half.

His second visit began in 1663, and he stayed eight years. No one has ever quite determined how politically advisable this last long stay may have been. His books offended the Puritans, whom he declared, together with all non-royalists, to be "perverse spirits," and contained effusive mentions of the "royal martyr," although the Josselyn family was known to have Puri-tan and Parliamentarian inclinations. Quite possibly, like others in high places at this time, Josselyn was merely playing it safe. In any case, as he stated at the end of one of his books, he finally returned home, and an-nounced himself "safely arrived in my Native Countrey, having in part made good the French proverb — Travail where thou canst, but dye where thou oughtest, that is, in thine own Countrey."

New-Englands Rarities Discovered was published in 1672 followed soon after in 1675 by *An Account of Two Voyages to New England. New-Englands Rarities* is dedicated to his "Honoured friend and kinsman, Samuel Fortrey,

Esq.," whose assistance enabled him to commence a voyage "to those remote parts of the world known to us by the painful discovery of that memorable Gentleman Sir Francis Drake." Josselyn therefore "adventured to
obtrude" upon his benefactor "these rude and indigested Eight Years Observations." They are the first "painful," or, as we now say, careful account
of all the flora and fauna of New England insofar as he was able to observe
them. Besides all this, he undertakes, as stated upon the title page, to give
us "The Physical and Chyrurgical Remedies wherewith the Natives constantly use to Cure their Distempers, Wounds and Sores" and a pretty little
poem to an Indian squaw made over, he says, from a poem to a gypsy.
And lastly, a chronological table of "the most remarkable passages in that
Countrey amongst the English."

Curiously enough, although less determinedly scientific in subject
matter, Josselyn's second book in order of publication, *Two Voyages to New
England*, was "most humbly presented" to the President and Fellows of the
Royal Society. This is a much less "painful" and more haphazard account-
ing of all sorts of interesting facts and events, interlarded with fuller descriptions of selected birds, beasts, fishes, trees and plants, a scattering of
recipes and remedies, descriptions of the Indians, and another chronological table. On the title page is the summary of the contents: "Wherein you
have the setting of a Ship, with the charges; the prices of all necessaries for
furnishing a Planter and his Family at his first coming; a Description of the
Country, Natives and Creatures; the Government of the Country as it is
now possessed by the English, etc. A large Chronological Table of the
most remarkable passages from the first discovering of the Continent of
America to the year 1673."

It is to Josselyn's two small but lively volumes that we owe our most extensive knowledge of what plants the early settlers used locally or grew
themselves. And, actually, John Josselyn is a splendid example of the "spare
kindred" whom Captain John Smith urged the landed gentry in England to
send over to help settle New England.

The final grand repository for all the earlier accounts, including those of
Captain John Smith, was, of course, that successor to Hakluyt's *Voyages*,
Purchas his Pilgrimes, a copy of which appears in a Salem inventory of 1647
beside "a bible." This was available to all who might wish to study it, or
indeed avail themselves of it, before writing their own accounts, as many
did. John Josselyn, however, appears to have preferred to be an inde-

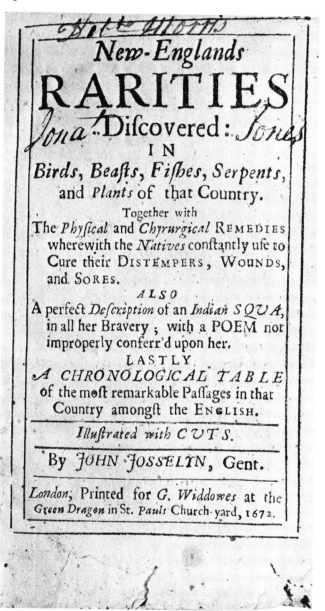

Title page of John Josselyn's *New-Englands Rarities*, 1672.

pendent observer whenever possible. His acknowledgments are infrequent. He quotes Captain John Smith by name upon the subject of "a certain berry called Kermes" which Josselyn searched for "as for a needle in a bottle of hay," and failed to find. He refers in the same context to his authority, Gerard, "our famous Herbalist" whose 1633 edition, as corrected and added to by Thomas Johnson, Josselyn elsewhere declares to contain many additions from the New World, as, also, he asserts, does Parkinson. He quotes "Mr. George Sands in his Travels" on the subject of these puzzling berries, for information on which Captain John Smith does not seem to have depended upon entirely reliable sources. Josselyn seems not to sense his own occasionally apparent indebtedness to William Wood. And yet it seems, from several direct acknowledgments by name to those who told him stories or whose accounts he read, that he meant to give credit wherever he felt it was due.

John Josselyn may be considered the most original observer among the early chroniclers. He shows himself to be a most attractive and intelligent individual, somewhat encumbered by overeager credulity but determined to tell all he knows about a subject of which most of the world of his time knew almost nothing.

To accept Josselyn as our most complete authority for what the early settlers had in their gardens, it is important for us to realize something of his personal world. Although decidedly out of sympathy with the "thorough-paced Independents and rigid Presbyterians" of the Massachusetts Bay Colony, he seems to share their respect for the Scriptures as final authority and to feel that God has provided a system for existence full of secrets for man to uncover. Everything, Josselyn assumes, has its purpose. Nothing is created in vain. Everything is, ultimately, for the use of man. He tastes, tests, experiments. He listens to descriptions of plants, seen though not brought back by those wandering in the wilds. He searches for some and finds them. He discovers some for himself, like the skunk cabbage and the pitcher-plant, which he is amazed, so "fantastick" is the flower, to find unrecorded. He exchanges remedies and plant uses with the Indians. He likes to describe as "trees of God," "the great trees that grow of themselves without planting." He notes the familiar English names which the English settlers have given to some of the plants of the New World and tries to judge the nature of the assumed relationship. Josselyn does not seem to be given to astrological interpretations concerning the usefulness of plants.

Nor does he seem to subscribe to the doctrine of signatures, a theory that the use of a plant might be indicated by a resemblance in its leaves or roots or blossoms to a feature of the disease it is meant to cure. But he was dependent upon those who did. He is a wholehearted Galenist and a natural meddler in medical remedies, vain of his skill but not pretending to knowledge except from experience. And it is to John Josselyn we owe our instructions for transplanting "Indian" plants and trees of which he finds the roots "but of small depth and so they must be set."

Josselyn recognized that Gerard and his editor, Johnson, and also Parkinson had purposefully included all the newly discovered plants from the New World. When Josselyn finds plants he believes not hitherto noticed, he produces his descriptions of them with all modesty, and adds his own rather clumsy drawings to illustrate his possibly original discoveries.

His journal for his first visit, which is contained in the second book to be published, is short, comprised of brief chronological entries detailing extraordinary tales told to him by other settlers, minor adventures of his own with wolves, wasps and snakes, and a touching story of a Negress slave appealing to him because she did not wish to be bred to a Negro man, also a slave, to increase her master's (Mr. Maverick of Boston) supply of Negroes.

The lists of supplies advisable to bring with one — food, clothing, tools, arms and utensils — make up the most interesting part of this document, which is otherwise the rather superficial account of a carefree young adventurer. There is only one entry which presages the later voyage's more serious involvement with the life of the country and gives an indication of the writer's personal tastes and requirements. After listing "the common proportion of victuals" from "pork or beef, fish, butter, cheese, pease, pottage, water-gruel, bisket and six shilling bear," he adds: "For private fresh provision, you may carry with you (in case you, or any of yours should be sick at sea) Conserves of Roses, Clove-Gillyflowers, Wormwood, Green-Ginger, Burnt-Wine, English Spirits, Prunes to Stew, Raisons of the Sun, Currence, Sugar, Nutmeg, Mace, Cinnamon, Pepper and Ginger, White Bisket, or Spanish rusk, Eggs, Rice, juice of Lemmons well put up to cure or prevent the Scurvey." And there is a remedy for seasickness which is also a forecast of those to come with the second voyage over twenty years later. "To prevent or take away sea-sickness Conserve of Wormwood is very proper, but these following troches I prefer before it. First make paste of Sugar and Gum-Dragagant mixed together, then mix therewith a reasona-

ble quantitie of the powder of Cinnamon and Ginger, and if you please a
little Musk also, and make it up into Roules of several fashions which you
may gild, of this when you are troubled in your stomach, take and eat
a quantity."

Josselyn's account of the second voyage is more ambitious, much longer,
and crammed with dates and information of all sorts. Even so, Josselyn's
style always has both dash and charm. In describing the plants — after having got in all possible animals — he says:

The plants in New England for the variety, number, beauty, style and vertues, may
stand in Competition with the plants of any Countrey in Europe. Johnson hath
added to Gerard's *Herbal* 300, and Parkinson mentioneth many more; had they
been in New England they might have found 1000 at least never heard of nor seen
by any Englishman before. Tis true, the Countrie hath no Bonerets, or Tartar-
lambs, no glittering coloured Tuleps; but here you have the American Mary-Gold,
the Earth-nut bearing a princely Flower, the beautiful leaved Pirola, the honied
Colibry. They are generally of (somewhat) a more masculine vertue than any of
the same species in England, but not in so terrible a degree as to be mischievous or
ineffectual to our English bodies.

In its context this warning about the "plants in new England" follows a
description of the climate, which he says is "reasonably temperate." A suggestion of the background information as to temperatures and humors
which was common knowledge and practice is contained in his very explicit
description of the climate and what it was supposed to do to the settlers.

Josselyn says it is, "hotter in Summer and colder in winter than with us,
agrees with our constitutions better than hotter climates, these are limbecks
to our bodies, forraign heat will extract the inward and adventitious heat
consume the natural. So much more heat any man receives outwardly from
the heat of the sun so much more wants he the same inwardly, which," he
concludes, "is one reason why they are able to receive more and larger
draughts of Brandy and the like strong spirits than in England without of-
fence." But, he says, "Cold is less tolerable than heat" which is a "friend
to nature. Cold is an enemy. . . . Many are of opinion that the greatest
enemies of life, consisting of heat and moisture, is cold and dryness; the
extremity of cold is more easie to be endured than extremity of heat; the
violent sharpness of Winter, than the fiery raging of Summer. To conclude, they are both bad, too much heat brings a hot Feaver, too much cold
diminisheth the flesh, withers the face, hollows the eyes, quencheth natural

heat, peeleth the hair, and procureth baldness." Which settles the climate, and explains why it was important to know the qualities of the plants to be used as remedies — some cold, some hot, and all in differing degrees. Some writers have suggested that John Josselyn must have had medical training, but there is nothing in anything he has written that would indicate this. These were the times when everyone had to know all he could about everything and be willing to act upon it.

Armed with his chief reference book, the 1633 edition of Gerard's *Herball*, as corrected and amplified by Thomas Johnson, Josselyn quite literally nibbled and tasted his way about the wilderness, finding several plants not noted before, drawing the first known pictures of the pitcher-plant, and trying to follow up any mentions of plants he does not know. Josselyn divided this book into several parts, of which plants are the subject of at least half, as can be seen in the following prospectus for "the proposed discovery of the Natural, Physical, and Chyrurgical Rarities."

In order these rarities are: (1) Birds, (2) Beasts, (3) Fishes, (4) Serpents and Insects, and (5) Plants. This section is divided into five categories as follows:

(1) such plants as are common with us, (2) of such plants as are proper to the country, (3) of such plants as are proper to the country and have no name known to us, (4) of such plants as have sprung up since the English planted and kept cattle there, (5) of such Garden Herbs (amongst us) as do thrive there and of such as do not, (6) of Stones, Minerals, Metals, and Earths.

John Josselyn will appear often in our references. His plant list will be given in full later, but here we may quote his references to gardens from his *Two Voyages*, which gives us our best idea of the importance of gardens and orchards in the new settlements.

"Governours Garden, where the first Apple-Trees in the Countrey were planted, and a vinyard . . ."

"Dorchester, a frontire Town pleasantly seated, and of large extent into the main land, well watered with two small Rivers, her body and wings filled somewhat thick with houses to the number of two hundred and more, beautified with fair Orchards and Gardens . . ."

"Roxbury, a fair and handsome Countrey Town, the streets large, the Inhabitants rich, replenished with Orchards and Gardens . . ."

"Boston . . . the South-side adorned with Gardens and Orchards."

"Dedham . . . abounding with Garden fruit . . ."

"Charles-Town . . . The market-place not far from the waterside is surrounded with houses, forth of which issue two streets orderly built and beautified with Orchards and Gardens . . ."

"New-Town . . . now called Cambridge where is a colleg for Students of late . . . the neatest and best compacted town, having many fair structures and handsom contrived streets, the Inhabitants rich, they have many hundred acres of land paled with one common fence a mile and a half long, and a store of cattle. . . ."

"Water-town . . . very fruitful . . . the Inhabitants live scatteringly. . . ."

"Mistick . . . On the West of this river is Merchant Craddock's Plantation where he impaled a park. . . ."

"Marble-head . . . Orchards and Gardens. . . ."

"New Salem . . . store of Meadow and Arable, in this Town are some very rich Merchants. . . ."

"Ipswich . . . store of Orchards and Gardens. . . ."

And, thankfully, we append also Josselyn's list of garden tools: "Wheelbarrow you may have there, in England they cost six shillings . . . broad howes . . . narrow howes . . . spades."

4

"Too Scanty a Stage"

Before we can replant their gardens upon which so much and so many depended, we must take the measure of the people themselves, and see how they, like their plants, survived in the new land. No seeds or roots or plants or scions could be wasted. Sometimes someone may have wondered if the people were not casting themselves upon barren ground.

When John Winthrop, Junior's exuberant and faithful friend, Edward Howes, wrote to him in New England that their mutual and most glamorous acquaintance, Sir Kenelm Digby, had suggested that Winthrop come home to England where he could do more good than he could upon New England's "too scanty a stage," Howes was reporting a concern which was being widely voiced in old England. Even as early as the 1620's, it began to be obvious in England that people who could be ill spared were leaving the kingdom. By the end of that decade such drastic official efforts were being made to stop them, even as they were embarking, that some of the greatest worthies came out of England in disguise and by subterfuge. As we have seen, Milton mourned that "free-born Englishmen . . . had been constrained to forsake their dearest home . . . whom nothing but the wide ocean and the savage deserts of America could hide and shelter from the fury of the Bishops." Still, one has only to read the accounts of these clergymen who had been closeted, however briefly, with the great Archbishop Laud to see why they were glad to leave their parishes like fugitives and to board a little ship bound for a new land, even when all they knew of their destination beyond its hardships was that there they would be free to have their own views of church worship. We must not forget that they knew they were good — not merely morally good, but valuable to the state. They came proudly, if sometimes precipitately.

England woke up too late to the realization that she was losing good men. But the good men had known this before they left. The thought appears in tracts and arguments, such as the one with which they were all familiar called *General Considerations for the Plantation in New England, with the Answer to Several Objections.* In this tract, widely quoted and variously attributed to several of the worthiest fugitives, the first stated objective is to carry the gospel to New England and by so doing to raise a bulwark against the Jesuits. The second objective is to find a refuge for those whom God means to save from what appears to be His coming judgment upon all the other churches in Europe. The third has to do with the "weariness" of the homeland with its inhabitants, so that children, especially those of the poor, are considered burdens rather than blessings. The fourth declares that the high cost of living and the difficulties of a man trying to "keepe saile with his equalls" have made it hard for the upright to live comfortably. Fifth, schools are corrupted by the evil examples of their governors, and are very expensive. Sixth, as the world is God's garden, why should any "stand starving here for places of habitation" when whole countries elsewhere "lye waste without an improvement." Seventh, what could be more nobly Christian than to support and raise a church in its infancy. And finally, "if any such as are knowne to bee godly and live in wealth and prosperity here, shall forsake all this to join themselves to this church, and runne in hazard with them . . . it will be an example of great use . . . and also . . . encourage others to join."

In the second half of the pamphlet, objections are raised and answered, and this last point is taken up first. It seems to have been the one which chiefly concerned many. To the objection that "it will bee a great wrong to our owne church and countrey to take away the best people," the modest answer is that there will still be plenty left. They were not conceited. The interesting point for us is that they knew their own worth.

This last is implicit in much of the correspondence after the settlers had proved their point, and it had become a great thing in England "to be a New England man," one of the arguments used by Edward Howes to coax John Winthrop, Junior, back. Howes's letters are the liveliest insight we have into the difficulties of correspondence in the early days of the colony, when there was even a need for writing some things in code for fear of some ships' captains reading the mails. Two subjects especially — alchemic experiments and sending money to New England — demanded secrecy. It is, moreover, Edward Howes's cheerful, gossipy, teasing letters that give us a measure of the younger Winthrop, interested in all that went on in the new science of chemistry but so far from it, eager for books and news and roots and trees and roses from his old friend, and yet quietly committed to continue in the New World which he knew he was bound to lead. One can only believe he often found himself incredibly lonely and cut off from the exciting and stimulating world he had known as a young man. Indeed, late in his life he admits this.

Many bright minds must have felt lonely and cut off. One marvels today at their steadfastness. One of the sources of their strength to stay the course was the extraordinarily strong relationship of that day between father and son, son and father. Responsibilities are handed on, without question. Without question, they are accepted. Expression of emotion was not difficult for the Puritans, whatever we have been misled to think. The senior Winthrop writes with deep gratitude to his "very loving son" for "the goodness I find in you towards me," for which he prays his son may be rewarded as God promised those "who give due honor to their parentes." We see the same strong feeling of love and honor in the three generations of Mathers, grandfather, father and son; and between — to introduce the other sex — Samuel Sewall and his daughter Betsy, Anne Bradstreet and her father, Thomas Dudley; and between Anne Bradstreet and her son Simon. All these examples of deep devotion and understanding are repeated in the marriage relationships. Between the elder Winthrops, husband and third wife, there was a bond whose strength reaches us today in their letters. One has only to read Anne Bradstreet's poems to her husband to recognize the vitality of their marriage. And from all the considerable estates willed to their wives to be managed by them, one senses the husbands' confidence in their wives' total abilities.

An age can be no greater than its average man. It is one of the phenomena of the Puritan settlement of New England that the intellectual content was so high, the moral propulsion so strong, and the foundations so solidly secured that there is no American today, however recently arrived and from whatever quarter of the globe, that does not share to some degree in their legacy. And this remains so in spite of their detractors, who have been legion. Criticism of Puritans quite literally knows no bounds. Fleeing from intolerance themselves, they are said to have become the most intolerant of men. Some critics even blame the Puritans' aversion to the theater for causing the age of Shakespeare to be so brief, ignoring the fact that many people in positions of responsibility in London took a far dimmer view than any Puritans of the goings-on in the playhouses across the river. While the Puritans, like most of the English of their time, do not seem to have read Shakespeare, they owned and read many of Shakespeare's source books, and, as with Hakluyt and Plautus, brought them with them to the New World. To realize the Puritans as human beings of diverse abilities and accomplishments is to recognize that what they created was truly great. And it is not to invite an anticlimax to say it would have been impossible for them to succeed without their gardens.

While the avowed purpose of the early Puritan settlers of New England was to found a City of God upon a Hill, New England was not the first site chosen. An early editor of the founders' papers for the Massachusetts Historical Society stated the case for choosing New England in terms less religious than practical. His impetuous and confused style is appropriate to the desperation of his subject. "Where," he asks, "could they fly? A religious war raged in France between the catholicks and protestants; a puissant protestant league of the States of Holland, Sweden, Saxony, and Denmark waged a bloody war against the Emperor; France embroiled with Austria, and England with France, Spain and Austria; Sweden with Poland; and Holland with Spain, while every petty state was drawn into the vortex as inclination promoted or necessity forced; deplorable must have been the prospect! But fortunate for the human race! These calamities laid the foundation in North America for establishing in Europe, the rights of men on the basis of reason."

This state of the world in no way detracts from New England's role in the great experiment, but it explains why the Puritans came so well supplied, especially for gardening. For, when the great wave of Puritan migrants began to arrive upon the New England shores, they represented a group of people doing something for the first time in the history of the world. Much as they liked to compare themselves with the Israelites, and hopefully as they liked to see a resemblance in this country to the Promised Land, the fact remains that they were doing what no one had ever done before. Although the Puritans were joined by people with varied motives, the mass and power of the move which carried them through the century was primarily religious. The Pilgrims, who were true Separatists and had been cruelly treated as such, had been here for nearly ten years before the great tide rose and sent waves of Puritans, some of the strongest minds and characters in England, to the shores of this surprisingly New World.

Although the Pilgrims seem to have come in something of a hurry, lacking forewarning of their needs or preparation for them, brought to the wrong place at the wrong time, and certainly backed by some very wrong people, the Puritan migration was deliberate, determined and carefully planned. The libraries of some of the early arrivals show us what they knew about their venture beforehand. Hakluyt and his successor, Purchas, left no doubt that Englishmen would do well to get away from "their own chimneys' smoke" and the "houses at their towns' ends," to travel over the seas and secure lands for the English before the Spaniards and the Portuguese and the French had seized them all. Captain John Smith urged them on. His book would appear to have been a "must." Various promoters stood ready to help them. They were told what to take with them for the voyage, and how to dress for it. The settlers of New England came under no illusions that the Indians would prove useful labor, a hope which had been dashed in Virginia. And they soon learned not to expect a beast of burden in the moose. They knew they had to bring with them everything they would need for a year — animals, servants, implements, seeds, roots, saplings, nails, "spirits." Only their own industry and God's help, if they deserved it, could be counted upon. They were brave, they were determined, and they were wise. Some of them were gifted. Many had left property behind them for others to tend. They were confident they could manage — the women as well as the men. It is when we search their letters, their sermons, their lists and diaries and papers, to find out what they grew, that we come to a full appreciation of their stature.

To take account of this stature, we have only to consider a handful of their leaders. The Winthrops, quite naturally, come first, each John, father and son. Puritans, pioneers, staunch husbands, willing amateur healers and keen gardeners. There was never any doubt about the quality of the senior Winthrop, who was the leader of his group from the time they were first "gathered" in England to consider the feasibility of setting up a more ideal state in a new world. A large landholder, a lawyer, as accustomed to London and business as to the countryside and its management, he was a good and loving husband, a kind father, a just judge and a particularly keen gardener. We have read his love letter when he was courting his third wife, whom he brought to the New World. We have heard him exclaim at the fragrance of the air blowing toward his ship from the new land, "like the smell of a garden," and we have seen him developing a rocky island in Boston Harbor into a place so fruitful that it became "the Governor's Garden" and a place for all ships to visit as they came in.

East Anglia, where the Winthrops and so many of the settlers of New England came from, bulges out into the North Sea and toward Holland like the archetype of all the woolsacks it must have symbolized to many sixteenth-century minds. Sheep raised upon these swelling uplands were the source of much of the warm clothing of Europe and England, after the wool had been spun and woven in the Low Countries. Stout little ships of native oak, framed so like the local barns that one could never look up into the roof of one or down into the hull of the other without observing their relationship, tackled the North Sea on any and all of its own terms. To take loads of fleece from these black-timbered barns across high seas to countless clacking looms and bring it back again as cloth for London tailors was the regular business of several centuries. When Andrew Boorde's *Book of Knowledge*, in which he instructs us in the people, customs and money of many nations, shows the typical Englishman striding along nearly naked, cloth on one arm, shears in the other hand, not having yet made up his mind how to be in the latest fashion, the little woodcut merely marks the peak, in mid-sixteenth century, of England's success in the wool trade. Silks and velvets, gold lace and tapestries, were the business and skills of Europeans. In a cold world, devoid of comforts, men needed wool. And England, especially East Anglia, furnished it.

This pretty land of mild contours, stout hedges, timbered houses and small clear rivers, washes its steep cliffs in a sea that is suddenly deep and strong. No great houses seem to command the landscape, or ever to have done so. Small holdings are evenly spaced. Towns are only gently crowded, the better houses sharing street frontage with the lesser or with their own barns and storage places. Gardens stretch back in fruitful little strips of orchards and patterned plantings near the house terrace. Churches are small and relatively simple. Even the graveyards have a quiet decorum and lack of individual ostentation. There are two explanations for this general gentle conformity — first, this was, for ages, a countryside where everyone was relatively prosperous, interdependent and cooperative, and secondly, there is no local stone.

In a countryside without stone, striking ostentation is almost impossible.

Church walls are made of a sort of mixture of small stones of varied shapes and sizes, too small to be shaped, lending themselves to being gathered up and molded into thick walls, the only larger cut stones being reserved for corners and coins and arches. Similarly the houses, whose lower walls are shaped up in the same way and then covered up with plaster, have thin bricks set diagonally between stout timbers framing their upper storeys. The roofs of the churches are almost inevitably thatched. It is a countryside of quiet self-sufficiency and wise improvisation, happily still preserved and recognizable as the sort of place from which one could contemplate crossing the great Atlantic Ocean to form a new England, bare-handed but with seeds and stores and trusting and well-trusted friends.

The other centers from which the Puritan wave gathered its forces are harder to realize as the stuff of memory for the early settlers. The rolling

A satirical portrait of the vain Englishman of the time of Henry VIII, so keen on being in the latest style that he strides along with a length of woolen cloth over his arm, unable to decide "what rayment" to wear. This is the first among the national portraits and characteristics portrayed in Andrew Boorde's book of travel. Boorde's scorn for the vanity of the English may well be an indication of how the Puritans felt.

moors and great hedges of the west of England, the gray stone of Yorkshire houses and walls, the lovely little moated castle of Lord Saye and Seele, where Puritan sympathizers met with Cromwell, are all still there. But even in this last handsome bastion where one source of family income was until very lately weaving "plush" for royal coronation robes dyed with local blackberries, time seems to have obscured the Puritan image. Indeed, Puritans were most opportunely forgotten when Lord Saye and Seele hastened to greet Charles II. But in East Anglia the old wool merchants' dark-timbered houses, some with gaily glazed coats of arms of their owners set into the diamond-paned casements, seem to have set the pattern for building in New England.

To say these men quitted England because they could not rise politically there, or achieve authority, is to miss the point. Similarly, to see the migration as primarily an economic movement is to judge them by our modern selves. Economics is today a greater force in the shaping of mankind than religious belief. It is hard for this age to understand a Washington or a Lafayette, let alone a group of determined Puritans set upon improving their church by starting afresh upon another continent. That they were asked to take with them, and willingly did so, many comparatively impoverished hangers-on does not make the venture any the less.

The weaving and interweaving of the small seventeenth-century world is astounding to us who are accustomed to feel our own world grow suddenly smaller if we unexpectedly meet a friend of a friend far from home. It seems as if everyone at the beginning of the English-speaking seventeenth century is bound sooner or later to cross the path of everyone else in it, inherit his property, marry his widow, sail in his ship, or copy his books. The dovetailing becomes very fine. Consequence and inconsequence are neatly fitted to form odd conclusions. Sir Kenelm Digby's flagship the *Eagle* served that royal pirate well during his two years of sailing about the Mediterranean, with Letters of Marque, courting royal favor and getting himself and his country into a peck of trouble until he captured a final plum of several treasure-laden ships. Within a year of his triumphant return the *Eagle* was purchased by the wise and provident Matthew Cradock and re-named the *Arbella*. And upon her — with the same captain she had in her pirate days — John Winthrop sailed. His provident but unsanctioned bringing of the charter to accompany the settlers instead of leaving it, as would have been usual, in England with the Board of Governors was one of the important influences in the formation of our government. To some of the governors in England it may have seemed all too much in keeping with the earlier spirit of the ship.

It was a smaller world, even in New England, than it is today. Everyone knew everyone else, or who they were. The little ships labored back and forth across the Atlantic, making crossing after crossing after crossing. It was well known who was the most attentive and obliging captain, and which crew could be counted upon to take the best care of passengers and animals and plants. Occasional loads of passengers appeared without sufficient provisions, and the already hard-pressed inhabitants of Plymouth or Salem became even harder-pressed, but on the arrival of such a load, they shared what they had. In one case, a whole group of indentured servants were freed to fend for themselves, because there was not food enough to meet the terms of their agreements to serve. A stiff price for some to pay for their freedom, but there is no record that they suffered for it. And a hard

One of the few plants illustrated in *The Countrie Farme*, and that not very well, nor even well defined. There was confusion among experts upon both plants and remedies.

loss for those who had ventured an average of fourteen pounds each to bring them over.

In the beginning the settlers had to send to England for all sorts of supplies — food, stockings, glass, seeds, nails, books and even sheepskins. But very soon, by mid-century, they were sending books to England to be published, and then beginning to publish their own here, arguments for and against all sorts of issues which today seem rather remote. Naked truth remained always the stated goal; to help others, the avowed reason for publishing.

With all this literary industry chiefly on the part of clergymen, it must have been surprising to everyone to have a modest poetess arise in mid-seventeenth-century New England and write as she pleased. Still, it is only when Anne Bradstreet writes love poems to her husband, a lament upon the burning of her house, addresses to her children — of whom she bore and raised seven — a dedication to her father, and an especially disarming little poem to her book — "Thou ill-formed offspring of my feeble brain" — that we are able to see her for herself.

Her magnum opus which she calls her *Four Times Four* deals with the four elements, four humors, four ages of man, four seasons, and ends with a monumental effort, aided by Sir Walter Raleigh, on four monarchies, somewhat disturbing her arithmetic. She shows that she is aware of that relatively recent discovery, the circulation of the blood, but we are especially indebted to her, as will be seen later in this book, for explaining in the first two sections of this epic effort, what was then common belief in the Aristotelian theories.

The remarkable thing is that this "gentlewoman" who was married in England at sixteen, and at eighteen came to the New World with Governor Winthrop and two future governors (her father, Thomas Dudley, and her husband, Simon Bradstreet) should have been able to write at all under any circumstances. Even in England she would have been considered singular, although there may have been other women who wrote poetry which never had the fortune to be taken, like Anne Bradstreet's, to a publisher in England by an admiring brother-in-law. Even he, in his preface to the little volume published in 1650, feels obliged to emphasize that the poems were written, not, he says, when she was neglecting her "discreet managing of her family occasions" but were the "fruit of some few hours, curtailed from her sleep and other refreshments." Without the apparently necessary defensive explanation that she took no time from her duties to write, it was important to her to show she realized her place as a woman. Even one of her admirers, Nathaniel Ward, ends praise of her with "Let Men look to't, least Women wear the Spurres." Anne Bradstreet discreetly averts such criticism in her prologue:

> Men have precedency and still excell
> It is but vain unjustly to wage warre
> Men can do best and women know it well. . . .

It is true in this same prologue that she shows her "gracious demeanor, her eminent parts, her pious conversation, her courteous disposition, her exact diligence in her place . . ." which her brother-in-law has commended in his introduction to her book.

"I am obnoxious," she says, "to each carping tongue who says my hand a needle better fits. . . ." and she acknowledges men's "Preheminence in all and each," begging only "some small acknowledgement of ours." And then she makes her crown.

> And oh ye high flown quills that soar the skies
> And ever with your prey still catch your praise,
> If e're you daigne these lowly lines your eyes,
> Give Thyme or Parsley wreath, I ask no bayes. . . .

A pity that she did not give us more of the thyme and parsley garlands from her own experience. For these we must search her *Meditations*, in prose, composed for her son Simon at his request. Apparently, Simon had felt, like us, that his mother had more to offer than her conventional poems and he asked her to set down something in writing to him to look upon "when you should see me no more." She has, she says, set down "nothing but myne owne." There are two observations having to do with the up-bringing of children which are as valid today as they were then.

Meditation X. Diverse children have their different natures; some are like flesh which nothing but salt will keep from putrefaction, some again like tender fruits that are best preserved with sugar; those parents are wise that can fit their nurture according to their Nature.

Meditation XXXVIII. Some children are hardly weaned, although the teat be rub'd with wormwood and mustard and bitter together; so it is with some Christians, let God embitter all the sweets of this life, that so they might feed upon more substantial food, yet they are so childishly sottish that they are still huging and sucking these empty breasts, that God is forced to hedg up their way with thornes or lay afflictions on their loynes, that so they might shake hands with the world before it bid them farewell.

Anne Bradstreet died in 1672 and no one knows her grave, although Cotton Mather avers that her "Poems, divers times Printed, have afforded a grateful entertainment unto the Ingenious and a Monument for her Memory beyond the stateliest Marbles." Somehow it defines the intimate convolutions of their world to note that her husband, four years after her death, married the widow of a fighter of the Indians killed storming the Narragansett Fort, another "Gentlewoman of very good birth and education, and of great piety and prudence, who was the daughter of Emanuel Downing, friend and brother-in-law to Governor Winthrop, Senior."

Anne Bradstreet was not alone in being a remarkable woman. Although she was certainly the most talented, there are other Puritan women to stand beside her. Two of these are clergymen's wives who had their own especial occupations as well as child-rearing and housekeeping. Eminent historians have been content to allocate the administering of herbal remedies among the early settlers to old women with a few simples and cures to give out for a small fee, like gypsies. Actually, when it was not the governors or the leading clergymen who undertook to administer healing, in the absence of or even in conjunction with, recognized doctors, it was frequently their wives. Of the two outstanding examples of this reliance upon capable women in the colony to use their talents outside their homes, one was Mrs. John Cotton and the other, Mrs. John Eliot.

When the second John Cotton was pastor of the church in Edgartown and receiving in addition to his salary an honorarium from the London Society for the Propagation of the Gospel in New England, his wife was paid the sum of ten pounds for her services for one year to the natives in "Physicke and Surgery." In this she seems to have succeeded "a Mrs. Bland" who, with two Indian interpreters and schoolmasters and Peter Folger, formed the staff of workers of Worshipful Thomas Mayhew, Esquire, in his first year as a missionary on Martha's Vineyard. To judge the value of Mrs. Cotton's monetary reward, it can be noted that Peter Folger's salary for one year was twenty-five pounds. Mayhew himself received twenty. What Mrs. Bland received, I do not find recorded, but the inference is clear, Mrs. Bland filled a recognized place in the missionary effort.

Another Puritan woman of repute and accomplishment was the wife of John Eliot, translator of the Bible for the Indians and our most indefatigable and best-remembered worker among them. His wife seems to have

fulfilled indeed that earliest requirement of a wife — "a help meet for him."
Engaged to marry him before he left England, she came out with friends
soon after and married him in the New World. As Cotton Mather puts
it: "He left behind him in England a vertuous young gentlewoman whom
he had pursued and proposed a marriage unto; and she coming hither
the year following that marriage was consumated in the month of October
A.D. 1632." Cotton Mather has much to say of her grace and godliness:
"Her name was Anne and gracious was her nature" . . . "a rich blessing,
not only to her family but to her neighborhood. . . ." He then excuses
himself from "further epitaphs upon that gracious woman," to devote
thereafter all his many, many pages to Eliot, who appears to have prayed
aloud as easily and almost as frequently as he breathed and to have been
a most engagingly enthusiastic worker, one to whom all things seemed
possible, as befitted anyone who could first give the Indians a written
language and then translate the Bible into it. Until, still in praise of Eliot,
we pick up one of the signs of Mrs. Eliot's grace.

"It was an extreme satisfaction to him," writes Mather of Eliot, "that his
wife had attained unto a considerable skill in physicks and chyryugery,
which enabled her to dispense many safe, good, and useful medicines unto
the poor that had occasion for them, and some hundreds of sick and weak
and maimed people owed praises to God for the benefit, which therein they
freely received of her. The good gentleman her husband would still be casting oyl into the flame of that charity, wherein she was of her own accord
abundantly forward thus to be doing good unto all; and he would urge her
to be serviceable to the worst enemies he had in the world. . . ."

While Cotton Mather seems to us to shower his credits rather unevenly,
Eliot's charitable impulses were also rather one-sided. On one occasion the
cloth in which Eliot was carrying home his recompense after service was
tied in knots to prevent his giving any away before he got home. Finding a
poor person in need of help and failing to untie the knots, Eliot gave away
the whole sum, since, as he said, the Lord appeared to intend it. Although
he had few enemies and would seem to have been fairly safe in urging his
wife to treat them all well, on one occasion he made a really bad enemy with
some remark he made in a sermon. The man "did passionately abuse him
for it, and this both with speeches and with writings." This reviler soon
after wounded himself dangerously and Mr. Eliot "immediately sends his
wife to cure him." This she did accordingly and refused any pay when,

cured, he came to thank her. Eliot, says Mather, as if this was not enough, insisted the man "stay and eat with him." And by this conduct "he molli- fied his reviler." A helpmeet indeed was Mrs. Eliot.

There must have been many women who deserved immortalizing in "this wildernesse," but even the most attractive reputations are usually fleeting and subject to varying interpretations by later and ever later historians. Yet, one of the oldest and most charming of reputations in this country — that of being the first lady who delighted to garden in Connecticut — has never been disputed or even dimmed by frequent quoting. It depends only upon a line in a letter from her husband, but it gives us a picture of a charming and clever person, and of marital cooperation and companionship in a task which was vital to the common welfare in a new world.

"We both," wrote George Fenwick, in the little colony of Saybrook, to John Winthrop in 1639, "desire and delight much in that primitive imploy- ment of dressing a garden. . . ." By "primitive" he meant man's earliest and oldest occupation. The seventeenth century was familiar with Genesis, "every herb" and "every tree" being created "for meat," and in seventeenth- century New England, gardens and orchards were the first necessities of life. To "desire and delight much" in "dressing" a garden, however, implies both skill and experience. And to "Lady Fenwick" belong the laurels for being the first recorded lady gardener in America. Curiously enough, this is one of the few firm facts about her, whose name was not even "Lady Fen- wick." Her grave has been moved to make way for a railway which has also been removed. The site of her house and garden are only to be surmised. Her date of death is not exactly known, and remains as roughly approxi- mated as her name on her tombstone. And yet her reputation is happily still with us as a reminder of the women who came with their husbands from easier lives overseas to cope with settling and holding a new world.

In the letter from which this line is taken, George Fenwick begins by thanking John Winthrop "kindly on my wife's behalf for your great dain- ties" and after saying that they both delight in gardening he adds that "the taste of so good fruits in these partes gives us good incouragement." There is another letter two years later in which he says, "I have receaved the trees you sent us for which I heartily thank yow . . . I am prettie well storred with chirrie and peach trees and did hope I had had a good nurserie of Apples you sent me last yeare but the wormes have in a manner distroyed them all as they came up. I pray informe me if you know any way to prevent the like mischiefe for the future." Which must surely place George Fenwick himself in one sort of first place among American gardeners.

John Winthrop the Younger had been commissioned for one year in 1635 by the English patentees for Saybrook to "endeavor to provide" a work force for "makeing of fortifications and building of houses . . . ," some to be "such houses as may receive men of qualitie which latter houses we would have to be builded within the fort." Saybrook was named for two of its aristocratic backers, Lord Saye and Seele and Lord Brooke. They were combining a business venture with assurance of a place to which they could flee if necessary because of their Puritan sympathies. Winthrop was also instructed to make sure that there was good ground of one thousand or fifteen hundred acres "reserved unto the fort for the maintenance of it." He seems to have demurred somewhat, apparently, about putting so many houses inside the fort, probably to his father first, who gently indicated he had better do as he was told by those "who imploy you."

Winthrop sent men to Saybrook to work through the winter and joined them in March 1636. In April there arrived by the ship *"Bachelor*, whom God save," Lion Gardiner who had been hired by the same group who engaged Winthrop. Gardiner had been employed at fort building for many years in Holland and brought with him much ironware, portcullis chains and all things he deemed necessary for fort building, of which a list was sent to Winthrop. In May of this year, George Fenwick came "suddenly" from London, as agent for, and himself one of, the patentees who had hired both Winthrop and Gardiner. He assured Winthrop that his coming was not to "dissolve" Winthrop's commission and urged him to "procure what shall be for your comfortable continuance." And he adds that, concerning Winthrop's "resolution to keep the fort intire within itself," he must not "care though it be displeasing to some." In spite of these assurances Winthrop appears to have felt his commission for one year was sufficient under the circumstances, and he left to settle himself in the future New London. Fenwick then sailed for England and left Gardiner quite literally to hold the fort and to extract what amusement he could from tricking the prowling and attacking Indians by leaving doors with nails driven through them pointing upward along the trails where the Indians sneaked to attack the fort. With Indians hanging drawn-and-quartered Englishmen beside the river as a warning to others, and Gardiner allowing an Indian to be split in half (he says it took twenty men) in retaliation, even what Gardiner considered light touches were rather grim. But it was back to this sort of situation that George Fenwick elected to return and bring his new wife in 1639.

Alice Apsley Butler Fenwick, the daughter of Sir Edward Apsley, was of the same class and type of English country gentlewoman as the wife of John Winthrop the Elder, used to life in a rural manor to which the surrounding countryside could look for advice, help and medicine. She had married John Butler, son of a future baronet whom he predeceased. However, our first Connecticut lady gardener appears in Thomas Lechford's *Plain Dealing or Newes from New England*, published in 1642, as "the lady Boteler," which is a rather haphazard designation for the wife of Master Fenwick but does imply a title which some seem to have loved even then.

Though Lechford did not visit Connecticut, he tries to write as if he had, and he appears well informed. Of Saybrook he says, in 1641, "and master Fenwike with the Lady Boteler at the rivers mouth in a faire house and well fortified, and one master Higginson a young man their chaplain. These

A "draught" of a knot from *The Countrie Farme*.

Plantations have a Patent; the Lady was lately admitted of master Hookers church, and thereupon her child was baptized." Master Hooker had but lately arrived to found Hartford after walking overland with the huskier members of his parish and their cattle from Charlestown in the Bay Colony, while the Reverend Whitfield had arrived on the same ship as the Fenwicks. George Fenwick had allowed him to take Alice Fenwick's share of cattle with him, since apparently Whitfield had none of his own and would need some to found relatively nearby Guilford. One wonders why Alice Fenwick went so far to join a church and have her child baptized. Seven weeks on the high seas must have proved trying in many ways.

Fenwick had appropriated some cattle from those consigned to Winthrop, who apparently owed him money for a loan which Fenwick then canceled. The Fenwicks would also seem to have had a less usual sort of stock. From the speeches made when her grave was moved in 1870, we learn, as quoted from other of her husband's letters, that she kept rabbits. She was said also to have a "shooting gun," which would seem to mean she had brought a gun to shoot game, of which there was a great abundance. However, as Lion Gardiner advised no one to go out shooting wild fowl without an escort to watch for Indians, the sport cannot have proved very entertaining or relaxing. We learn, also, from these speeches, that she cultivated fruits and flowers and — best of all — that she was "cheerful."

And that is all we know about her. Her two small daughters went back to England with their father after she died, probably in 1645, and was buried "within the fort." By 1648 George Fenwick was a colonel fighting on the Parliamentary side in England. He had sold the interests of the patentees of Saybrook to longer- and better-entrenched colonies upstream, but he remembered Connecticut in his will when he died in 1657, in England after remarrying and serving the public cause there with distinction.

The shores of the Connecticut river above the little headland look as wild today as they did when "Lady Fenwick" dressed her garden and tended her fruit trees on the point of land between the two small coves whose neck could be so conveniently fortified. Nothing man-made remains except her tombstone and her bright reputation.

And here may be the place to give an idea of the size and scope of the establishments some of these early-settling ladies might expect to preside over. We have seen Endecott appropriating the "Greate House" which had belonged to the "Old Planters" headed by Conant. References to a "greate house" in each settlement are frequent. We have a good description of what comprised such a "Greate House" in seventeenth-century New England from the will of Richard Cutt made in 1675 at Portsmouth, "in Piscataqua." Richard Cutt and four others had petitioned the General Court at Boston in 1653 on behalf of fifty or sixty families for a definite township to be called Portsmouth instead of the plantation then named "Strabery Banke" — "accidently so called, by reason of the banks where straberries was found in this place." The General Court appears to have had misgivings "because of Mr. Mason's claim on the land," but these were overcome and the township of Portsmouth was that same year declared to reach from the "sea by Hampton lyne to Wynnacot river, leaving the proprietors to their just right." The petitioners had asked for "the necke of land beginning in the great bay at a place called Cotterill's delight," and "soe running to the sea." It is to be hoped that the place with the charming name was included in the final decision. In any case, by 1675 Richard Cutt, one of three Welsh brothers who had arrived before 1646, felt secure enough in possession of the Great House, which had been built in 1631 as a part of Mason's property, to make careful disposition of it in his will.

In 1644 it was lived in by an assistant to the Governor, both appointees of Mason, but, the assistant having left for Port Royall with a quantity of arms and stores belonging to Mason, the house passed into the possession of Richard Cutt in 1646. Originally Mason's plantation had consisted of the Manor, the Great House, and the Saw-Mill, three separate establishments, and apparently, to judge by Saybrook, a typical seventeenth-century set-up for an early settlement. The Great House appears in detail in Richard Cutt's will, bequeathed to his "beloved wife Elenor Cutt," as his "new dwelling house, with the bake house, brew house, barn, and all housing thereunto belonging; with the log warehouse and wharfing, my stone warehouse only excepted" (this is kept out to go to a daughter), "together with my garden, orchard, and all the land in fence in the home field, adjoining to my house; as also my corn mill, with my house and barn up at the creek, with all the upland and meadow thereunto belonging, excepting the tanyard and building thereunto belonging so far as home . . . furthermore . . . all my plate, brass, pewter, iron, beding, utensils belonging to the house, together with all my stock of cattle . . . and the five Negro servants." Richard Cutt died in 1676. His brother John Cutt died in 1681 leaving his widow, Ursula, a second wife, five hundred pounds and the "Liberty to dwell" in his house until his son should need it, or the "use of that land at ye Pulpit" — a rocky point up the river — "and build upon it as she pleases." Apparently, she did please, although this land lay two or three miles away. In 1694 she was surprised there by Indians while three men were haying and she herself was occupied near the house with her maid, preparing to make all ready before she should leave because of warnings of Indians at Dover. The maid escaped, but the hayers and Madam Ursula were all shot and scalped. Her hands were severed and carried away since the Indians could not remove the rings.

It was not for self-aggrandizement that "Great Houses" were made so all-encompassing and self-sufficient. Each was quite a province for a New England housewife to manage as a widow, although the old wills show the husbands' total faith in their competence. Seeing the housewives coping with the garden and still-rooms and kitchens and bake and brew houses must have reassured them. Like the forts and garrison houses and the closely settled small towns, they might have to bear the impact of sudden gatherings when their gardens and orchards and "house fields" would have to serve as refuge and sustenance for half the countryside.

And where among these varied folk do we put handsomely bewigged, ascetic soul-searcher, Cotton Mather? Born in the New World into a family of great distinction, son and grandson of distinguished scholar clergymen who were, like him, the very epitome of Puritan divines, his life reached into the next century where they, as well as he, were beginning to be discredited. Dedicated religious historian of all the facets of the founding and founders of New England, he kept scientific pace with his times, subscribing to the newest literature on land management, "keeping saile" with the Royal Society in London, and — most fore-reaching of all in his time — introducing inoculation against smallpox, learned from his Negro slave. His conviction of the reality and danger of witchcraft was as real as his appreciation of the menace of smallpox. He foresaw the relationship between the ills of the mind and those of the body. And, if he was never a practicing gardener himself, he certainly depended upon the industry in others. Eager to present the wonders of the New World to the Old, he corresponded regularly with the Royal Society on every possible subject. His presentation of six new plants, found in New England but not mentioned in herbals, ranks him with John Josselyn as one of our early botanists, howbeit his interest in plants was primarily medicinal.

This handful of characters could be duplicated again and again, although these few are the most helpful to us in replanting their gardens. From the younger Winthrop, unashamed to be seen hoe in hand, to the gentle Fenwicks, absorbed in garden planning, to the dynamic Cotton Mather, writing down all his remedies for body and soul in his *Angel of Bethesda*, these are all great people, on this "too scanty a stage."

5

"In Equal Tempers"

While the great Puritan migration was gathering force to sweep across the Atlantic with its unique accumulation of able and determined individuals united by one overriding purpose, other fresh currents and tides in exploration and experiment were rising and swirling in new patterns in the English and European world. Their just due has always been given to monasteries for preserving European culture; but the barbarians have long lacked credit for contributing enlightenment in matters other than theological and philosophical. In fact, the rebirth of European man was most ably assisted by these non-Christians who had been busily advancing man's scientific knowledge of himself and his universe while the monks guarded what seemed to them all the knowledge worth preserving. Hazards of travel and barriers of language had strengthened the assumption in Europe that knowledge was a relatively stable commodity needing only the protection of a privileged few. With increasing facilities in printing and publishing, and the translation of books from Greek and Latin into English, it became apparent to many that there was a great deal of thinking still to be done and that every man capable of thought was free to have a try. Through new translations, the wisdom of the ancients was made available, as well as their myths and errors, and joined to all this were the contributions of "barbarians" in the then barely recognizable infant sciences, alchemy among them.

It was a rich and promising world for original thinkers and experimenters from which the Puritans sailed away to form a perfect state in the eyes of God. They were as intellectually alive and curious as any of those left behind them. They may, indeed, have been more truly vital, since their convictions and surmises had carried them so far and so fast in what was, in fact, one of the great practical experiments of the modern world. Secure in their belief that the Bible was God's directive to mankind, and armed with their interpretations of the knowledge of the ancient world as part and parcel of God's will, they left behind them a world ayeast with questioning of hitherto accepted facts and athirst for fresh revelations of man's true relationship to his environment. Cut off by the still very formidable ocean, hardened by the difficulties of survival, marooned intellectually without the easy polishing of one man's mind by another's of a different texture, the Puritans strove to remain true to their original purpose while the rest of the world moved on. That they accomplished their purpose is apparent even today as proof of their originality and power. Those among them who retained the curiosity and ambitions of renascent man could hardly be expected to be able to keep up with all that was being experimented with elsewhere. The wonder is that they were able to accomplish so much that was new in the whole world then and has endured to this day in our systems of government and universal education.

Their accomplishment is all the more remarkable when one considers the burden of ancient and outmoded thinking with which they were encumbered. This is not to say that the New England Puritans alone were caught in time. Many people in the England of the seventeenth century were unable to think beyond medieval concepts of man and his world. Because the enormous and sustained difficulty of their undertaking froze our country's forebears into the rigidity of thought for which they are reproached today, it becomes all the more incumbent upon us to explore their fixed pattern of beliefs in trying to understand them and estimate their great contribution. Even to be able to reconstruct their gardens, it is essential to make, however cursory, an observation of their background beliefs.

Reliance upon the Bible is implicit in the very fact of the settlers being here at all, to build their "City upon a Hill." That God created the world for the use of man is the basic assumption not only of the colonists but also of all the herbalists and explorers and chroniclers. The additional dependence upon the lesser beliefs of the time is harder for us to understand than their faith in a paternal Almighty whose slightest wish for them it was their duty to ascertain and fulfill. These other beliefs seem to us now so diverse in import and origin as even to conflict with one another, but to those who held them, they seemed completely compatible and parts of God's original plan for the universe and man's place in it.

After the universal acceptance of the idea that the world was created by God for the use of man, the next most widely held belief of the whole nature of man's place in the universe stemmed from those whom the most eminent herbalists described as "the ancient heathens," the Greeks. Through Galen, a second-century Greek physician in Rome, Aristotle's theory of the universe had become common knowledge. Aristotle had seen the universe as composed of matter impressed by form to produce the four elements of fire, air, water and earth. Each of these elements had two of the four primary qualities, moist, dry, hot and cold. However, as two of each of the four qualities are opposites and cannot be coupled, the four possible combinations are: hot and dry, assigned to fire; hot and moist, assigned to air; cold and moist, assigned to water; cold and dry, assigned to earth. One quality dominates in each combination. Each element can pass into another through the quality they share in common. Thus, fire can become air through heat. Two elements can become a third by removing a quality from each, if the two remaining qualities are not identical or contrary. Thus, fire and water, by giving up the dry and cold qualities and retaining the hot and fluid, can become air. The basic matter, from which the elements are made, always remains the same.

From this system is derived the theory of the "Four Cardinal Humours," or the four chief fluids which supposedly make up the body: blood, phlegm, choler and melancholy or black choler. By their relations and proportions, an individual's physical and mental qualities were determined, his "animal temperament." A preponderance of any one would produce a bodily habit or temperament, called, in the same order: sanguine, choleric, phlegmatic and melancholic. These four states, or "tempers" were, in turn, complicated by being hot or cold, moist or dry. It was a fortunate man who could maintain them in due proportion. In fact, to help his patient achieve that balance was the province of every physician.

Galen, upon whose edicts the settlers so completely depended, was familiar with the schools of Greek philosophic thought, had studied medicine and traveled widely. He settled finally in Rome where he was court physician to Marcus Aurelius and made part of his practice tending the wounds of the gladiators. He acknowledged his debt to Hippocrates and to the anatomists of Alexandria, that seat of learning and experimentation, and today is himself regarded as the founder of experimental physiology. A firm believer in botanical remedies, he dominated medical thought and practices for nearly two thousand years, enjoying a great resurgence in popularity when his works, long preserved chiefly in Arabic and by the Arabs, were translated back into Greek and Latin in the fifteenth century. The Latin translation continued to be studied in our medical schools through the eighteenth century.

From the custody of the Arabs to the custody of the New England Puritans appears now to have been no jump at all for either the ancient herbalists or the Puritans. The beginnings of chemical medicine had also been nurtured by the Arabs, but these were not to have wide acceptance until long after their introduction by Paracelsus. Galen as traditional physician became almost universally accepted, a household habit.

How much these ancient philosophical theories were part of the life of the New England Puritan can be judged by quotations from Anne Bradstreet, the "Tenth Muse." In her *Four Times Four* she presents as her characters the four elements and the four humors, declaring:

> These are of all the Life, the Nurse, the Grave,
> These are the hot, the cold, the moist, the dry,
> That sink, that swim, that fill, that upwards fly,
> Of these consist our bodies, Cloathes and Food,
> The World, the useful, hurtful, and the good.

She goes on to explain:

> My first do show their good and then their rage,
> My other four do intermixed tell
> Each others faults, and where themselves excell:
> How hot and dry contend with moist and cold,
> How Air and Earth no correspondence hold
> And yet in equal tempers, how they 'gree
> How diverse natures make one Unity. . . .

She describes the elements contending for the first place and makes reference to alchemy, our next consideration in the background beliefs which so influenced the Puritan settlers.

"Fire" says:

> And you Philosophers, if 'ere you made
> A transmutation it was through mine aid.

and

> Ye Paracelsians too in vain's your skill
> In Chymistry, unless I help you still.

"Earth" speaks of:

> The rare found Unicorn,
> Pysons sure antidote lies in his horn.

Which justifies Governor Endecott for having offered a horn to Mrs. Winthrop when whe was ill and John Winthrop, Junior, for possessing one himself, just in case.

"Earth" also addresses merchants who send forth well-manned ships "where sun doth rise," and adds:

> Ye Galenists, my drugs that come from thence,
> Do cure your Patients, fill your purse with pence;
> Besides the use of roots of hearbs and plants,
> That with less cost at home supply your wants.

"But how," Earth says, "my cold dry temper works upon the melancholy constitution."

"Water," after preening herself on her charms, including "the Dolphin loving music," admits to:

> What ill there in me lies.
> The phlegmy constitution I uphold.
> All humours, tumours which are bred of cold.

"Air" says that, although she is considered last, she knows her place is first:

> The ruddy sweet sanguine is like to air
> And youth and Spring, Sages to me compare,
> My moist hot nature is so purely thin,
> No place so subtilly made, but I get in.
> I grow more pure and pure as I mount higher,
> And when I'm thoroughly rarified turn fire
> So when I am condensed, I turn to water,
> Which may be done by holding down my vapour,
> Thus I another body can assume
> And in a trice my own nature resume.

Alchemy might not seem relevant to a study of the gardens of New England Puritans, and yet this combination of infant chemistry and ancient residue of suspect magical arts had a direct bearing upon the everyday life of the average New England housewife, endeavoring to survive and succeed in "this wildernesse." It was on her knowledge, conscious or not, of alchemy that she would have to draw to make her salves and syrups, her waters and plasters and "Oyels" and fermentations.

While John Winthrop, Junior, dedicated alchemist, was lending some of his vast library of alchemic books to the son of Governor Brewster of Plymouth (who was seeking to find the philosopher's stone in the wilds of Connecticut), he was also corresponding, sometimes rather wistfully, with his friends of experimental bent in Old England. On a simpler schedule of experiments, the Puritan housewife was working with her own rudimentary equipment and whatever maids or friends she had to help her in her well-named "still-room." When she made her dyes for the stuffs she spun and wove, she had to depend upon vegetable and plant concoctions, fixed with mordants tested by her own experimenting with metals and minerals. In fact, a book listed in a Boston inventory of 1660 as *Baker on Gardening* turns out to be actually a treatise on "the wonderful hid secrets of nature touching the most apt formes to prepare and destyl Medicines for the conservation of health and Quintessence, Aurum potabile, Hippocras, Aromatical wynes, Balmes, Oyles, Perfumes, garnishing waters and other manifold excellent confection. Whereunto are joyned the formes of sondry apt Fornaces and vessells required in the art. . . ." In fact, just another facet in the life of the New England housewife.

Baker on Gardening can only be by George Baker (1540–1600) surgeon, and a member of the Barber Surgeon's Company of which he was elected master in 1597. He was retained in the household of the Earl of Oxford, and his first tiny book had an enormous title which promised to tell how to make a "pretious oil" called Oleum Magistrale, and to include the whole third book of Galen which would show a method of curing wounds and how wrong everyone else was on the subject. After preparatory praises he gives the recipe for Oleum Magistrale, "formerly known as Aparice." Here would seem to be the place to hand it on. So: Into a quart of the best white wine and "Oil Olif," pound varying quantities of both the leaves and flowers of "Hipericon," "Cardus benedictus," "Valerian" and "Sage." Boil them all in a copper kettle until the wine is consumed, strain, add Venice

¶ The newe Iewell of Health, wherein is

contayned the moft excellent Secretes of Phificke and Philo-
fophie, deuided into fower Bookes. In the which are the beft ap-
proued remedies for the difeafes as well inwarde as outwarde, of all the partes
of mans bodie : treating very amplye of all Dyftillations of Waters, of Oyles,
Balmes, Quinteffences, with the extraction of artificiall Saltes, the vfe and preparation of Antimonie, and potable Gold. Gathered out of the beft and moft ap-
proued Authors, by that excellent Doctor *Gefnerus.* Alfo the Pictures, and maner
to make the Veffels, Furnaces, and other Inftruments therevnto belonging. Faithfully corrected and publifhed in Englifhe,
by George Baker, Chirurgian.

Printed at London, by Henrie Denham.
1 5 7 6.

The embattled Alchymia among her limbecks and furnaces. A title page from one
of the books comprising George Baker's *Jewell of Health.*

turpentine, boil again, and repeat with "Ohbanum," "Mirrah" and "Sanguis draconis." Let it cool, bottle it and set it in the sun a few days, then put it away, ready for use for practically anything. I am not identifying the plants here for reasons that George Baker goes into in his next book, published in 1576, and several times thereafter, and because all plants will be defined together later in my Appendix.

Baker's next book is the *New Jewell of Health*, "wherein is contayned the most excellent secretes of Phisicke and philosophie. Faithfully corrected and published in English." It is, in fact, the *Evonymuns* of Conrad Gesner,

❡ The fourth Booke of Dyſtillations,
conteyning many ſingular ſecrete
Remedies.

A title page from George Baker's *Jewell of Health* which suggests the magical associations with alchemy.

Swiss father of German botany. In his preface Baker defends himself for turning it into the vulgar English tongue, but he has made a reservation, like mine above. "As for the names of the simples," he says, "I thought it good to write them in the Latin as they were, for by the searching of their English names the reader shall very much profit, and another cause is that I would not have every ignorant asse to be made a chirurgeon by my book for they would do more harm than good."

It is interesting to note here that when he speaks of "Philosophie" he means what is called "the chemical art" or distillation. He is sure that distilled medicines exceed all others in power and value. He says, "for three drops of sage doth more profit in the palsie, three drops of oil of coral for the falling sickness, three drops of oil of cloves for the colicke than one pound of these decoctions not distilled."

Alchemy as a study, an art or a practice is as old as man himself. Sometimes referred to as the "Egyptian art" in medieval Europe, it had associations with black magic and was so popularly suspect that many of the researchers found it prudent to disguise their operations and materials under a whole set of curious terms, recognizable only to the initiate — as witness when Jonathan Brewster writes to John Winthrop, Junior, of "head of the crowe" or "raven's head," which is lead, and of "Virgine's Milke," which I will guess is silver. Unfortunately, such practices must have confirmed the idea of alchemists as being close to medieval spellbinders, a view quite obvious in the illustrations of even the most serious works on the subject, where cuts depicting alchemists at work in laboratories seem to derive straight from the pointed-hatted, star-embroidered, heavily gowned, bearded ancients we recognize as magicians. Although the chief aim of alchemy in the popular mind was to change base metals into gold, it had a far deeper significance for its dedicated practitioners. This was the search for the Alkahest, the Philosopher's Stone or universal panacea. The Puritans were able to sympathize with one of the basic tenets of alchemy, though its import as a philosophy trying to turn into a science might escape them. Only the pure in heart, the alchemists had always believed, could truly succeed in finding the Philosopher's Stone.

Alchemy can be said to have arrived in Europe after spending the Dark Ages being hatched into science by the Arabs, who had had it from the Alexandrians of the early Christian era, who had in their turn earlier received it from the Greeks.

In Alexandria all the then-known scientific crafts, such as dyeing, glass-making and metal-working, were combined to become the source of alchemy. Called also "the Hermetic Art" (from which we derive today's "hermetically sealed"), alchemy recognized as its founders a small confusion of three original heroes, all named Hermes. According to Arab chroniclers, Hermes I, called Trismegistus, was a grandson of Adam, existed before the Flood, and was the first man to wear sewn clothing (which is one up for the seventeenth-century Anglican reproached in meeting for being heard to ask a neighbor how Adam and Eve kept the leaves together). Hermes I also wrote books, built pyramids, and was the patron of mathematics and science. Hermes II came after the Flood, lived in Babylon and taught Pythagoras. Hermes III lived in Egypt, and was a town planner, a physician, philosopher, and author of books on poisonous animals and on alchemy.

Another character of importance in alchemy of interest to both early Puritans with their devotion to the Old Testament and to modern housewives today is "Maria the Jewess," said to be the sister of Moses. She discovered that liquefaction produced by a mild heat is more successful as a solvent than calcination produced by violent heat. This theory was also advocated by Sir Francis Bacon, whom the New England Puritans read, but the name of the equipment for this process is what immortalizes this first lady scientist for us — the *bain-marie*, still in use today.

In the light of modern science, it is interesting to realize that early alchemists were convinced that transmutation of base metals into gold could be effected if only they could get a sufficiently high temperature. Bellows are prehistory, but dampers were not invented until the fifteenth century in England by an alchemist, Thomas Norton of Bristol. And in that century, a chimney to blow over the fuel was introduced by Glauber, another alchemist, immortalized because his name was given to the salts *he* hoped were the Alkahest. Before Glauber, a chimney's sole function had been to let out smoke. Only the century before that, Paracelsus, whom Cotton Mather enjoys quoting, was the first to introduce to medicine the use of chemical compounds. In fact, alchemists make up a distinguished, unexpected and often unsuspected company. James I practiced alchemy in Scotland before he became King of England. Sir Walter Raleigh, to improve his time during his imprisonment in the Tower, ordered in an extensive collection of alchemical equipment. And that bright inconstant star,

Sir Kenelm Digby, who keeps turning up in various contexts all through the seventeenth century, has his alchemical overtones, as evidenced in his writings and by his friendship with Edward Howes, the lively and enquiring friend of John Winthrop, Junior.

John Winthrop, Junior, must rank alone as the leading alchemist in the new world. His initial shipment of alchemic glasses and paraphernalia is as long and as carefully worked out as his list of seeds. His library was heavily weighted on the side of alchemical textbooks, and he liked to use, as well as his signature, his "mark," the odd little figure that had been the monadic emblem of the great Dr. Dee, alchemist above all others in sixteenth-century England. Combining all the alchemical symbols of the seven basic metals, it looks like a squatting little round-faced man wearing a small, turned-up hat. In his "case book" where he records his treatments and his patients, John Winthrop, Junior, uses the symbols of the metals separately. These symbols also appear in the little manuscript "Dispensatory" of Edward Taylor, a Connecticut poet and divine, who was also physician to his back country parish. Without trying to make more than is necessary of the beliefs and goals of alchemy, it must be recognized that this pseudoscience in its wider reaches sought to establish the relationship between visible and invisible, spiritual and material, the macrocosm of the universe and the microcosm of man's world.

Considering this vast interrelationship which the Puritans tried to consider and rationalize, it should be small wonder to us today that witches occasionally played a part. In the search for physical conditions in the invisible and spiritual world, and vice versa, the possibility of a spirit or power in inanimate objects could not be ruled out. This animism, such as one still finds in primitive peoples, may have been handed down from the earliest Greeks. From them, too, may have come the idea of the sun as male and gold, the moon as female and silver, and the planets as various metals. The Romans substituted their own names, and we have Venus represented by copper, Mercury by mercury, Mars by iron, Jupiter by tin, Saturn by lead. If we feel amusement to see how long it took primitive beliefs to perish utterly, and to wonder how the Puritans seemed to find them easy to absorb, we have only to say over to ourselves now, thoughtfully, the days of our week.

Benjamin Tompson's "Funeral Tribute" to John Winthrop, Junior, spells out Winthrop's deep involvement with alchemy and its recognition by his Puritan peers. Here are mentioned the tools, the dedicated secrecy, the interruption only by death of the effort to discover the elixir of life, or the Philosopher's Stone. "Hermetically" in Tompson's poem refers to the methods of the art, and "anatomise" means to explain all, not to perform an anatomy or "give" one, like Giles Firmin, as mentioned later in this book.

John Winthrop, Junior, died on April 6, 1676. A few short excerpts from the tribute to "John Winthrope, Esq. a Member of the Royal Society and Governor of Connecticut Colony in New England . . . who expired in his Countreys Service," suffice to show his alchemical bent.

> Three Colonies his Patients bleeding lie,
> Deserted by their great Physicians eye. . . .
>
> Projections various by fire he made
> Where Nature had her common Treasure laid
> Some thought the tincture Philosophick lay
> Hatcht by the Mineral Sun in Winthrops way.
> And clear it shines to me he had a Stone
> Grav'd with his name which he could read alone. . . .
> His common Acts with brightest lustre shone,
> But in Apollo's Art he was alone.
> Sometimes Earths veins creeping from endless holes
> Would stop his plodding eyes; anon the Coals
> Must search his Treasure, conversant in use
> Not of the Metals only but the juice.
> Sometimes the Hough, anon the Gardeners Spade
> He deigned to use, and tools of th' Chymick trade.
> His fruits of toyl Hermetically done
> Stream to the poor as light doth from the Sun.
> The lavish Garb of silke, Rich Plush and Rings,
> Physicians Livery, at his feet he flings.
> One hand the Bellows holds, by t'other Coals
> Disposes he to hatch the Health of Soules
> Which Mysteries this Chiron was more wise
> Than unto Ideaots to Anatomise. . . .
>
> To treat the morals of this Healer Luke
> Were to essay to write a Pentatuke. . . .
>
> Now Helmonts lines so learned and abstruse
> Are laid aside and quite cast out of use. . . .

Maiſon Ruſtique,

OR

THE COVNTRIE
FARME.

Compiled in the French tongue by
Charleſ Steuens and *Iohn*
Liebault Doctors of
Phyſicke.

And tranſlated into Engliſh by RICHARD
SVRFLET Practitioner in
Phyſicke.

Alſo a ſhort collection of the hunting of the Hart,
wilde Bore, Hare, Foxe, Gray, Conie ; of
Birds and Faulconrie.

The Contents whereof are to be ſeene in
the Page following.

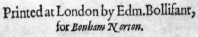

Printed at London by Edm. Bollifant,
for *Bonham Norton.*

1 6 0 0

Title page from John Winthrop, Junior's own copy of *The Countrie Farme*, a book translated from the French which became extremely popular. Copies existed in Plymouth and Boston.

This last refers to van Helmont of whom Cotton Mather was also aware and to whom he would have left all laurels undisturbed, but Tompson is eulogizing. "Balsome of his Countries Health" is another of his names for Winthrop who, incidentally, did indeed not "anatomize unto idiots" any of his remedies. He did not even divulge the ingredients of his favorite remedy "Rubilia" which was finally discovered and analyzed by modern science from a few grains of powder still on one of his papers.

Astrology, insofar as it concerns us in this book, may be defined as the practical application of astronomy as an art or science to determine the influence of the stars upon the affairs of men, more particularly here in reference to his plants and their sowing and reaping.

Astrology as the art or science of divining the fate of human beings from a study of the positions of heavenly bodies at their birth, came from the Babylonians to the Greeks to the Romans, and by the beginning of the Christian era was able to join the other pseudosciences in its full plumage. Indeed, the Greeks had considered that astrology embraced all the other then known sciences. Under its wings were included botany, chemistry, mineralogy, zoology, anatomy and medicine. Affected by the sun, moon, the five planets, the more prominent stars, and the signs of the Zodiac, were all things animate and inanimate — metals, stones, minerals, plants and different parts of the human body. Again we see evidences of this documentation in the little "Dispensatory" of Edward Taylor, our best example of a Puritan as a man of his world who would try to make use of all it offered or he could grasp.

Though the early settlers in New England refused to name the days of the week by the names of the planets, preferring to confound their contemporaries and especially their later-date readers by insisting the week begin with the Lord's Day and simply numbering the days after that, astrology was woven into their lives. It appears in all three of their chief herbals, and in the book of that practical gardener, Leonard Meager. While Bacon and Sir Thomas Browne in the Puritans' libraries ridiculed the astrologers of their day, Milton, the good Puritan whom Cotton Mather complimented by drawing upon, repeatedly refers to planetary influence. And Nicholas Culpeper, derided by his fellow physicians and repudiated by the College of Surgeons, appears to have lost no Puritan readers by ascribing every plant to its planet.

So, after Galen and the four humors, and alchemy, and astrology, any

A retort illustrated in *The Countrie Farme.*

other of the interwoven beliefs and practices of the seventeenth-century New England Puritans seem child's play. And the so-called doctrine of signatures, is indeed, exactly that. This is another of the beliefs commonly held and somehow assimilated by the Puritans without in the least conflicting with their religious convictions. Actually, the doctrine of signatures stemmed directly from these last. Besides creating herbs for the use of man, God was thought to have left little clues in the leaves of the plant, or its flowers or roots or juice, as to the disease or the organ for which it was intended, as see liverwort's leaf-shape and the spots on the leaves of lungwort, poppies for hemorrhages and agrimony for jaundice, the root of the mandrake for sterility, birthwort for diseases of the uterus, and so on. Popularized by an Italian botanist named Porta in the sixteenth century in Italy, the doctrine of signatures has been called a part of the earliest folklore about plants, and is a theory of "resembling tokens."

Another primitive or even prehistoric belief which governed the choice of plants to be cultivated and cherished was a conviction that whatever the affliction, the remedy would be found close by. So for rheumatism, aggravated by living in damp and swampy areas, there was the willow bark, a discovery which has paid off, since the aspirin of today contains as a main ingredient the chemical substitute for the salicylic acid found to be a relief in such cases. Digitalis, used in some heart diseases, is said to have been discovered by sufferers in England who plucked it to relieve their distress. Poison ivy, an affliction of the new country, was supposed to be cured by rubbing on the affected part whatever grew nearest to the ivy, so that even today we have people exposed to poison ivy picking dogbane or the nearest common weed and assuming it to be the antidote. Involuted thinking created some special "vertues." Stonecrop keeps off lightning because it grows on roofs, so it must be there for a purpose, and what better one than to keep off lightning? And for afflictions with unpleasant odors, one could eat a plant smelling in much the same way and hope for relief, so that even Gerard and Culpeper fall into this pit and recommend plants with most unsavory odors for troubles which are similarly ill-scented.

And, of course, to ward off witchcraft, there were plants no one could be without — plants of such well-known virtues that it was folly to dispense with them. One of these was rue, which supposedly prevented anything evil from growing anywhere near it in the garden. Another was snakeroot, only a piece of whose root would protect one against snakebite or provide a cure if one were bitten. And angelica was considered truly miraculous as well as being delicious when candied.

But magic plants were the least of the many preoccupations of the New England gardener of the seventeenth century. There was no room for superstition with so much good, solid, ancient, trustworthy information available. And the very art of collecting these valuable materials was mostly described as "simpling."

But although simpling was the common word for gathering medicinal plants in Old England, its definition in William Cole's little book on *The Art of Simpling*, published in London in 1657, includes just about everything one had to know, except about astrology, which he disdained. His subtitle is "An Introduction to the Knowledge and Gathering of Plants." According to Cole, "Simpling is an Art which teacheth the knowledge of all Druggs and Physical Ingredients but especially Plants, their Divisions, Definitions, Differences, Descriptions, Places, Names, Times, Vertues, Uses, Temperatures, and Signatures."

6

For Meate . . .

When the Puritans in New England sang together from the Bay Psalm Book:

> For beast hee makes the Grasse to grow,
> Herbs also for man's good:
> That hee may bring out of the earth
> What may be for their food.
> Wine also that man's heart may glad
> And oyle their face to bright;
> And bread which to the heart of man
> May it supply with might.

they were singing about their own lives. For, as everyone who lived in either England, Old or New, in the seventeenth century knew well, after God had created man and woman on the sixth day he said to them, "Behold, I have given you every herb bearing seed which is upon the face of all the earth and every tree in the which is the fruit of a tree yielding seed . . . to you it shall be for meat."

It was this assurance which made opening up a new world, disclosing so much more of the "face of all the earth" than had been known before, of such vital importance to the health and welfare of mankind, since there would obviously be revealed a great new store of herbs "for meate and medicine." If New England proved to be lacking in "Spanish bliss," or gold, it could make up for that lack with commodities which could be turned into gold, such as fish and timber and turpentine, and in the possibilities of new "herbs for the service of man," as the 104th Psalm phrased it before it was turned into the jingle of the Bay Psalm Book.

What did the early settlers find and how did they eat, after they had established themselves? How much did they depend upon their own garden plants grown from seeds and roots and bulbs and scions brought from England? How much upon skills learned there, and how much from those learned from the Indians in the new land?

Facts are hard to come by. Non-truths have charm. From that well-known painting of the Pilgrims wandering through snowy woods to church derives our national image of our forebears, forever cold, forever fearful. Celebrations of our own lavish Thanksgiving have transformed what was first declared as a day for fasting and prayer. We accept the often-repeated statement that the importance of pepper, in the search for which so many serendipitous discoveries were made, was due to the lack of refrigeration and the necessity of somehow swallowing tainted meat. And we allow yet another commonly accepted untruth to blur our understanding of our country's founders — that a lot of our "escaped" wildflowers arrived accidentally in feed for cattle and packing for furniture. All so sadly false, as is our general estimation of the Puritans. Brave they are allowed to be, but not bright, attractive, and good company. And if one does not like a people, one seldom likes their food. Although the early Puritans are the source of some of our most important basic institutions and national traits — our government, our system of education, and the compulsion to help others less fortunate than ourselves — scholars have never considered the early New England diet tempting, even for research. And yet to find out what they ate may be one way to find out what our founders were really like.

Mention of food, its lack and then its plenty, is frequent in early accounts, but there are few details. It is left for us to pick up crumbs. Even references to what they disliked may reveal as much to us as what they liked. For instance, when Bradford writes in his chronicle, *Of Plimouth Plantation*, that on one occasion they had only lobsters and water to offer new arrivals, we cannot be properly sympathetic until we realize that lobsters were mainly what Indians used for catching better fish, like bass and cod. Indian women had to dive for lobsters off the rocks and what the braves did not take for bait, the women smoked and dried. William Wood in his *New Englands Prospect* in 1634 says of lobsters "their plenty makes them little esteemed and seldom eaten."

On the whole, the early settlers seem not to have been responsible for our national predilection for shellfish. Clams were rated low as either a staple or a delicacy and pigs were run upon the clam flats. William Wood claims the pigs grew to know when it was ebb tide and would rush to the clam flats as if driven. John Higginson of Salem reports that "an honest man" said grace over a dish of clams. He "invited his friends to a dish of clams and at the table gave thanks to Heaven which had given them to suck the abundance of the seas and of the Treasures hid in the sands." An amusing tale to show only the wit and godliness of the host.

On the other hand, if they appear to have overlooked plentiful shellfish as a blessing for other than Indians, they were repeatedly grateful for all other fish and for a plant called the "ground-nut." As we have seen, no less an authority than Governor Bradford in his *History of New England in Verse* gets fish and ground-nuts into the third line:

> Famine once we had (wanting corn and bread)
> But other things God gave us in full store
> As fish and ground-nuts to supply our strait
> That we might learn on Providence to wait
> And know by bread man lives not in his need
> But by each word that doth from God proceed.

Any single plant which can rank with fish, the quantity and variety of which we know astounded them, must have seemed of great importance. It is a pity that we have no exact description of just which plant they meant by ground-nut. John Josselyn lists a plant he calls the "earth-nut" which he says is "of divers kinds, one bearing very beautiful flowers." Again he refers to the "earth-nut" with a "princely flower." Edward Tuckerman, editor of a reprint of Josselyn's *New-Englands Rarities* in the mid-nineteenth century, thinks Josselyn means, of our two plants called "ground-nut," the wild bean, or *Apios tuberosa*. In any case, it all goes to show that the early settlers were quite willing to grub for food and to eat the Indians' vegetable diet, even if they did not seem to care much for their shellfish.

The Pilgrims were greatly assisted also by the corn they found stored in baskets in Indian ground-cellars, and immediately appropriated. Early advocates of settling in the new land were eloquent about the Indians' corn and its phenomenal yield per planted kernel, and they were eager to learn from the Indians how to grow it (in hills with four plants to a hill and a manuring of dead fish). Francis Higginson, who had arrived in Salem in 1629, explains in his posthumously published book *New England's Plantation*, what "great gaine some of our English Planters have had by our Indian

Corne." He says they sowed it with such increase that they sold their harvest to the Indians for beaver skins and made a great profit, "where you may see how God blesseth husbandry in this land." "There is not such great and beautiful eares of corne I suppose anywhere else to be found but in this Countrey," he continues, "being also of various colours, as red, blew, yellow, etc. and of one Corne there springeth four or five hundred."

Roger Williams in his *Key into the Language of the Indians of New England* printed in London in 1643, says corn is called "Ewachimneash" and that "there be divers sorts of this corn and of the colours, yet all of it is either boiled in milk or buttered, if the use of it were known and received in England (it is the opinion of some skillful in physic) it might save many lives in England, occasioned by the binding nature of English wheat, the Indian corne keeping the body in a constant moderate looseness."

John Josselyn, like Roger Williams, is interested in the practical uses of corn. "It is light of digestion," he says, and, speaking of the New Englanders as still English, "the English make a kind of Loblolly of it to eat with Milk which they call Sampe; they beat it in a Morter and sift the flower out of it; the remainder they call Homminey, which they put into a Pot of two or three Gallons with Water and boyl it upon a gentle fire till it be like a Hasty Pudden; they put of this into Milk, and so eat it. Their Bread also they make of the Homminey so boiled, and mix their Flower with it, cast it into a deep Bason in which they form the loaf, and then turn it out upon the Peel, and presently put it into the Oven before it spreads abroad; and the Flower makes excellent Puddens." (A peel is a baker's shovel upon which the bread is thrust into the oven, which has been heated beforehand with coals.)

Another locally used grain noticed by Josselyn he calls "Naked Oats, there called Silpee, an excellent grain used instead of Oat Meal. They dri it in an Oven, or in a Pan upon the Fire, then beat it small in a Morter." With this, he says, they make one of the "Standing Dishes" of New England. (The other he mentions is pumpkin pie.) Into a "Pottle of Milk," or the equivalent of two wine bottles in quantity, "when the Milk is ready to boil" they put "about ten or twelve spoonfuls of this Meal, so boil it leasurely stirring of it every foot least it burn too; when it is almost boiled enough they hang the kettle up higher, and let it stew only in short time it will thicken like a Custard; they season it with a little Sugar and Spice and so serve it to the Table in deep Basons, and it is altogether as good as a

Spelt Corne.

White-pot. It exceedingly nourisheth and strengthens people weakened with long sickness. Sometimes they make a Water Gruel with it, and sometimes thicken their Flesh Broth either with this or Homminey if it be for the Servants."

A "white-pot" is a traditional Devonshire dish made from cream and flour and eggs and spices, also called "white pudding." The grain Josselyn has described is said by Tuckerman to be *Avena muda*, also called pillcorn or peelcorn because the ripe grains drop from the husks. This sounds a likely identification and as if the housewives had brought with them the seeds of labor-saving grain described in Gerard's *Herball*, 1633 edition, Josselyn's authority. Gerard notes this grain as the favored oatmeal of

Norfolk and Suffolk housewives, because it can be used without the aid of a mill. The *Herball* scorns the East Anglian housewives as being the sort "that delight not to have anything but from hand to mouth" since "while their pot doth seeth" they may "go to the barne and rub forth with their hands sufficient for that present time," not providing for the morrow but willing to "let the next day bring with it."

Pumpkins and squashes made the other great staple besides corn whose cultivation the Pilgrims learned from the Indians. According to Higginson, when the corn season was over, pumpkins and squashes became the Indians' chief food. Everyone remarks upon the "Pompions" and the "Isquontersquashes" of the new land, but John Josselyn, again, is our chief authority. Of the "Squashes, but more truly Squontersquashes," he says they are "all of them pleasant food boyled and buttered and seasoned with Spice." Of pumpkins, whose earlier English name he uses, he speaks more fully as he gives his second "standing dish." "The ancient New England standing Dish," he says, again in his *Rarities*, "is made from Pompions, there be several kinds, some proper to the Countrey, they are dryer than our English Pompions, and better tasted; you may eat them green. But the Housewives' manner is to slice them when ripe and cut them into dice, and so fill a pot with them of two or three gallons, and stew them upon a gentle fire a whole day, and as they sink, they fill again with fresh Pompions, not putting liquor to them; and when it is stewed enough, it will look like bak'd Apples; this they dish, putting butter to it and a little vinegar (with some Spice, as Ginger, etc.) which makes it tart like an Apple, and so serve it up to be eaten with Fish or Flesh. It provoketh Urin extremely and is very windy." Considering that Josselyn felt compelled to say also that some kinds of squashes give the eaters worms, he does not appear to have felt kindly toward that vegetable family.

As we have seen, the earliest explorers and seamen coming to fish had heralded the profusion of fruits and berries and nuts. Captain John Smith, dusting his hands of indigent Virginians and sailing along a rugged coast to name it New England, had forehandedly planted his own little garden for "sallets" on a rocky island on the coast of Maine. Careful observer of the profusion of berries growing upon a bank of the Piscataqua River, he named it "Strawberry Banke." As one who had carried the Virginia settlement through its first year by coaxing corn from the Indians, he providently noted hilltops cultivated with "Indian Gardens." Seeing all this promise of fertility, he made a note to ask the English gentry for their "spare Kindred" to form up this new settlement — no more idle aristocrats and ne'er-do-wells like those who had made things difficult in Virginia.

Of all the crops, strawberries were the most abundant. "This berry," says Roger Williams in his *Key into the Language of the Indians in New England*, published in London in 1645, "is the wonder of all the fruits growing naturally in these parts. It is of itself excellent, so that one of the chiefest doctors of England was wont to say that God could have made, but God never did make, a better berry. In some parts where the natives have planted, I have many times seen as many as would fill a good ship within a few miles compass. The Indians bruise them in a mortar and mix them with meal and make a strawberry bread." Later he reiterates, "They also make great use of their strawberries, having such abundance of them, making strawberry bread, and having no other food for many days; but the English have exceeded, and make a good wine both of their grapes and strawberries also, in some places, as I have often tasted."

We are fortunate to have a recipe for making strawberry wine from no less a personage than Sir Kenelm Digby in his *Closet*, a book of his secret recipes published, in 1669, by his manservant after Digby had died. This is only one of the many small debts we owe Digby, definitely the seventeenth century's bright, inconstant star.

Of wine made from strawberries, Digby says:

Bruise the strawberries and put them into a Linnen-bag which hath been little used that so the Liquor may run through more easily. You hang in the bag at the bung into the vessel before you put in your Strawberries. The quantity of the fruit is left to your discretion for you will judge there to be enough of them when the colour of the wine is high enough. During the working, you leave the bung open. The working being over, you stop your vessel. Cherry wine is made after the same fashion. But it is a little more troublesome to break the Cherrystones.

Roger Williams does not say he made wine himself, but Josselyn said, "It was not long before I left the Countrey that I made Cherry wine, and so many others, for there are good store of them both red and black."

Digby gives us instructions for the Countess of Newport's cherry wine.

Pick the best cherries free from rotten and pick the stalk from them. Put them in an earthen pan. Bruise them by griping and straining them in your hands and let them stand all night. On the next day strain them out (through a napkin, which if it be a course and thin one, let the juice run through Hippocras or gelly bag, upon a pound of fine pure Sugar in powder, to every gallon of juyce) and to every gallon put a pound of Sugar and put it into a vessel. Be sure your vessel be full or your wine will be spoiled; you must let it stand a month before you bottle it and in every bottle you must put a lump (a piece as big as a Nutmeg) of Sugar. The vessel must not be stopt until it hath done working.

Digby's currant wine must have been fairly tart.

Take a pound of the best Currants, clean picked, and pour upon them in a deep straight-mouthed earthen vessel six pounds or pints of hot water, in which you have dissolved three spoonfuls of the purest and newest Ale-yeast. Stop it very close till it ferment, then give such vent as is necessary and keep it warm for about three days, it will work and ferment. Taste it after two days to see if it be grown to your liking as soon as you find it so, let it run through a strainer to leave behind all the exhausted currants and the yeast, and so bottle it up. It will be exceedingly quick and pleasant and is admirable good to cool the liver and cleanse the blood. It will be ready to drink in five or six days after it is bottled. And you may drink safely large draughts of it.

Josselyn reports another drink. "Our fruit Trees," he says in *Two Voyages*, "prosper abundantly, Apple-trees, Pear-trees, Quince-trees, Cherry-trees, Plum-trees, Barberry-trees. I have often observed with admiration that the Kernels sown or the Succors planted produce as fair and good fruit without grafting as the Tree from which they were taken; the Countrey is replenished with fair and large orchards. It was affirmed by one Mr. Woolcut (a magistrate in Connecticut Colony) at the Captain's Messe (of which I was) aboard the Ship I came home in, that he made Five Hundred Hogsheads of Syder out of his own Orchard in one year. Syder is very plentiful in the Countrey, ordinarily sold for Ten shillings a Hogshead. At the Tap-houses in Boston I have had an Ale-quart spic'd and sweetened with Sugar for a groat, but I shall insert a more delicate mixture of it. Take of Maligo-raisins, stamp them and put milk to them, and put them in an Hippocras bag and let it drain out of itself, put a quantity of this with a spoonful or two of Syrup of Clove-Gilliflowers into every bottle, when you bottle your Syder and your Planter will have a Liquor that exceeds passada the Nector of the Countrey."

A "Hippocras bag" is a woolen cloth strainer of a long and tapering pattern said to be copied from and named for the sleeve of that earliest and greatest Greek physician, Hippocrates. It would seem to have been an essential of every kitchen and every stillroom in the seventeenth century (see illustration).

John Josselyn included an account of a way to make syrup of "clove-gillyflowers." Gillyflowers (pronounced Jilly-flowers) are said to be so called because they bloom in July. Some see a derivation from the French *"girofle,"* the name for cloves. Anyway, clove pinks were a most popular concoction for flavoring and as a cordial. "Gilliflowers," says Josselyn in *Two Voyages*, "thrive exceedingly there and are very large, the Collibuy or humming bird is much pleased with them. Our English dames make a syrup of them without fire, they steep them in wine till it be of a deep colour, and then they put to it spirit of Vitriol, it will keep as long as the other."

John Josselyn reports upon our native blueberries very simply. Of "Bill-berries," he says, there are "two kinds, Black and Sky coloured, which is

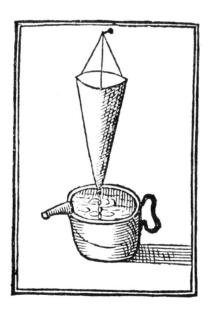

A "Hippocras bag" used for fine and slow straining, named for the great early physician Hippocrates, after whose sleeve it was fashioned. This illustration is from George Baker's *Newe Jewell of Health*, 1576 edition.

more frequent." He finds that, "they are very good to allay the burning heat of Feavers and hot Agues, either in Syrup or Conserve." He calls them, as the settlers use them, "a most excellent Summer Dish." "The Indians dry them in the Sun and sell them to the English by the Bushell, who make use of them instead of Currence, putting of them into Puddens, both boyled and baked, and into Water Gruel." He says the settlers "usually eat of them put into a bason, with Milk and sweetened a little more with Sugar and Spice, or for cold Stomachs, in Sack."

Of richer concoctions made from other fruits, Josselyn refers to the "Quinces, Cherries, Damsons," which "set the dames at work. Marmalad and preserved Damsons is to be met with in every house." Describing Boston, he has used marmalade again in another and probably familiar sense. "On the South there is a small but pleasant Common where the Gallants a little before Sun-set walk with their Marmalet-Madams, as we do in Morefields, etc. till the nine a clock Bell rings them home to their respective habitations, when presently the Constables walk their rounds to see good orders are kept, and to take up loose people."

And marmalet, marmalade or marmulate is made exactly as we do today, even as couples still walk on Boston Common.

Salads or "sallets" would appear to have been a regular accompaniment of seventeenth-century meals, even before or without their exposition by the great John Evelyn. We have seen the early explorers like Champlain and Captain John Smith eagerly planting their own salad beds. We see reference to "Oyle of olives" in supplies sent or brought or requested. Many of the plants grown or collected in the wilds were for salads, since everyone appears to have been well aware of the need for green stuffs in the diet; but, except for the allusions to the pots or waters in which salads have been cooked, we have no first-hand description of how their salads were made up. One assumes that, like water, they were considered dangerous to eat raw. There is, however, a reference to overdoing the idea of salads which we cannot bear to leave out, a quotation from Burton's *Anatomy of Melancholy*, a book we know was owned and read by the Puritans. Burton is quoting from Plautus, a Roman dramatist whom learned Puritans read in the original Latin. For, whatever they have reputedly thought about play-going, the Puritan attitude seems to have been much the same as ours today toward dog racing — more opposition to its frequenters and social consequences than to the sport itself.

Robert Burton, 1577–1640, a wit and scholar, wrote his *Anatomy* as a philosophical treatise on man's lot. A "variety of much excellent learning," it was admired by Johnson, adapted by Milton, and remained popular reading for the scholarly and witty for two hundred years. The salad reference which concerns us is from a chapter on "causes of melancholy" which Burton lists in order as: God, bad-angels and devils, witches and magicians, stars, old age, parents, and bad diet. Burton declares that salads "are windy and not fit therefore to be eaten of all men raw, though qualified with oyl, but in broths, or otherwise." Some, Burton says, "are of opinion that all raw herbs and sallets breed melancholy blood, except bugloss and lettice." Curious how everyone has always sworn by the borage family. "Borrage brings courage," is perhaps the oldest saying about flowers there is, and bugloss (pronounced as in bugle) belongs to the borage family.

The Romans, Burton tells us, always began their meals with a salad until the second century, when, woe to them, the custom was not continued, and we all know what happened to Rome. But Burton wishes to have salads cooked or "sodden," and he delights to quote Plautus whose "scoffing cook" says,

> Like other cooks, I do not supper dress,
> That put whole meadows on a platter,
> And make no better of the guests than beeves,
> With herbs and grass to feed them fatter.

We know John Winthrop, Junior, brought many seeds for salads with him. We do not know if he also cooked them.

When the senior Winthrop writes from "this strange land, where we have mett many troubles and adversities" to his "lovinge and dutiful sonne" in England, in March of 1631, he wishes his son might delay the voyage to join his father and come in Mr. Pierce's ship, since Mr. Pierce takes such good care of his passengers, and the younger Winthrop is bringing with him his young stepmother and several children. In any case, "Bring no provision with you," says the father . . . "but meale and pease, and some otemeale and Sugar, fruit, figges and pepper, and good store of saltpeeter, and Conserve of redd roses. . . ." And then, after "mithridate," an antidote against poison taken as an electuary, which means with honey or in a syrup, he lists at great length such items as calves' skins, shoes, stockings, hats, sheepskins with wool "dyed redd," woolen clothes of "sad" colors, millstones, shoemakers' supplies, wine vinegar, pitch, suet or tallow, chalk, chalk line, compasses, linen and birdlime. We can only conclude that the "no provision" advice means to bring no staple supplies for the voyage. If they are not to be under the care of the kind Mr. Pierce, they will need these little extras for their comfort. What word would cover the substantial shipment of supplies we do not know.

While the conserve of red roses may sound charmingly frivolous to us today, it was indeed one of the necessary comforts for a rough and long voyage in a very small ship. We are fortunate, again as always, to have Sir Kenelm Digby's suggestion.

After a lengthy and complicated set of directions for a "Dr. Glissons" method of making conserve of red roses, Digby gives us a relatively simple method related to him by a Doctor Bacon who had it from "a Roman Apothecary," which certainly places it in the medicinal class, festive though it sounds. Briefly, Dr. Bacon, or the Roman apothecary, clarified twelve pounds of the best sugar with whites of eggs and spring water, which, after boiling and skimming, came to about nine pounds of clarified sugar. Boiling this to a syrup, about halfway in the process he began to "beat his rose leaves in a mortar," the "rose leaves," by which he means the petals, having been carefully picked and the "nails," or places where they join the bloom, cut off beforehand. The juice of two lemons squeezed over them while they are beaten brings out their color. If the roses are allowed to stand after the beating they will turn black, so they are put quickly into what is by now, hopefully, a "high Syrup." After stirring them and letting them boil gently, skimming the while, for about a quarter of an hour, a drop upon a plate should show you it is "of due consistency." Then put it all into pots which can be left open for ten or twelve days to get their tops candied and so naturally sealed, and tie paper covers over them. For good measure, "Dr. Bacon useth to make a pleasant Julep of this Conserve of Roses by putting a good spoonful of it into a large drinking glass or cup, upon which squeeze the juice of a Limon; and clip in unto it a little of the yellow rinds of the Limon; work these well together with the back of a spoon, putting water in little and little, til you have filled up the glass with Spring-water; so drink it. He sometimes passeth it through an Hypocras bag and then it is a beautiful and pleasant Liquor."

But the conserve need not be strained. As the more complicated of the recipes for the conserve of roses given by Digby states firmly, "The colour both of the Rose-leaves and the Syrup about them will be exceedingly beautiful and red, and the taste excellent, and the whole very tender and smoothing and easie to digest in the stomack without clogging it. . . ." We can be glad of the older Winthrop's concern and forethought as, obviously, this is not a remedy that can be run up at a minute's notice.

But all cooking anywhere in the seventeenth century, of meat or for medicine, was never intended to take place quickly, notwithstanding Gerard's condemnation of the East Anglian housewives. Cooking in seventeenth-century New England required what Sir Kenelm Digby liked to call "discretion and experience." It had to take place in and around the large central chimney of what would be the largest room in the house, where, as likely as not, the huge four-posted and handsomely curtained bed of the owners would stand in one corner, their chests against the walls nearby. The wide hearth and deeply set fireplace under the huge open chimney would be, in itself, the kitchen. Even when there might later be separate rooms for eating and sleeping, the generous hearth with overhead cranes set into the center of the chimney, the oven on one side of the back, constituted the whole cooking area. This is why we do wrong to imagine roaring fires with great logs burning furiously, because the housewife, in order to cook, would have to be able to step or lean into the fireplace herself. And she needed to have several fires going at the same time, at least three in different degrees of burning: one hotly flaming, one of glowing embers, and one of hot ashes with "coals" on top. With these and hooks on the cranes to hang pots at different heights above the heat, and the oven at the side and back to be heated with hot coals before using, she would be able to follow even the most elaborate operations. Double boilers could be made by putting hay in the water at the bottom of a large iron kettle so a smaller kettle could rest inside. Reflecting ovens, "spiders," or frying pans on legs, spits for roasting meat, toasting forks, long-handled dippers, earthenware dishes, glass bottles and copper kettles — a formidable array of tools and utensils — equipped her to work minor miracles of cookery. The great beam stretching across the front of the fireplace, almost above her head, would have a smoothly worn hollow where she rested one hand to balance herself while she stirred with the other. The inner side of this beam, close to the fire, could become charred and constituted one of the commonest hazards in seventeenth-century cooking. The many remedies for burns in all the old records show that layers of heavy woolen garments often caught fire from being too near to the flame. For with everything cooking at once and in different ways, close watching was essential lest all those concoctions made with meal or flour or sugar should "fire too" or burn on.

Sugar, used to an extent which would appall us today, was so plentiful since it came from Barbados. Josselyn suspected it as the cause of people's being pitiably "tooth-shaken." The number of remedies for toothache would seem to bear him out, and the desperation of some of the remedies — such as rubbing the jaws with gunpowder — indicate the pain must have been severe. This would also explain the prevalence in gardens of poisonous plants, a little of which could be taken for a numbing effect; but a discussion of this belongs elsewhere.

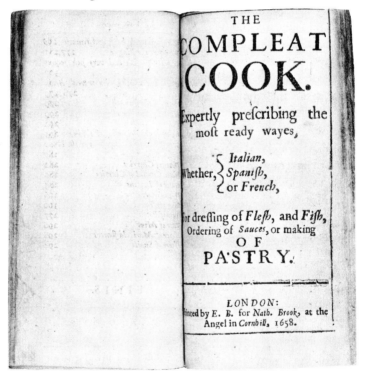

Title page of *The Queens Closet* introducing the section on cookery, last part of the book.

Adept as she had to be in the kitchen, in the stillroom the Puritan house-wife exercised even more skills. The stillroom was usually a little room at one side of the kitchen, slightly lower to ensure coolness, or it was a sepa-rate little house. To operate in it, the housewife had to understand and practice what are now the rudiments of modern chemistry. Here she made her cheeses, started her wines and sugar-fermented drinks, brewed beers, set dyes with mordants of her own concocting suited to each plant and the stuff it was to dye, and made all her plasters, salves, oils and waters . . . in fact, she ran what was a combination brewery, dairy and apothecary's shop. Fortunately for her then, even as it is fortunate for the modern American housewife, the men of the seventeenth century took an active interest in cookery as well as the making of all kinds of drinks. In fact, it is due chiefly to the men that we are able to tell what was made and how. It was not for another century that women felt the urge or were encouraged to explain what they did with all their time at the fire and in the stillroom and the gardens.

It is thanks to this male interest in food that we know how one could manage to feast on a venison pasty long after warm weather had set in. In his Journal, as we have seen, John Winthrop tells of his arrival in Salem Harbor and being feasted by "Mr. Endicut," the Governor of the Salem Colony on "a good venison pasty and good beer."

The "venison pasty" which Winthrop found good was served in June. Perhaps the deer were being killed whenever it was convenient to do so, but Digby has furnished us with a useful recipe for storing venison. Having cooked it, he says, in the oven in pots sealed with pastry, you drain off the broth, let pot and meat cool, and then repack the pot with the meat, over which you pour "butter very well clarified" to a height of two inches above the meat. The next day "binde it up very close with a piece of sheeps Leather so that no air can get in. After which you may keep it as long as you please."

For the pasty, Digby says, quoting My Lady of Newport, "Line the dish with a thin crust of good pure paste but make it pretty thick upward toward the brim, that it may be there Pudding crust. Lay then the venison in a round piece upon the Paste in the dish that must not fill it up to touch the Pudding but lie at ease; put over the cover and let it over-reach upon the brim with some carved Pasty work to grace it, which must go up with a border like a lace growing a little way upwards upon the cover, which is a little arched up and hath a little hole in the top to pour into it unto the meat and the strong well-seasoned broth that is made of the broken bones, and remaining lean flesh of the venison. Put a little pure Butter or Beef-suet to the Venison before you put the cover on unless it be exceeding fat. This must bake five or six hours or more as an ordinary Pasty. An hour or an hour and a half before you take it out to serve it up, open the Oven and draw out the dish far enough to pour in at the little hole of the cover the strong decoction of the broken bones and flesh. Then set it in again, to make an end of his baking and soaking."

"Puff paste," Digby says, crediting no one but himself, "is made by tak-ing a Gill of cold water, two whites of eggs, and one yolk, and to a quart of flour one pound of butter. So rowl it up but keep out of the flour as much as will rowl it up." From elsewhere we know he means by this last to roll it out, dot with butter, roll up and repeat many times, as we do today.

Pastry seems to have been fairly common, to judge by the poem about the women of Boston beginning a fortification on Boston Neck when they feared an Indian attack. It is written in a frivolous vein belying both the times and the subject, but has great charm.

> A grand attempt some Amazonian Dames . . .
> Contrive. . . . A Ruff for Boston Neck of mud and turfe.
> Their nimble hands spin up like Christmas pyes,
> Their pastry by degrees on high doth rise . . .
> A tribe of female hands, but manly hearts
> Forsake at home their pastry-crust and tarts
> To knead the dirt their samplers down they hurle,
> Their undulating silks they closely furle. . . .

and they have come supplied against fainting fits:

> They want no sack nor cakes, they are more wise.

And in the end, of course, "Maie stronger hands" take over:

> But the beginners well deserve the praise.

An engaging picture of Boston dames in a crisis, toward the end of the seventeenth century.

The beer served Winthrop by Endecott must have been as like the beer of Old England as possible, made with malt and yeast brought from the homeland. The hops could have been native, since we know early voyagers up Maine rivers rejoiced at the quantity of hop vines luxuriously covering the banks on either side. Malt we know John Winthrop had brought with him and we may presume that Mr. Endecott had done so also. And although we know that "Beer" was made of almost anything handy in early New England, malt was preferred. One of Winthrop's young fellow voyagers wrote a desperate letter to his father asking him to send over a hogshead of malt because they were in such sore straits as to be drinking water. So we know the English grains they had planted were not yet flourishing as they did later, when, as we know from Bradford, all sorts of English grains were doing well. Of the yeast, the "ale-yest" of Digby's *Closet* and Josselyn's eel

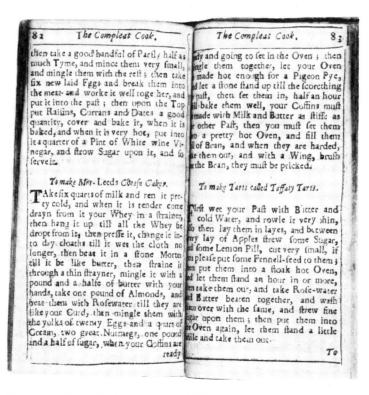

Two pages from the last section of *The Queens Closet* telling how to make cheesecakes and tarts.

dish (of which more later), we have no direct accounts. It would seem to have been one of the essential commodities, so precious as to be assumed, perhaps, as every housewife's special concern. But of what they did to keep it "alive" on shipboard or to make it always new and fresh after they landed, I have found no mention. Remote homesteads in our own Far West today have a similar problem to keep the yeast working.

Josselyn tells us that in Maine beer was made of molasses, which would seem to have put it rather on the rum side. To the molasses they added chips of sassafras root, a little wormwood, water and bran (elsewhere he refers to "Molasses Beer"). He does not mention yeast, and perhaps the sugar in the molasses furnished the fermentation. In any case, the seventeenth century was a great one for concocting drinks of most unlikely materials, since water was obviously so dangerous.

The New England settlers were under no illusions about water after years of experience with it in England. No wonder the fresh waters of New England amazed and delighted the newcomers. When Dr. Andrew Boorde of sixteenth-century England advised that moats be kept moving to ensure the health of the household, he was not necessarily referring to moats surrounding great castles for defense but to those dug for drainage and convenience around many well-planned small, medieval English houses. Burton, in *The Anatomy of Melancholy*, cautions against using the water from these moats even for cooking and denies that, as some have said, it makes the best beer, although he concedes it may certainly make the strongest. So we can understand the lyric celebration by Captain John Smith of the springs near every New England house, and the staunch recommendation by William Wood of the waters of New England as being better than bad beer, though of course not to be preferred to good beer. Although he says of New England water, obviously astounded, that "those that drinks it be as healthful, fresh and lustie, as they what drink beere." That water in New England could be drunk with no ill effects whatever also astounded the delicate and hence doubly cautious Higginson.

On the whole, ten or twelve years after Winthrop landed the settlers were not doing too badly. Johnson's *Wonderworking Providence of Sion's Saviour in New England* lists among the blessings: gardens, orchards, "fair Houses" and good eating. Of the year 1642 he says:

First, to begin . . . the increase of food . . . you have heard in what extreme penury these people were at first . . . when ships came in it grieved some Master to see the urging of them by people of good rank and quality to sell bread unto them. But now take notice how the right hand of the most high hath altered all . . . and now good white and wheaten bread is no dainty, but every ordinary man hath his choice, if gay clothing and a liquorish tooth after sack, sugar and plums lick not away his bread too fast . . . there are not many towns in the country but the poorest person in them hath a house and land of his own and bread of his own growing, if not some cattel; besides, flesh is now no rare food, beef, pork, and mutton being frequent in many houses, so that this poor wilderness hath not only equalized England in food, but goes beyond it in some places for the great plenty of wine and sugar, which is ordinarily spent, apples, pears, and quince tarts instead of their former Pumpkin Pies, Poultry they have plenty, and great rarity, and in their feasts have not forgotten the English fashion of stirring up their appetites with variety of cooking their food. . . .

Edward Johnson's reports are borne out by John Josselyn, whose recipe for cooking eels proves this last point. But before we quote this as a fitting end to this chapter on food, let us turn to Digby again, if only to show the variety of plants expected to be ready to hand. What the variety of cooking was, we can judge best by firsthand accounts, but there is much to be learned also from the instructions they had in their hands and minds from the old country itself.

Sir Kenelm Digby, so highly respected, although one cannot help wondering why, by such worthies as the younger Winthrop and Cotton Mather, mentions such a variety of "ingredients" in his recipe for metheglin as may give us pause, especially when we realize most of them came straight from the garden.

Metheglin, called, Digby says, "The Liquor of Life," was an alcoholic drink made from fermenting honey, and was particularly favored in Wales. A considerable number of the settlers came from near the Welsh border. Oddly enough, the drink was used as recently as the 1890's in the United States where it was flavored with sarsaparilla and activated or charged with carbonic acid gas, which would certainly have startled Sir Kenelm. But sarsaparilla would have seemed too simple a flavoring for him. "Take," he says, "Bugloss, Borage, Hyssop, Organ, Sweet-Marjoram, French cowslip, Coltsfoot, Thyme, Burnet, Self-heal, Sanicle a little, Betony, Blew-buttons, Harts-tongue, Meads-sweet, Liverwort, Veriander two ounces, Bistort, Saint John's wort, Liquorish, Two ounces of Carraways, two ounces of Yellow-saunders, Balm, Bugle, Half a pound of Ginger, and one ounce of Cloves, Agrimony, Tormentil-foots, Cumfrey, Fennel-roots, Clowns-all-heal, Maiden-hair, Wall-rew, Spleenwort, Sweetoak, Paul's betony, Mouse-ear. . . ."

For two hogsheads of metheglin you take two handfuls apiece of each of the above herbs except sanicle, of which he says, "you need only half a handful." All in all, it makes for quite a garden for the Puritan wife to tend.

But of course this is nowhere near the end of the metheglin "receipt." Perhaps we had better start fresh with another set of directions Digby gives as taken from a Mr. Pierce, whom we can only hope is the kind captain that John Winthrop recommended to his son.

In the directions for Mr. Pierce's "Excellent White Metheglin" he first boils three hogsheads of the best water and throws in four handfuls of sweetbrier leaves, the same of eyebright, two of rosemary, the same of sweet marjoram and one of "Broad-thyme." After a quarter of an hour's boiling he strains out the herbs and lets the water cool to blood-warm when he puts the honey to it, about one part to four of water, and mixes it all well for at least an hour. A test for a "strong" enough mixture, is to float an egg in it so that "a Groat's-breadth, or rather but a threepence, of the Eggshel must swim above the Liquor." Mr. Pierce makes his test with a whole dozen eggs, to "make a medium of their several emergings." (This because some eggs are fresher or rounder than others.) The egg test passed, the concoction goes back into the "Copper" to boil again and be skimmed. "Turn up an hour Glass and let it boil well a good hour." Just at the end of the boiling, a pound of ginger is added and the whole is allowed to cool. Then comes the operation of pouring it from a height into a "pottle of New-ale-barm," and when all the liquor is in and uniformly mixed with the barm, or yeast broth, it will begin to "work." There is more skimming before it is put into two hogsheads which have served for Spanish wine, where it works more until it finally runs clear, when it is gradually sealed, first with paper dipped in yeast and then with a cork. Three weeks after broaching, "which is best not done till a year be over after making it," Mr. Pierce flavors it with cinnamon and cloves, although he sometimes leaves out the cloves. After a year, if it is put into bottles, it will keep for several years. In any case "This Metheglin is a great Balsom and strengthener of the Vicera and is excellent in colds and consumption."

After all this literally rather strong medicine culled from the records of domestically accomplished and eager gentlemen, we are fortunate to have discovered, very late in the day for this book and by one of those minor miracles that take place at dinner parties, a copy of a book considered unobtainable. This is a small volume intended chiefly for the housewife, several copies of which appeared in Boston almost immediately after being printed in London. Its title and frontispiece outlines the general competence expected of the seventeenth-century housewife: *The Queen's Closet Opened. Incomparable Secrets in Physick, Chyrurgery, Preserving, Candying, and Cookery.* This combination of skills is something we hope has been made obvious, but it is reassuring to see it all so succinctly expressed. It is also reassuring for us, having made so much of how everyone turns up everywhere over and over again in the small seventeenth-century world, to find that one of the "most Experienced persons of our times" who divulged their secrets to the Queen, consort of Charles I, was our friend Sir Kenelm Digby.

"W. M.," one of the Queen's "late Servants," has made a much gentler compilation, in the "Receipt Books" of Queen Henrietta Maria, than Digby's servant served up from Digby's own collection. However, the material is obviously based upon much the same common practice. Digby and the Queen preserve venison in much the same way and make the same pastry. The Queen leans towards puddings, "white pots" and jellies, and has a very good and what we might now call "traditional" white cake. The waters she distilled from roses and later used in cooking and creams are much what we might expect. Her "receipts" were given her by as motley a group of friends as Digby's own, and, like his, include both countesses and ordinary misters and mistresses. The Queen does not, however, appear to have known any sea captains.

And now we come to John Josselyn's own recipe for cooking eels, which compares very favorably with anything we have seen in either Digby's closet or the Queen's. In this, curiously enough, he mentions several herbs which he would seem to have considered so essential as not to need listing with the other plants he has carefully remarked upon as growing well in this country.

"The Eal," he says, "is of two sorts, salt water eels and fresh water Eales; these again are distinguished into yellow bellied Eale and silver bellied Eale; I never eat better Eals in no part of the world that I have been in, than are here. They that have no mind or leasure to take them may buy of an Indian half a dozen silver bellied Eals as big as those we usually give 8 pence or twelve pence a piece for at London for three pence or a groat. There are several ways of cooking them, some love them roasted, others baked, and many will have them fryed; but they please my palate best when they are boiled; a common way is to boil them in half water, half wine with the bottom of a manchet, a fagot of parsley, and a little winter savory; when they are boiled they take them out and break the bread in the broth, and put to it three or four spoonfuls of yest, and a piece of sweet butter; this they pour to their Eals laid upon sippets (a small loaf of white bread) and so serve it up. I fancie my way better which is this: After the Eals are fley'd and washed I fill their bellies with nutmeg grated and cloves a little bruised and sow them up with a needle and thread; then I stick a clove here and there in their sides about an inch asunder, making holes for them with a bodkin; this done I wind them up in a wreath and put them in a kettle with half water and half wine vinegar, so much as will rise four fingers above the Eals; in midst of the Eals I put the bottom of a penny white loaf and a fagot of

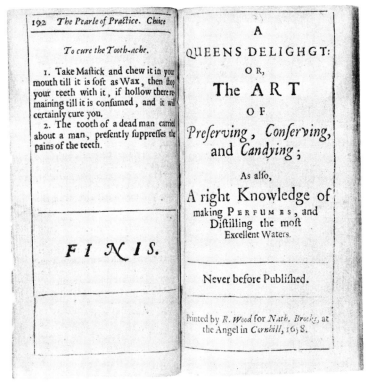

Title page from the section on preserving in *The Queens Closet.*

three yolks, and tie two pieces together,
and lay them in water an hour, and take
as much sugar as will fill up your mould,
and boil it in a *Manus Christi*, then pour
it into your mould suddenly, and clap on
the lid , round it about with your hand,
and it will be whole and hollow , then
colour it with what colour you please,
half red, or half yellow , and you may
yellow it with a little Saffron steept in
water.

Touching Preserves and Pomanders.

To make an excellent Perfume to burn between two Rose leaves.

TAke an ounce of Juniper , an ounce
of Storax , half a dozen drops of
the water of Cloves , six grains of
Musk , a little Gum Dragon steept in
water,

water, and beat all this to paste , then
roll it in little pieces as big as you please,
then put them betwixt two Rose leaves,
and so dry them in a dish in an Oven,
and being so dried , they will burn with
most pleasant smell.

To make Pomander.

Take an ounce of Benjamin, an ounce
of Storax , and an ounce of Laudanum,
beat a Mortar very hot , and beat all
these Gums to a perfect paste; in beating
it , put in six grains of Musk , four
grains of Sivet ; when you have beaten
this to a fine paste with your hands
with Rose-water, rowl it round betwixt,
your hands, and make holes in the beads,
also string them while they be hot.

To make an Ipswich water.

Take a pound of fine white Castle
soap, shave it thin in a pint of Rose-
water, and let it stand two or three
days; then pour all the water from
and put to it half a pint of fresh
water, and so let it stand one whole
day, then pour out that , and put half
O   a

Remedies against unpleasant odors. From the section of *The Queens
Closet* on preserving, conserving and candying.

these herbs following, Parsley one handful, a little sweet Marjoram, Penniroyal, and Savory, a branch of Rosemary, bind them up with a thread and
when they are boiled enough take out the Eals and pull out the thred that
their bellies were sewed up with, turn out the Nutmeg and Cloves; put the
Eals in a dish with butter and vinegar upon a chafing dish with coals to
keep warm, then put into the broth three or four spoonfuls of good ale-
yeast with the juice of half a lemon; but before you put in your Yeast beat
it in a porringer with some of the broth, then break the crust of bread very
small and mingle it well together with the broth, pour it into a deep dish
and garnish it with the other half of the Lemon, and so serve them up to
the Table in two dishes."

7

And Medicine

Practice of medicine among and by the early settlers in New England was, from the beginning, almost entirely their own affair. Although efforts were made by financial underwriters and council members in England to send a "physitian" or a "chirurgeon" or, at the very least, a "barber-chirurgeon" with every group, there were not enough physicians and surgeons in seventeenth-century England to be well spared, or who would wish to risk lives and practices in the new wilderness. One who did "come over," like Giles Firmin of Ipswich, a skilled chirurgeon who performed the first "anatomy" in New England, might easily find himself, as Giles Firmin did, unable to make a living by his practice and forced to return to England. Another might be like Samuel Fuller, a physician who came with the Pilgrims and was impressed into other kinds of services. Fuller was used to being loaned out by the Pilgrims, as when he made one day's work of bleeding twenty people in Mattapan. But when he was called upon to help with illness in Endecott's colony in Salem, and found he lacked sufficient drugs, he achieved immortality by giving advice on how to separate from the church in England. As time went on, ambitious physicians might find patients grown independent in medical practices, as in all else. Later in the century, when the affluent and influential Samuel Sewall watched beside his ill wife and other members of his Boston household, he had no less than three doctors in attendance, but he felt competent to sit in judgment upon their treatments, which consisted mainly of opiates and did not greatly impress him.

Obstetrics, in New England as in Old, remained primarily in the hands of midwives. Here even Samuel Sewall felt no need to take charge, though we see him in his diary escorting the midwife to her home, "her stool upon my arm." Trained by other midwives in honorable succession, the midwife represented a profession and estate to which any competent neighborly wife or widow might aspire.

An "anatomy," or the dissection of a corpse with accompanying explanation, was rare even in England and France in the seventeenth century. It depended entirely upon a skilled practitioner coming upon, or by, a suitable corpse. One could wait a lifetime for an opportunity to see such a demonstration and everyone was welcome and eager to attend. Giles Firmin's one chance, as a practicing physician, to "anatomize" was long remembered. It would have been folly to forgo being present upon any such occasion. The knowledge gained could prove convenient. A dead Indian or a drowned Negro might afford amateurs a chance to investigate the unknown, but it was difficult to profit if no one knew what he was looking at. See again Samuel Sewall, laughing gently into his diary after an anatomical exploration of an executed criminal during which a friend held up the heart and called it the stomach. At least they were all trying to learn. Some even hoped to contribute to science, as when John Josselyn, conscientious describer of New England's ills and remedies, announced, after treating a wounded Negro, that he thought the reason for the color of Negroes' skins was an extra skin underneath.

Bloodletting, that long highly considered remedy for nearly everything (it was Washington himself who asked to be bled in his last illness) seems to have remained within the preserve of barber-surgeons. As we have noted, during the ages we egotistically call Dark, the Arabs were forging ahead in the sciences and had taught the early European practitioners of medicine a relatively merciful method of bleeding by scarifying an area and submerging it in warm water. Until then bloodletters had cut the most accessible vein and let it run. But, while New England Puritans sometimes refer to "cupping," or placing the mouth of a warm glass over an affected area, anything to do with a scalpel and lancet they felt was not for amateurs. The founding fathers, however, considered themselves reasonably able in all other medical matters, even when most of their medicinal material had to come from their own gardens.

The technical equipment of the amateur practitioners included only a few simple tools, apart from the equipment for distilling and brewing, and what was needed in the rare alchemical experiments. A silver tongue-scraper could prove useful, as when Winslow from Plymouth was fortunate enough to be able to cure the Sachem Massasoit by scraping his tongue and treating him with tinctures of sassafras and strawberry leaves. Winslow was acting as a good neighbor, but it was undoubtedly as a grateful patient that Massasoit warned him of impending Indian attack. Besides the tongue-scraper, another popular gadget of the times, for use in almost any emergency, was the "antimonial cup," made of antimonial glass and capable of turning any wine into an emetic. One of the earliest references to this aid is in a letter to John Winthrop in 1636 from Matthew Cradock, early and staunch sup-

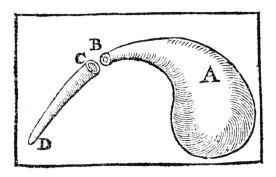

"Antimonial cup" from *The Countrie Farme*.

porter of the Puritan venture. Matthew Cradock writes that it was by the good Captain Pierce (everyone's favorite captain) he "had full purpose to have sent you an Antimonial Cupp which I make doubt whether I shall gett to send by him. If I be not misinformed the use thereof (I fear immoderat) was an occasion of shortening Sur Nathaniell Riches dayes who hath made exchange of this life for a better." So they were warned. Yet when a professional physician on Cape Cod died in the last half of the seventeenth century, his inventory listed what may well have been the mainstays of all hopeful practicers of physicke — an antimonial cup and a copy of Culpeper's *English Physician*.

On the whole, the practice of "physick" would seem to have depended chiefly upon anyone who felt the urge and had a flair for healing or was endowed with special authority in quite other matters. Cotton Mather spoke eloquently of the "angelic Conjunction" evidenced in so many divines being medically inclined, though in no case that we know of were they medically trained. It has been assumed that those bright young men with Cambridge degrees who came to help purify the Church of England and set up a true City of God in New England, *must* have had some sort of training in medicine while in college, since they so willingly gave medical advice and remedies to ailing and loyal parishioners here, but there is nothing to prove that they did.

In the world of seventeenth-century England from which the settlers had come, medicine in small town and rural areas was largely presided over by the lady of the manor and other housewives of distinction and warm hearts. New England was fortunate in having among its Puritan women many who had been raised in the gentle tradition of helping the less fortunate even in considerable numbers. With the clergy and the governing officials to give advice and set the tone, the leading housewives to furnish nursing skills and substantial supplies from their gardens and stillrooms, the colony seems to have felt reasonably secure without professional guidance.

A practitioner among the early divines whose reputation for cures was greater than his training for them was the poetic and hypochondriac Michael Wigglesworth. As a boy, his first year in this country was spent sheltering in a leaky hole in the ground. Moody and hysterical as a youth, possibly as a result of this experience, he studied for the ministry, and lived a long life of ever-failing health. Obituary poems and sermons lauded his healing power and industry, but are not specific. He is the author of that early best-seller for nearly a century, *Day of Doom*, where he endears himself to us for condemning unbaptized infants to the "easiest room in hell." Still, his patients loved him.

Another popular and medically practicing seventeenth-century divine was Edward Taylor of Connecticut, an able early poet of greater skill and range than Wigglesworth, and directly influenced by George Herbert whom he surpasses in charm. Taylor was an indefatigable copier of books he needed in his library and could not come by otherwise. He left among them a fat little manuscript volume called his "Dispensatory" in which he lists nearly four hundred plants and their properties, the various ways of treating them for use, and some alchemical data. For good measure he includes a list of unusual animals, some information about stones and minerals, and remedies quoted from a Jesuit priest where, unfortunately, the pages are torn. In any case, Edward Taylor, to whom we owe some of our best lyrics, felt it incumbent upon him, like the other clergymen of his time, to care for the health of his flock as well as its souls.

There was a sound tradition for these clergymen to practice physic on the side. Sir William Osler remarked in an essay that "the parsons of that day

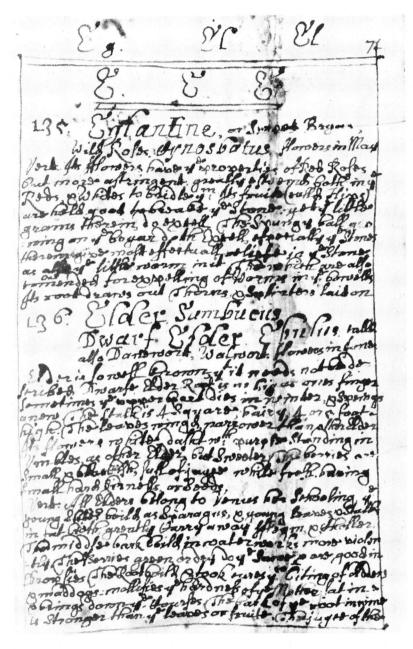

A page from the little manuscript volume of Edward Taylor's "Dispensatory" with which he took care of the ills of his parishioners.

had often a smattering of physic." Osler said they illustrated what Cotton
Mather called an angelic conjunction, and he quoted Mather: "Ever since
the days of Luke the Evangelist, skill in Physick has been frequently pro-
fessed and practiced by Persons whose declared business was the study of
Divinity." But there was a good seventeenth-century authority for this
from a source much admired by the same clergymen — George Herbert,
English poet and divine who himself must have been medically inclined.

In his *Country Parson*, Herbert stays his parson's hand from nothing. No
wonder the Puritans loved him. In his chapter entitled "The Parson's Com-
pleteness," he states that "the Country Parson desires to be all to his parish;
and not only a Pastor, but a lawyer also, and a Physician." He disposes of
the legal aspects fairly quickly, advising the reading of "initiatory treatises,"
conversing at length with lawyers while carefully allowing them to do most
of the talking, and in cases "of an obscure and dark nature," having recourse to the law. The role of physician he finds easier to fill if here the
parson may expect the help of a wife. (Failing her and with no skill or
knowledge himself, he might have to call a doctor.) Herbert's problemati-
cal parson is fortunate in his helpmeet. "If there be any of his flock sick, he
is their physician — or, at least his wife, of whom instead of the qualities of
the world, he asks no other but to have the skill of healing a wound, or
helping the sick." On the whole, he says ". . . it is easy for any scholar to
attain such a measure of physics as may be of much use to him both for
himself and others. This is done by seeing one anatomy, reading one book
of physic, having one herbal by him." The parson may even be able to
practice economy, provided his wife is a good gardener. "In the knowledge
of simples, wherein the manifold wisdom of God is wonderfully to be seen,
one thing would carefully be observed; which is to know what herbs may
be used instead of drugs of the same nature, and to make the garden the
shop. For home-bred medicines are both more easy for the Parson's purse
and more familiar for all men's bodies. So where the apothecary useth
either, for loosing, rhubarb, or for binding, bolearmena, the Parson useth
damask or white roses for the one and plantain, shepherd's purse, knot-
grass for the other, and that with better success." As for his wife, she "seeks
not the city, but prefers her garden and fields. . . . And surely hyssop, valerian, mercury, adders tongue, yarrow, melilot, and St. John's wort made
into a salve; and elder, camomile, mallows, comphrey and smallage made
into a poultice, have done great and rare cures."

Caltha Sylvanica.

It is the greatest Vulnerary in the World. Make a Balsam of the Green Leaves with Oil-olive, it cures ye most grievous pains of the Nerves, and Cuts of all sorts. ... with an admirable Efficacy & Celerity. Ordinary wounds are cured in four or five hours with ye accomplishment of ye patient; and extraordinary ones in a few dayes or forthwith. The Indians are extremely venturous in their Surgery, if this ... be near ym.

Consolida ...

For Internal Wounds, it will do wonders. It relieves oppressions from clotted Blood in ye Stomach. ... of Blood it affords so solid a remedy to be parallel'd. —

Ophioglossa.

Bruise it, and lay ye cataplasm to ye part bitten by ye Rattlesnake; it infallibly & immediately fetches out ye harmful poison, which else would in a few hours bring a miserable death upon ye patient. It is a marvellous gift of Heaven to ye countreys, which has ... venemous vipers in ym. For ...

Cotton Mather's listing of the plants he felt the Royal Society in London should know about, with their uses. He had not seen these listed before in any herbal.

Although we have considered the two distinguished parsons' wives who were medically inclined, we have no record of their cures. Mrs. John Cotton was handsomely paid for her practice of physic and surgery among the Indians on Martha's Vineyard. Mrs. John Eliot, who seems to have practiced to oblige her husband, must have worked literally for love. Their lack of note-taking is regrettable. Women were early admonished not to act so beyond their stations as publicly to take notes on sermons, so the lady practitioners may have felt that ostentatious record-keeping upon any subject might not become them. Even Cotton Mather, inveterate note-keeper himself, did not urge women to keep records, although he approved of women in medicine and discussed training his daughter to become a doctor. He had considered studying medicine as a young man because his stuttering seemed to unfit him for the ministry. However, his affliction being "most marvelously lifted from him" at a fairly early age, he made medicine a side interest and recorded his observations and practices meticulously in a still unpublished volume called "The Angel of Bethesda." It is in one of these chapters that he urges gentlewomen to keep their gardens full of helpful plants and their closets ready stocked with "several harmless and useful (and especially external) remedies for the help of their poor neighbors." But not to keep records.

After clergymen and their wives, magistrates seem to have been resorted to — the more important, the more importuned. The elder Governor Winthrop and his contemporary, Governor Endecott of Salem, were both great gardeners, but that cannot wholly account for Governor Winthrop's being called upon for advice by ailing colonists, or for Governor Endecott's urgently sending Governor Winthrop a most extraordinary assortment of advice, comfort and remedies, including a unicorn's horn, when Madam Winthrop was having "a fit of the mother." A "mother-fit" is a Saxon word for hysterics. We must think with sympathy of Mrs. Winthrop, so used to ministering to others, now being in the position of being cared for by the two foremost male governors in the colony. Endecott's letter reads:

Worthy Sir. I am sorrie to heare of your affliction in this visitation of God, though you know that whom he loveth he chastiseth let that comfort you. . . . I have sent you all I have or what I can get; viz. Syrup of Violetts, Sirrup of Roses, Spirit of Mints, Spirit of Annis, as you may see written on the several vialls. I have sent you Mrs. Beggarly her unicorns horn and Beza stone I had of Mr. Humfrey who is sorry also for your exercise. I have sent you a Bezoar stone and mugwort and organie if you should have need of it. They are both good in the case of your wife, and also I have sent you some galingall root, Mrs. Beggarly knows the use of it. If the fitt of the mother comes very violently as you write, there is nothing better to suppress the rising of it than sneezing, a little powder of tobacco taken in her nose I think is better than Hellibore. If I knew how or which way in this case to do her good I would with all my heart, and I would now have gone to you but I am altogether unskillful in these cases of women . . . please to tell Mrs. Beggarly that all her family are well. . . .

Endecott, indeed, was sending all he had or could get including Mrs. Beggarly's own unicorn's horn, a remedy we have seen Anne Bradstreet allowed to be an antidote to poisons. John Winthrop, Junior, had a horn and there was one kept handy at Windsor Castle far into the next century, although the great French surgeon Paré published a book doubting the existence of such an animal. Bezoar stones were easier to come by, as everyone had goats and this is the stony residue which forms in the stomach of ruminants. The best bezoar stones were supposed to come from Peru, since high mountain herbs were the stronger, but New England goats must have been able to oblige with this early and highly charged vitamin pill.

The younger Governor Winthrop was more nearly an accredited physician in his own right than either his father or Governor Endecott. A dedicated and industrious alchemist as well as an expert gardener, although he accepted useful roots and plants from his father (who seems to have prescribed botanical remedies only), the younger Governor became famous for some of his chemical and mineral prescriptions. After being importuned to come and live in several communities, even Manhattan, he settled finally in Connecticut where he kept a medical journal. From his "cases," all duly listed, we can judge his fondness for his secret powder, "rubilia," lately analyzed from a smudge to be a concoction of antimony, niter and tin. But he also relied upon the tried and true plants the settlers had brought with them: elecampane, elder, wormwood, anise and plantain root. The younger Governor Winthrop is rated by medical historians as among our early medical men, although they are unable to say when he studied medicine and where. Still, as we know from Dr. Andrew Boorde of the century before and from Sir Kenelm Digby, to travel at all anywhere was to pick up medical information. The younger Winthrop had traveled widely before deciding to follow his father to the New World.

Some writers have suggested that John Josselyn must have had medical training, but there is nothing in anything he has written that would so indicate. Armed with the 1633 edition of Gerard's *Herball* as corrected and amplified by Thomas Johnson, he literally nibbled and tasted his way about the wilderness.

Even so, Josselyn in his role of doctor does not feel he lacks authority.

The Diseases that the English are afflicted with, are the same that they have in England, with some proper to New England, griping of the belly (accompanied with fever and ague) which turns to the bloody-flux, a common disease in the countrey, which together with the small pox hath carried away abundance of their children, for this the common medicine amongst the poorer sort are Pills of Cotton swallowed, or Sugar and Sallet-oyle boiled thick and made into Pills, Alloes pulverized and taken in the Pap of an apple. I helped many of them with sweating medicine only. Also they are troubled with a disease in the mouth or throat which hath proved mortal to some in a very short time, Quinsies, and Imposthumations of the Almonds, with great distempers of cold. Some of our New England writers affirm that the English are never or very rarely heard to sneeze or cough as ordinarily they do in England, which is not true.

Pleurisies and Empyemas are frequent there, both cured after one and the same way; but the last is a desperate disease and kills many. For the pleurisie I have given Coriander seed prepared, Carduus seed, and Hartshorn pulverized with good success, the dose one dram in a cup of Wine.

The stone terribly afflicts many, and the Gout, and Sciatica, for which take onions roasted, peeled and stampt, then boil them with neats-feet oyl and Rum to a plaister and apply it to the hip.

Head-aches are frequent, Palsies, Dropsies, Worms, Noli-me-tangeres, Cancers, pestilent feavers, Scurvies, the body corrupted with Sea-diet, Beef and Pork tainted, Butter and Cheese corrupted, fish rotten, a long voyage coming into the searching sharpness of a purer climate, causeth death and sickness amongst them.

Josselyn's list of the diseases which did "most painfully afflict" the settlers seems to be much the same as those which do most painfully afflict us today with a few notable exceptions. But Josselyn fancies himself as a doser.

"Take," Josselyn recommends, "for cough or stitch upon a cold, Wormwood, Sage, Marygolds, and Crabs-claws boiled in posset-drink and drunk off very warm." This, he says, "is a sovereign medicine." Crabs-claws were left to the housewife to find in her garden. In Josselyn's reference book, Johnson-upon-Gerard, "crabs-claws are known also as 'Water-houseleek.'" One wonders what the Puritan housewife did to supply this one. What the housewife came up with was probably the regular houseleek which Cotton

Mather also recommended at the end of the century. All this would have been considered common procedure, simpling indeed.

With disease defined as an "unknown guest," one can understand the violent measures to which practitioners often had recourse, depending upon what nature had already started to do. Emetics were administered to those who were vomiting and purges to those with a flux. Provoking a sweat and inducing sneezing were considered invariably beneficial. And it was not unusual to try everything at once, which accounts for the popularity of herbs which would purge "both upwards and downwards." If the patient recovered, it proved that the "imbalance" had been got rid of. If the patient found recovery beyond his powers, God had intended him to exchange this world for a better.

In those days the West Indies were nearer to New England than they are today, even with today's airplanes. They were an integral part of the total scheme of the move to the New World. Sea traffic was normal and land travel hard, as in the Old World. The West Indies, called the "Salt Islands" for the ease with which salt could be made from seawater there, in the seventeenth century contained the largest English settlement in the New World. It was not unreasonable for New England to expect to procure supplies without undue difficulty or delay. Sugar, lemons, oranges, limes, almonds, medicinal plants and occasional slaves were not unusual in Boston. Cotton Mather mentions a fair number of West Indian remedies in his medical writings. Samuel Sewall went courting with West Indian fruit and nuts. New England rum, which supported more ships than all else except the seas, was entirely derived from the West Indian sugarcane. And one of the worst epidemics of smallpox in seventeenth-century Boston, brought by a ship from the West Indies, precipitated the first official municipal effort at quarantine in the New World, and perhaps anywhere. So reference to cinchona bark (from South America) and jalap (from Mexico) and aloes (from the West Indies) need surprise us no more than the continuing argument as to whether the virtues of domestic or "monks' rhubarb" could ever equal that imported from China by land route or sea since time immemorial.

But however near exotic remedies, and however much the most highly placed individuals in church and state were resorted to as a matter of course, most material for home cures had to be grown and prepared at home by the Puritan women. Dunton, the gay and unreliable bookseller, says of a Dr. Bullivant in Boston in 1686, "He does not direct his patients to the East

Indies to look for drugs when they have far better out of their gardens."
Gardens with "a variety of flowers" were cultivated by housewives for the
general well-being. Ladies who arrived from manor houses in England well

tigate the burnings, out of the fame Auctour.

The felfe fame Auctour, both otherwyfe prepare the oyle,
which worketh ftronger and to greater purpofe in the abouefayd
burning. Take on Mydfomer eue, the toppes, flowers & leaues,
with which let a new potte be fylled, hauing in the bottom a lytle

hole, and let the mouth of the vpper pot be
dyligently ftopped, which fet into ye mouth
of another pot ftanding vnder, the mouth
of which lute round about with the other,
that no ayre breath forth: this done, fet the
pots fo deepe into the earth that they may
wholy be couered and buried in the earth,
after let them ftande for a whole yeare in
the grounde: at the ende of which tyme,
drawe the pottes forth, and you fhall finde

in the neather potte a cleare oyle, which by the heate of the fumo-
fities of the earth, is drawne forth from the Henbane. This maner of inftruction is founde perfiter, in the difcription of the oyle
of yuie berries (where is otherwyfe lefte in the earth for fixe
monethes) with this are members labouring and fore payned
with dayly fluxes falling to them, annoynted.

A compounde oyle out of Sædes, procuring flæpe: Take
of the Sædes of Lollij, of Henbane, of the whyte and blacke
Poppie, of the Lettuce and Purcelane fædes, of eache fowre
fmall handfulles, of the fædes of Faba inuerfa, which is Tele-
phium, two fmall handfulles: let all thefe be diftylled togyther:
of this diftylled, mynifter two fcruples wayght at a tyme, with
a lytle or fmall quantity of Opium.

A simple method of extracting oils from plants by burying. From George Baker's *Jewell of Health.*

versed in the arts of distilling and fermenting and brewing, seething and drying, and making waters and spirits and pills and powders, found themselves as busy as any apothecary in England. They would accept the efficacy

The Italians haue inuented another maner and way of Dyſtilling waters in the Sunne, which wyth them is often vſed after this maner. They take two Glaſſe Bodies wyth narrowe

neckes and mouthes, the one being emptie, and the other filled with Herbes o? Flowers. Thys Glaſſe ſo filled, they cloſe o? ſtop with a fine Lynnen cloth (bounde about) th?ough which the lycour may aptly paſſe o? dyſtill. After that, they th?uſt the necke of this Glaſſe, into the necke of the emptie Glaſſe ſtanding vnder, and then diligently ferment and ſtop the paſſages and wayes rounde about, with Lute o? Potters Claye, o? other lyke matter, to the ende, that no vapour no? vertue of the ſubſtance may b?eathe fo?th: This done, ſet theſe two Glaſſes on ſuch wyſe iopned and bounde togyther in the beames of the Sunne, after ſuch maner, that the ſame Glaſſe which contepneth the Herbes o? Flowers, maye ſæne to be aboue, and the other whpch is emptie, to ſtande vnder, fo? to recepue the lycour which is heated and decoded by the Sunnes fo?ce, that ſo dyſtilleth downe into the Glaſſe. And on ſuch wpſe, doe the women of Bononie in Lumbardie, p?epare and purchaſe the water of B?emble flowers, fo? the benefit and ſingular comfo?t of the epes. As touching another maner o? wape of Dyſtilling in the Sunne, reade hereafter in the p?oper place taught.

A simple method of distilling waters from plants in the sun. See Rosemary in the Appendix. From George Baker's *Jewell of Health.*

Of the compounde waters, efpecially of leaues, flowers, rootes,
feedes, fruite, herbes, and trees, lycours,
gummes, and woode.
A water for the eye fight.

The.Lxxxvi.Chapter,

A Water defending and preferuing the fight for a long time,
and purging the eyes of all fpottes : Take of the beft and
pleafanteft white wyne, twelue pyntes, of newe breade light

wrought and well wafhed, fower poundes, of Fennell, Celon-
dine, and of the heades of the fquill onpon, of eche fower ounces,
of Cloues fower drammes : thefe mire diligently togither in a
M.j.   glaffe

Distilling in large quantities. From George Baker's *Jewell of Health.*

of growing a little poison on the side, to be used in moderate amounts as a
soporific or pain-killer. After all, a lady of the manor in England, with some
deadly poison she happened to have on hand, had been able to finish off a
whole group of enemy soldiers quartered upon her. Handy poison is something for us to remember as we continue to grow monkshood in our gardens today. This is the plant that killed so many in Belgium in a salad in

the fifteenth century that the party has never been forgotten. Dioscorides' only antidote was swallowing whole a mouse which had just eaten the root.

Some of the books listed as "Markham" on seventeenth-century Boston booksellers' import lists may have included Gervaise Markham's *English Housewife*, as well as his work upon horsebreeding. We refer to *The English Housewife* here as expressing the social truism of its time. Markham's treatise on the good wife rivaled Leonard Meager's *English Gardener* in running through edition after edition in seventeenth-century England. In his list of accomplishments to be expected of the housewife, Markham sets first a knowledge of medicine, then cooking, gardening and distilling, in that order. If we ever wonder who really dispensed all these remedies made from the gardens, or who tended the gardens from which all the remedies came, or who made up all those syrups of clove-gillyflowers or waters of angelica and mint and roses and strawberries, here is our answer. It was the English housewife, whether she was at home in Old England or recreating her indispensable garden in "this wildernesse."

Thomas Tusser was another domestic authority of the times. A well-born and well-educated farmer and poet who lived in East Anglia in the sixteenth century, he left a great output of rhymed advice on farming and gardening, the ordering of a pious country household, and wise land use. First published in 1557, his *Five Hundred Points of Good Husbandry* was a much quoted reference book which went into many editions. He advises housewives on what they should have on hand:

> Good huswives provide, ere an sickness do come,
> Of sundry good things in her house to have some.
> Good aqua composita and vinegar tart,
> Rose water and treacle to comfort the heart.
> Cold herbs in her garden, for agues that burn,
> That over strong heat, to good temper may turn,
> White endive and succory, with spinage enough,
> All such with good pot herbs, should follow the plough.
> Get water of fumitory, liver to cool,
> And others the like, or else go like a fool.
> Conserves of barberry, quinces and such
> With sirops, that easeth the sickly so much.

Tusser does advise consulting a "Medicus" before giving medicine, but he does not appear to include the above in that category.

Since it was chiefly from their gardens that the early settlers had to take care of their medical needs, with their plant lists to guide us we can have a fairly good idea of what they hoped to do in "physick." So interrelated were "physicking" and gardening that we can see how a book on alchemical remedies by the great English physician Baker could be listed in a Boston seventeenth-century inventory as *Baker on Gardening*.

Surprisingly, considering their origins, the Puritans brought few plants with them for purely "magic" or superstitious reasons. To disregard the existence of witches when they were mentioned in the Scriptures would have been unthinkable in seventeenth-century New England. The devil's existence, of course, was a foregone conclusion, and to many he was as real as the Indians. Samuel Sewall, we know, felt the devil was at the bottom of the witch troubles. Sewall is lauded in our history today as the justice who repented. Actually, the doubt and fear expressed in his famous "Bill" seem to stem from the possibility that he may have been sitting upon a bench not legally qualified to try persons for witchcraft. He confides to his diary some years after the trials that, when he had been viewing the coffins of six of his children in the family tomb, he could not but wonder, like any good Puritan, what he had done to deserve such misfortune. He had had doubts, he says, about accepting a place on the court of Oyer and Terminer, so hastily created by Governor Phips to end the witch troubles. The legality of the court was in public dispute. "A great silence," when he thought about it then, had seemed to warn him not to go. His "Bill" which was read in church, while he stood to acknowledge it, says only that he, "sensible of the reiterated strokes of God upon himself and family," and being, "sensible that as to the guilt contracted upon the opening of the late Commission of Oyer and Terminer at Salem (to which the order of the day relates)," he desires to "take the blame and the shame of it. . . ." There is no doubt, however, that he believed in witches — "woe, woe, woe, witchcraft," as he said in his diary.

With universal acceptance of the reality of an unseen world governed by devil-directed spirits, it is a wonder that the thinking of New England Puritans escaped being encumbered with even more useless lumber than their already heavy load. An example of the sort of background knowledge they relied upon is contained in the little dog-eared medical reference books from an old house in Newbury, Massachusetts. *A Breviary of Health* and its sequel, *Extravagantes*, had been written in sixteenth-century England by Dr.

Andrew Boorde. When still a very young man he became a Carthusian monk, but he soon asked to be excused from the "rugerosities" of the order, and studied medicine, traveling throughout Europe and England. He spied for Thomas Cromwell on the continent to discover the feeling there for Henry VIII, was called in consultation during that king's illness, wrote several books on travel, medicine and house-building, and was mysteriously executed in mid-career. He was well known as a benefactor of the

A woodcut made to serve as a portrait of the great Dr. Andrew Boorde, although it is a century earlier in costume and feeling. Boorde's name has been let down into one corner of the cut, which is indeed of an obviously wise and kindly man. From a reprint of Boorde's book of travel, "declarynge the properties of all the regions, countreys and provinces, the whiche he did travel thorow."

poor through his eagerness to practice medicine wherever possible. *A Breviary of Health* and *Extravagantes* were published several times in England in the last half of the sixteenth century, changing with the times from Gothic black letter to Roman type. Leavened with common sense and wit, they are still weighted with caution about another, invisible world, and they illustrate for us the background of approved medical practice of the time. They give us a useful springboard from which we can take off to observe the medical practices of the Puritans with more perspective and sympathy than we might otherwise be able to muster.

In fact, we can see how, in the century before the great migration, common sense hardly dared prevail. In his chapter on the nightmare, Boorde announces that although it is supposed to be a spirit which infects those sleeping, and in spite of Incubus and Succubus and something very odd shown him by "the Anchress at St. Albans here in England," he believes "the Mare" to be "some vaporous humour rising out of the stomach to the brain coming from surfeiting and drunkenness . . . bringing a man sleeping into a dream to think that the which is nothing is somewhat." And yet he cannot forbear to add that "if it be a spirite" he has read of a "herb named Furga Demonum, called by the English 'St. John's wort' " whose "vertue" it is "to repell such malificiousness of spirites." Our roadsides bloom yellow with this herb today, but there is nothing to show it was ever used against witches, unless a remark of John Josselyn's could be counted as such — that cheeses put up in it would keep at sea.

Considering this background we should admire how common sense did seem to prevail among the settlers and how much nonsense they left behind them. For instance, Sir Kenelm Digby's secret remedy, called his "powder of sympathy" seems never to have crossed the Atlantic, although the younger Governor Winthrop knew of it. This was a development of the ingenious theory that a wound could be cured by treating the weapon which made it. Digby worked his cure by dipping the injured man's bandage in his concoction, and his friend's wound healed. We have no account of anything like it being attempted here.

While the settlers soon gave up the old idea that boiling oil, treated with herbs, did a fresh wound good, and for wounds and bruises fell back upon "simples," those old reliable herbs which could be used alone, generally they appear to have favored package deals. Called "portmanteau" at a later date these cures comprised whole lists of well-known and trusted herbs all done up together, so that one or two at least could be expected to do the trick. We have an "excellent water for the eyes," for instance, from Captain Lawrence Hammond, who was deputy to the General Court from Charlestown and a Middlesex County recorder and militia officer, which would have required a good garden: "Take sage, fennel, vervain, betony, eyebright, celandine, cinquefoil, herb of grass [he means herb of grace, or rue], pimpernel. Steep them in white wine one night, distil all together, and use the water to wash the eyes." For a simple, he says, "the juice of eyebright is excellent for the sight," but he adds another cure for poor eyesight which can be taken internally. It too contains eyebright, but this time infused in white wine, seethed with rosemary, and drunk often.

A set of directions for the care and cure of measles is copied out in an Ipswich storekeeper's account book, along with some other good ideas for pains in the back and "wind collick," written upside down from the accounts to save paper. The measles cure seems to be a hand-me-down from the first medical broadside by the Reverend Thacher which he had copied from the great English physician Sydenham. It refers to the patient as "he" throughout and warns against a cold drink lest "he find death in the pot." The back and colic remedies are portmanteau cures. For "wind Collick," take summer savory, angelica, sweet tansy, elecampane. For pain in the back, a syrup is made of yarrow and either borage or "comphrey," and brandy and gunpowder are added. While it may surprise us to see gunpowder taken internally, even after various practitioners such as Cotton Mather had come out publicly against the swallowing of bullets as beneficial for "a kink in the guts," we know that John Josselyn was willing to report favorably upon gunpowder for toothache, "being put into the hollow."

But toothache was not the only affliction for which gunpowder was used. Although we know that the settlers had books upon animal husbandry, such as Gervaise Markham's book about the horse, and although we see Gerard occasionally giving hints about treating horses and cattle, and John Josselyn recommending hyssop for hens and so on, there is never any mention of what to do for an ailing dog. Which is all the more remarkable when we consider that, due to the prevalence of wolves, settlers in Essex County were encouraged to keep a dog for each family, a mastiff for those whose financial standing was five hundred pounds or over, and "a hound or a beagle" for those of lesser means. So we are pleased to find, in a rather unlikely

Title page of *The Queens Closet*, a mid-century book for the housewife, with her
potential skills listed as "Physick, Chyrurgery, Preserving, Candying and Cookery." This was a popular item with Boston booksellers.

place, directions for treating a mangy dog. *The Queen's Closet Opened* includes, among such mild remedies as distilled marigold water applied to a
lady's forehead for a headache, a very drastic treatment for a dog with the
mange. In winter he is to be confined to a bed of deep, fresh straw. In
summer he is to be tied in the full sun to help the treatment speed its cure.
To four ounces of tar, fresh grease is added until the whole substance is
easy to apply and to spread. To this then is added half a spoonful each of
brimstone and gunpowder and two spoonfuls of honey. Three treatments
of being smeared with this remedy will cure him. We know the Winthrops
brought mastiffs. We know Josselyn set dogs upon wolves on the coast of

Maine. And the Indians had dogs, at least after the settlers had arrived. So it is good to know that there was a note upon their medicine among all the many "receipts" for their masters.

The Indians do not appear to have been of much help in medicinal matters, legend to the contrary. What the Puritans learned from the Indians in medicine does not seem to have amounted to much, although we have an attractive picture from William Wood's *New Englands Prospect*, in 1634, of the Indian women in the Englishwomen's kitchens for a fair exchange of women's chat. It was these visits which so disturbed the Indian men who considered that Englishmen spoiled their own wives and were afraid good

16   *The Pearle of Practice. Choice*

~ *To make water of Life.*

Take Balm leaves and ftalks, Burnet leaves and flowers, Rofemary, red Sage, Taragon, Tormentil leaves, Roffolis, red Rofes, Carnation, Hyfop, Thyme, red ftrings that grow upon Savory, red Fennel leaves and roots, red Mints, of each one handful; bruife thefe hearbs and put them in a great earthen pot, and pour on them as much white Wine as will cover them, ftop them clofe, and let them fteep for eight or nine dayes, then put to it Cinnamon, Ginger, Angelica feeds, Cloves and Nutmegs, of each one ounce, a little Saffron, Sugar one pound, Rayfins Solis ftoned one pound, Dates ftoned and fliced half a pound, the loines and legs of an old Coney, flefhy running Capon, the red flefh of the finnews of a leg of Mutton, four young Chickens, twelve Larks, the yolks of twelve Egs, a Loaf of white bread cut in fops, and two or three ounces of Mithridate or Treacle, and as much Baftard, or Mufcadine as will cover them all. Diftill all with a moderate fire, and keep the firft and fecond waters by themfelves,

Phyfical and Chyrurgical Receipts. 17

felves; and when there comes no more by diftilling, put more Wine into the Pot upon the fame ftuff, and diftill it again, and you fhall have another good water. This water muft be kept in a double glafs clofe ftopt very carefully: it is good againft many infirmities, as the Dropfie, Palfie, Ague, Sweating, Spleen, Worms, Yellow and Black Jaundies; it ftrengthneth the Spirits, Brain, Heart, Liver, and Stomach. Take two or three fpoonfuls when need is by it felf: or with Ale, Beer, or wine mingled with Sugar.

Dr. Atkinfons *excellent Perfume againft the Plague.*

Take Angelica roots, and dry them a very little in an Oven, or by the fire; and then bruife them very foft, and lay them in Wine Vinegar to fteep, being clofe covered three or four days, and then heat a brick hot, and lay the fame thereon every morning; this is excellent to air the houfe or any clothes, or to breath over in the morning fafting.

To

How to make "water of Life." Except for acquiring the larks, the New England housewife could have accomplished this concoction, so much of it depending upon her garden. From *The Queens Closet*.

labor might be lost. There was a good deal of trying to find plants similar to those the English women knew in England, which might be employed to like ends. For instance, both sides so to speak, used mint for stings, and perhaps, for all we know, like the Greeks, as a contraceptive. And of course there were new plants of which too much was hoped, such as sassafras and sarsaparilla. The sixteenth-century discovery in South America of cinchona bark — "Jesuit's bark" — which became so important in the treatment of malarial fevers, made hopes run high for an equally dramatic and valuable discovery in North America for yet another scourge, the "pox," or venereal disease, which the Spaniards would appear to have exchanged for tropical fevers.

Sarsaparilla and sassafras at first excited great expectations, but by the time New England Puritans were employing them they had begun to occupy the relatively simple places they hold in pharmacopoeias today. The beautiful blue lobelia, given us by the Indians, has its supposed curative powers commemorated in its botanical name, *Lobelia syphilitica*. Cotton Mather, who had learned the technique of inoculation against smallpox from a Negro slave, introduced another hopefully valuable discovery from another primitive man, a cancer cure with which a "famous" Indian had great success. Cotton Mather wrote to the Royal Society about the plant which the Indian used and sent them a root. He called it "Fagiana" and said he had had experience of it himself only in drinking a tea made from it, which he had found vastly refreshing, but the "famous Indian" did "very strange cures upon cancers by a decoction of it taken inwardly" and a "Cataplasm of the boiled plant . . . upon the place affected." The plant he is referring to is the parasitic "beech drops" or *Epifagus virginiana*, and it is hard not to assume that the "famous Indian," like many searchers for cures before him, had assumed that what was of parasitic growth itself might be the remedy for a disease of like nature.

All the popular herbals, upon which everyone depended, linked "meate and medicine" in their portentous introductions. Gerard's *Herball*, in 1567, after exclaiming, "What greater delight is there than to behold the earth apparalled with plants, as with a robe of embroidered work," continues: "The delight is great but the use is greater, and joined often with necessities. In the first ages of the world they were the ordinarie meate of men, and have continued ever since of necessarie use both for meates and to maintain life, and for medicine to recover health." Parkinson in his *Paradisus* in 1629 says we have learned from Noah, who had it from Adam, "the knowledge both of what Herbes and Fruits were fit, eyther for meate or medicine, for Use or Delight," and mentions what many of his plants will cure. Culpeper, himself a Puritan, translated Galen into English for the benefit of the poor who could not afford professional help, and offended his fellow physicians in London by deliberately making fun of them in all their robes and rings, and, he says, ignorance. He added insult to injury by being so extremely astrological. Culpeper's remedies are herbal to a degree, to the last root, leaf and flower.

Sometimes the knowledge upon which all these men so depended seems to us today inextricably ill-assorted. Still, it charms us as we think of Samuel

10 *The Pearle of Practice. Choice*

A Cordial Syrup to cleanse the blood, open Obstructions, prevent a Consumption, &c.

    Take Rosemary flowers, Betony, Clove-gilly-flowers, Borrage, Broom, Cowflip flowers, Red-rose-leaves, Melilot, Comfrey, Clarey, Pimpinel flowers, of each two ounces, red Currans four pounds: infuse all these into six quarts of Claret Wine, put to it fourteen pounds of ripe Elder berries, make the Wine scalding hot, then put in the Flowers, Currans, and Elder berries, cover the pot, and paste it very close, set it in a kettle of warm water to infuse fourty eight hours, till the vertue of the ingredients be all drawn out, then press it out hard, and put to every pint of the liquour one pound and three quarters of powder Sugar, boil and scum it till you finde the Syrup thick enough, when it is cold bottle it, and keep it for your use. Take two spoonfuls in a morning, and so much in the afternoon, fasting two hours after it.

                        A

Physical and Chyrurgical Receipts. 11

A Medicine for a Dropsie approved by the Lady Hobby, *who was cured her self by it.*

    Take Carowayes, Smallage, Time, Hysop, Water-cresses, Penniroyal, Nettle tops, Calamint, Elecampane-roots, of each one little handful, Horse radish two pounds, boil them in six quarts of running water, until half be consumed, then strain it, boil it a new with a pottle of Canary Sack, Liquorish twelve ounces, sweet Fennel-feed, one ounce bruised, and a quarter of an ounce of Cumin-feed bruised; boil all these above half an hour, then strain it, and keep it for your use; nine spoonfuls in the morning fasting, and as much at three or four a clock in the afternoon, use it for some time together. This the Lady *Hobby* proved by her self.

Dr. Adrian Gilberts *most soveraign cordial Water.*

    Take Spearmint, Broom-mint, Mother of Time, the Blossome tops of Garden Time, red Penniroyal, Scabious, Celandine, Wood Sorrel, Wood Betony,
                         An-

Page from *The Queens Closet*, showing the plants customarily used in home medicines.

Sewall, our only Pepys, setting out to look for herbs and being pleased to find yarrow, that ancient wound-herb approved by Galen and said to have first been recommended to Achilles by his tutor, Chiron, a centaur.

The medical practice of the early settlers, however, did not depend merely upon a few plant-filled books for reference. Their backlog of knowledge which enabled them to proceed with such confidence represented a fair summary of the useful medical knowledge of mankind from the beginning of recorded time up to and through the discovery by Harvey of the circulation of the blood. But while scholars like Cotton Mather and John Winthrop, Junior, tried to keep up with medical advances in the Old World and add to them, it was primarily upon knowledge handed down from the early

190   *The Pearle of Practice. Choice*

good to fill hollow Ulcers with flesh and apply a Plaifter on the top of it.

7. Balfam two ounces good for all forts of green wounds, being put in warm.

A Receipt of the Oyl of Johns-wort.

Take a quart of the beft white Wine infufe therein pickt flowers of Saint Johns-wort, then ftow thofe flowers very dry, and put in more into the fame Wine, infufe them again, fo long till the Wine be very ftrong and red coloured with the Saint Johns-wort, then ftrain out the Wine clear from the flowers, put thereto a pint of the beft Sallet Oyl, a quarter of an ounce of Cinnamon bruifed, a quarter of Cloves bruifed, one race of very good Ginger fliced, one good handful of the yellow flowers of Saint Johns-wort pickt very clean; boil all thefe on a very foft fire till the Wine be all evaporated, when it is almoft boiled, put in one good fpoonful of pure Oyl of Turpentine, let it boil in it a little; fo keep it for your ufe the elder the better.

Phyfical and Chyrurgical Receipts. 191

A Receipt for an extraordinary wafting of the Back, and for the Stone and Shrangury ufed by Juftice Hutton.

Take Plantain and Ribwort, diftill them in an ordinary Rofe ftill, when you have occafion to ufe it, take Pippins and roaft them, and take away the skin and coat, and put them into the water, making thereof a lambfwool as thick as you pleafe, fweeten it with fome Loaf fugar, the fweeter the better, take thereof half a pint when you go to bed, and this do nine or ten nights together, efpecially when you feel an heat in the Back.

For the Teeth.

If you will keep your teeth from rotting, or aching, wafh the mouth continually every morning with juyce of Lemons, and afterward rub your teeth with a Sage leaf, and wafh your teeth after meat with fair water.

To

How to make home remedies from the garden. From *The Queens Closet.*

Greeks and Romans that most of the Puritans looked for their medical information, or what passed for it.

Aristotle's theory of the four elements — fire, air, earth and water — being affected by the four qualities — hot, cold, dry and moist — to form the human temperaments of sanguine, phlegmatic, choleric and melancholic, was a basic assumption in every household with social and civic responsibility. All the herbals, after recording the "vertues" of each plant, recorded also its qualities of temperature which would aid anyone applying this knowledge to the practice of "physick" — hot and cold, moist and dry, and in what degree.

To those trying to restore the balance or "temper" of an individual, be he melancholic, phlegmatic, choleric or sanguine, with a distillation of some plant, its roots, seeds, leaves or flowers, it was essential to know each quality in each particular degree.

Although realizing their debt to the "ancient heathen" for discovering part of the divine plan, the New England Puritans were, like most of their sixteenth- and seventeenth-century English countrymen, chiefly dependent upon the assurance in Genesis that everything growing upon the earth was there to serve men. On the third day they knew, as land appeared, the earth brought forth grass and "the herb yielding seed, and the fruit tree yielding fruit after its kind." On the fourth day came "lights in the firmament of Heaven," the sequence upon which opponents to astrological interpretation base their claims that plants cannot be dependent upon planets, as plants came first. But, as Gerard said, "Let us look not up at the planets but down at the plants."

But what the Puritans principally depended upon, of course, was what took place on the sixth day when God gave man and woman "every herb bearing seed which is upon the face of all the earth, and every tree, in the which is the fruit of a tree yielding seed; to you it shall be for meat." Confirmed in the 104th psalm, "He causeth the grass to grow for the cattle, and herb for the service of man," it was rendered into jingling verse in the Bay Psalm Book:

> For beast hee makes the grass to grow,
> Herbs also for mans good. . . .

and appeared again within the century as a hymn.

Grass for our cattle to devour,
He makes the growth of every field;
Herbs for man's use, of various power,
That either food or physic yield.

So Galen and the herbals and the Old Testament are all joined by the Puritans to help them take care of everyone in the New World with the plants there provided, grown and combined with those from the Old.

With all this ancient wisdom buoying them up, no wonder Puritan housewives felt able to deal with their troubles and those of their neighbors directly from their gardens. Based upon plants brought from England in seeds, roots and cuttings, with "Indian plants" added, preserved or dried in their attics and stillrooms with their knowledge of and skill in the pseudoscience of alchemy, the practice of medicine in seventeenth-century New England became the first flourishing home industry.

8

"A Pleasing Savour of Sweet Instruction"

In seventeenth-century Middlesex County Court records, a witness reports that the young man on trial had come to his house to call upon him and reached up over the fireplace to take down the "great herbal," to see if he could find confirmation there for his question about juniper berries having qualities to cause a mare to "slink" her foal. While this story placed the young man in a questionable position in a case involving paternity and murder, for us, however sympathetic we may feel after this interval of time, it places the herbal. When inventories mention "a herbal and a bible" we know where they were kept. When inventories give us long lists of distinguished books, like the libraries of the Mathers and John Winthrop, Junior, we can be sure of a "study" with shelves; but when the books seem to have been part and parcel with the pewter and consisting only of those no one could live without, they went into a place within quick and easy reach. Curiosity having led us to inquire into the properties of juniper berries in the three herbals we know were owned by the founding fathers, one realizes the young man had merely hearsay behind him.

John Josselyn has mentioned only that women dried sheets over juniper bushes and that a good hedge could be made from planting eglantine and junipers close together. The herbals are united in equally innocent reports. Johnson-upon-Gerard, in 1633, gives juniper many "vertues," including that it makes a good varnish with linseed oil to "beautifie pictures and painted tables" and to make iron "glister," but he does not mention what the young man was looking for. John Parkinson, in 1629, does not put juniper into either his pleasure garden or his kitchen garden. Nicholas Culpeper, who died in 1654, praises the juniper bush and advises eating a dozen berries raw every morning as a handy way of extracting the "chymical oil" which is so good for colic. One might think that the young man had no recourse, if all the herbals were so reticent; but, actually, Gerard names several plants with this power, one so strong that he has put sticks about it in his garden lest any woman so much as step over it. So all we have in the end is a certain kind of sympathy for the young man and assurance of the prominent position of the herbal in the household.

There were three great herbal authorities upon which the seventeenth-century settlers of New England mainly depended. Noted in inventories, listed by booksellers, mentioned in libraries and referred to in texts, these are: Gerard's *Herball*, 1633 edition, done over by Thomas Johnson, Parkinson's *Paradisi in Sole, Paradisus Terrestris*, published in 1629, and Culpeper's *Herbal* which burgeoned into many editions in the last half of the century. Parkinson's second book, *Theatrum Botanicum* published in 1640 was intended to supplement his first book but it became "a Herbal of large extent" and so encyclopedic as to serve us only as a final check on the comments familiarly served up in his *Paradisus*. A fourth herbal, that of Dodoens, first published in Flemish in 1554, was owned by Governor Brewster and was probably brought to this country on the *Mayflower*. This was the basic herbal upon which much of Gerard depended, so we are not quoting it separately. Presumably Brewster had Lyte's English translation from the French edition, and the translation Gerard is supposed to have used was a later one. In any case, no herbal of the seventeenth century could depart far from the bulk of constantly done-over knowledge they all contained.

And yet these three main seventeenth-century authorities are utterly dissimilar, each from each. Gerard's *Herball* sets out to be encyclopedic and achieves such vigor, charm and style as place it in the forefront of quotable books in English literature. It has both spark and salt. Neither of the others has either. In his *Paradisus* Parkinson is every inch and every word the gentleman, discriminating, full of taste, wishing to be constantly instructive, careful of his facts and credits, a gardener above all. Culpeper has no engaging qualities whatever, unless one admires a sharp temper. He is brusque and dogmatic. He goes out of his way to scorn his detractors and recognizes no equal. He is determined to enable people to cure themselves and to this end he uses all convenient authorities — especially, for instance, Gerard — without deigning to credit anyone since Galen. He plunges ahead, allocating to each heavenly body its due influence upon each plant. Every plant is tagged with its proper temperature. His examples of others' errors and omissions are blood-chilling. Gerard's illustrations in the same vein are funny and often sly. Parkinson's asides are always wise and good, never very amusing except when he is expressing a mild distaste for something, as, for instance, the Jerusalem artichoke. (Unless, of course, one accepts Paradisi in Sole, or Park-in-Sun, a pun on his own name, as wit.) Gerard is pithy and witty; Parkinson constructive and kind; Culpeper bristling with scorn for those who do not believe in him. And in the end they all add up to much the same sort of information about much the same plants. Still, it is better to consider each one separately, since it was upon each one, as an individual, that so many of the settlers and so much of their gardening depended.

The earliest of these herbalists is John Gerard (1545–1612) who describes himself upon the title page of his book as a "Master of Chirurgerie" in London where he also acted as superintendent of the gardens of Lord Burghley, to whom the herbal is "presented" for his Lordship's "protection." In 1597, when the herbal was first published, Gerard was Warden of the Company of Barber-Surgeons in London. He became its Master in 1607. Though gardening had always been the nursemaid of the medical sciences, Gerard excelled all others in being at the same time a practicing surgeon, an active and indefatigable gardener, a plant collector and a dedicated herbalist. While he superintended his Lordship's garden he maintained also a garden of his own in Holborn, "the little plot of myne own especiall care and husbandry," from which he made a famous list of garden plants at the end of the century. He had a wide acquaintance among plant collectors and gardening friends to whom he was happy to give credit. In fact, he is so generous to individuals whose only recorded notices are his immortalizing references to them, that it is hard for us to see Gerard as blackly plagiaristic as do Canon Raven and other modern critics. Gerard's corrector and amplifier of the second edition, Thomas Johnson, also saw him somewhat darkly and at closer range. But to call the herbal an "adaptation" of Dodoens, as Gerard's sharpest critics enjoy implying, is to deny the life that Gerard gives to every item. Here, for instance, is what he has to say of elecampane: "It groweth in meadows that are fat and fruitful; it is oftentimes found upon mountains, shadowie places that be not altogether drie; it groweth plentifully in the fields on the left hand as you go from Dunstable to Puddle hill; also in an orchard as you go from Colbrook to

Portrait of John Gerard from the 1598 edition of his *Herball* which shows him holding a specimen of the Virginia potato.

Ditton ferry, which is the way to Windsor, and in sundry other places, as at Lidde, and Folkstone, neere to Dover by the sea side." Even if this entry was sent him by a friend — as many were — the direct friendliness of it is disarming. Elecampane is that favorite remedy for throat ailments still used in New England for horses.

At his very least, John Gerard evokes the English countryside, the flowers and the people, as no one else did. Gerard's contemporary, the great Flemish botanist De l'Obel, his editor, Thomas Johnson, who corrected and amplified the *Herball* for its second edition in 1633, and the late Canon Raven are all agreed upon Gerard's charm, if also upon his wholesale appropriation of the works of others, faults and all. However, when Milton borrows from Gerard, practically verbatim, his description of the tree chosen by Adam and Eve to clothe them, no one thinks the less of Milton. Nor did Cotton Mather offend by borrowing from Milton his picture of hell's torments.

It is unfair to interdependent gardeners to let literary borrowers go free. Let us consider at least Milton's indebtedness to Gerard's description of the "Arched Indian Fig Tree":

This rare and admirable tree is very great, straight, and covered with a yellowish bark tending to tawny; the boughs are many, very long, tough, and flexible, growing very long in short space, as do the twigs of Oziars, and those so long and weak, that the ends thereof hang downe and touch the ground, where they take root and grow in such sort that those twigs become great trees; and these being growne up unto the like greatnesse, doe cast their branches or twiggy tendrils unto the earth, where they likewise take hold and root; by means whereof it cometh to passe, that of one tree is made a great wood or desert of trees, which the Indians do use for coverture against the extreme heate of the Sun, wherewith they are grievously vexed; some likewise use them for pleasure cutting downe by a direct line a long walke, or as if it were a vault, through the thickest part, from which also they cut certaine loope-holes or windows in some places, to the end to receive thereby the fresh coole aire that entereth thereat, as also for light, that they may see their cattell that feed thereat, to avoid any danger that might happen unto them either by the enemie or wilde beasts; from which vault or Walke doth rebound such an admirable echo or answering voice; if one of them speake unto another aloud, that it doth resound or answer againe foure or five times, according to the height of the voice, to which it doth answer, and that so plainly, that it cannot be known from the voice it selfe; the first or mother of this wood of desert of trees is hard to be known from the children, but by the greatnesse of the body, which three men can scarcely fathom about; upon the branches whereof grow leaves hard and wrinckled, in shape like those of the Quince tree, greene above, and of a whitish hoary colour underneath, whereupon the Elephants delight to feed. . . .

Of this Milton makes the following use, when Adam and Eve are searching, upon Adam's advice:

> Some tree whose broad smooth leaves together sowed,
> And girded on our loyne, may cover round
> Those middle parts. . . .

They go together

> . . . into the thickest Wood, where soon they chose
> The Figtree, not that kind for Fruit renowned,
> But such as at this day to Indians known
> In Malabar or Decan spreds her Armes
> Braunching so broad and long, that in the ground
> The bended Twigs take root, and Daughters grow
> About the Mother Tree, a Pillard shade
> High overarch't, and echoing Walks between;
> There oft the Indian Herdsman shunning heate
> Shelters in coole, and tends his pasturing Herds
> At Loopholes cut through thickest shade; Those Leaves
> They gathered. . . .

Which seems to me to leave Gerard well ahead, as his whole *Herball* stands out above all others.

Gerard, after his splendid and flattering flourish to the "right Honourable and my singular good Lord," his patron and employer, Lord Burghley, begins his book: "Among the manifold creatures of God that have all in all ages diversely entertained many excellent wits, none have provoked mens studies more, or satisfied their desires so much as Plants have done, and that upon just and worthy causes: For if delight may provoke mens labor, what greater delight is there than to behold the earth apparelled and garnished with great diversities of rare and costly jewels?" He extemporizes upon these delights of the outward senses and then announces, "but the principall delight is in the minde, singularly enriched with the knowledge of these visible things, setting forth to use the invisible wisdome and admirable workmanship of almighty God. The delight is great, but the use greater, and joined often with necessity. In the first ages of the world they were the ordinarie meate of men and have continued ever since of necessarie use both for meates and to maintain life, and for medicine to recover health." He goes on to say that he has worked for his patron for twenty years and, "to the large and singular furniture of this noble Island," he has added "from forreine places all the varietie of herbes and Floures that I might in any way obtaine." It charms us to see that with all that "varietie" he had

The layout of a simple and workable garden in sixteenth-century England, with raised beds and room for walking and talking between them.

his portrait done holding a leaf of the "Virginia potato." He has labored "with the soile to make it fit for the plants, and with the plants, that they might delight in the soile," with what success he leaves it to those to judge who have seen his Lordship's garden and "the little plot of myne own especial care and husbandry." Realizing that gardens are private and, under negligent successors, come soon to ruin, he has resolved to make his labors "common" and has brought this report or history of plants to a "just volume," "richer than former Herbals," and so presents it to his Lordship.

He has little more to add in his following address to the "courteous and well willing reader" except a greater embellishing of the "excellent art of simpling, which is neither so base nor contemptible as perhaps the English name may seem to intimate" but "a study for the wisest, and exercise for the noblest, a pastime for the best." After listing the various heroes whose names are commemorated in their special plants, he propounds a question which has a fresh validity for us today — "Who," he says, "would therefore look dangerously up at the Planets that might safely look down at Plants?''

Thomas Johnson, whose edition of Gerard is most frequently quoted and was the edition upon which John Josselyn depended, also addressed the reader, at greater length and with less charm than Gerard. He says he feels no need to tell more about the importance of plants than that "God in his infinite goodness and bountie hath by the medium of Plants bestowed almost all food, clothing, and medicine upon man." He then gives a running history of all those who have written upon plants since men could write at all, speaks at some length upon his own labors in correcting and amplifying Gerard's volume, acknowledges most especially his own indebtedness to Mr. John Goodyer of Maple Durham, his "onely assistant," for his observations of plants he has grown and seen in gardens, and to Mr. George Bowles of Chiselhurst in Kent for his knowledge of wild flowers.

Thomas Johnson, born in 1604 in Pontefract, English stronghold of the licorice root, was a practicing apothecary to Snow Hill, a Free Brother of the Society of Apothecaries, and one of a group of ten apothecaries who used to go on simpling trips, or botanical excursions, into the surrounding counties. His skill in medicine and in botany earned him the degree of Bachelor of Physic and, four months later, Doctor of Physic. He joined the London Royalists at Oxford and was wounded in 1644 in the shoulder. He died within a fortnight, from a fever which had set in after he was wounded. Johnson was a man of great ability to whom Gerard may well owe much of his fame. He revised Gerard's *Herball* in one year's time, to get it published ahead of Parkinson whom he knew was just finishing his *Theatrum Botanicum* for Johnson's own publishing house. Cutting throats was as usual then as wholesale borrowing.

The herbal which John Josselyn carried about with him, and to which he so hopefully wished to add, was the 1633 edition of *John Gerarde of London, Master of Chirurgerie, very much enlarged and amended by Thomas Johnson, Citizen and Apothecarye of London.* This is probably the edition that turns up in the last half of the century in the lists of Boston booksellers.

John Parkinson's *Paradisi in Sole, Pardisus Terrestris* would seem to be titled most forbiddingly, even considering it was intended as a lighthearted pun in an age that loved punning. Parkinson's *Earthly Paradise* might be the equivalent today. A dazzling and complicated frontispiece of Adam and Eve running happily about a Paradise where all the trees and flowers are much the same size, establishes the tone of the book, where words like "delight" abound.

Parkinson begins by offering "this speaking Garden" to the Queen, "that it may informe you in all particulars of your store, as well as wants, when you cannot see any of them fresh upon the ground." Always the reserved and modest instructor, he addresses the "Courteous Reader" at some length, setting out what everyone then knew to be true.

Portrait of John Parkinson holding what appears to be a primula. From a reprint of the 1629 edition of his *Paradisus.*

Although the ancient Heathens did appropriate the first invention of the knowledge of Herbes, and so consequently of physicke, some unto Chiron the Centaure, and others unto Apollo or Aesculapius, his son; yet we that are Christians have out off a better Schoole learned that God, the Creator of Heaven and Earth, at the beginning when he created Adam, inspired him with the knowledge of all naturall things (which successively descended to Noah afterwards, and to his Posterity); for as he was able to give names to all the living Creatures, according to their severall natures: so no doubt he had also the knowledge, both of what Herbes and Fruits were fit, eyether for Meate or Medicine, for Use or for Delight.

After a great deal more of the earliest possible background to gardening, he laments the loss of Eden and all that knowledge and then injects a charming line of his reasoning in writing this book which, perhaps intentionally, may serve as Parkinson's own epitaph.

That as many herbes and flowers with their fragrant sweete smele doe comfort, and as it were revive the spirits, and perfume the whole house; even so such men as live vertuously labouring to doe good and profit the Church of God and the Commonwealth by their paines or penne, doe as it were send forth a pleasing savour of sweet instructions, not only in that time wherein they live, and are fresh, but being drye, withered and dead, cease not in all after ages to doe as much or more.

John Parkinson, botanist, apothecary, prominent in the Society of Apothecaries, and, after the publication of this book in 1629, First Botanist to Charles I, cannot have been a Puritan. And yet his statement to the reader could not have seemed more sound to the Puritans who read him. His books appeared first in New England in the library of Leonard Hoar, president of Harvard College, who had hoped to institute a chemical laboratory and a garden and orchard for students "addicted to planting." After Hoar's death Parkinson's *Paradisi in Sole, Paradisus Terrestris* and *Theatrum Botanicum* were given to the Mathers, Increase and his son Cotton. Samuel Sewall was executor of Hoar's estate. Mrs. Hoar remarried and became Mrs. Usher. The Mathers, who had suffered a bad fire in their own library when Increase Mather's house burned in 1676, were invited to choose what books they would like from Hoar's library. It is interesting that they chose Parkinson, or did Mrs. Usher and Samuel Sewall press the two volumes upon them? Van Helmont's *Ortus Medicinus* was a more likely choice and was also among the Mathers' books from this source, later to be much quoted by Cotton Mather.

In his first section of *Paradisi in Sole, Paradisus Terrestris*, Parkinson discusses "The Ordering of the Garden of Pleasure," in nine short chapters. This is followed by "The Garden of Pleasant Flowers" which has one hundred and thirty-three chapters, each on a different sort of plant "and the kinds," and takes up rather more than the first two-thirds of the book. "The Ordering of the Kitchen Garden" follows with seven chapters and "The Kitchen Garden" with sixty-three chapters. These share the last third of the hundred and thirty-three chapters, each on a different sort of plant "and the kinds," and takes up rather more than the first two-thirds of the book. "The Ordering of the Kitchen Garden" follows with seven chapters and "The Kitchen Garden" with sixty-three chapters. These share the last third of the book with "The Ordering of the Orchard," first useful and then ornamental trees, which "beare not fruit fit to bee eaten," and finally some additional plants remarkable "more for their raritie." The book ends with a chapter on the "Virginia Sumach" and one on "The Virginia Vine, or rather Ivie." Three indexes assist us; one of the Latin names; one of the English, and "A Table of the Vertues and Properties." The entire intent of the book is to be easily useful, and never to aggrandize the author, delighted as he sometimes is to have acquired and cultivated rare plants "to be seen but with a few."

Parkinson's purpose in his *Paradisus* is primarily to introduce us to a garden where "beautifull flower plants, fit to store a garden of delight and pleasure" are "severed" from those that are "wilde and unfit." This aesthetic intent was then quite new in the history of gardening in Europe and England. Even Parkinson cannot quite break away from the traditional idea of gardening as primarily "for meate and medicine," and he hastens to assure us that he will "set down the Vertues and Properties of them in a briefe manner, rather desiring to give you the knowledge of a few certaine and true than to relate as others have done, a needless and false multiplicitie." However, this is his fourth and last point, his first being to show "all the chiefest for choice and fairest for show" flowers and their kinds; the second to name them as soundly as he can and show where others may have erred: the third to "embellish this Worke" with the "figures of all such plants and flowers as are materiall and different one from another, but not as others have done, that is, as number of the figures of one sort of plant that have nothing to distinguish them but the colour." And the fourth we know was useful and medicinal.

The second original contribution that Parkinson makes to gardening is

The title page of John Parkinson's *Paradisus*, the 1629 edition.

his calm and just consideration that man may have to live in a house not exactly of his choice, and that his garden may also have to be the best he can manage under the circumstances. This is a radical departure from the contemporary European standard of excellence in garden design which takes off from heights of grandeur and sophistication to immortalize the name of both owner and designer. Parkinson was gardener to royalty, but he is always aware of the ordinary mortal to whom his very first sentence is disarmingly directed to give comfort: "The severall situations of mens dwellings are for the most part unavoidable and unremovable. . . ." That being so, he advises that the garden be placed before the "fairest" parts of the buildings so they will form a shelter for it, and it may be enjoyed as conveniently as possible. The rooms "abutting thereon shall have reciprocally the beautiful prospect into it, and have both sight and sent of whatsoever is excellent and worthy to give content out of it, which is one of the greatest pleasures a garden can yeeld his Master." After facing facts about men's houses, Parkinson allows that, as far as the design and plan go, "every man will please his own fancie," so Parkinson will show only "the severall formes that many men have taken and delighted in, let every man chuse which him liketh best."

He illustrates with a set of six elaborate geometrical patterns on his first page of illustrations and then announces that every man may invent more as it shall please him, but that the important thing to remember in designing a garden is that "the fairer and larger your allies and walkes be, the more grace your garden shall have, the less harme the herbes and flowers shall receive, by passing by them that grow next unto the allies sides, and the better shall your weeders cleanse both the beds and the allies."

And then he goes on to his plants: those for edging, those for hedging, "out-landish flowers," and those that are "called usually English flowers," how to plant them, how to care for them, and their pests. And his very last chapter in the section on "Ordering of the Garden of Pleasure" announces firmly that there is "not any art whereby any flower may be made to grow double that was naturally single, nor of any other seat or colour that it first had by nature. . . ." So we know exactly where we are in time. We see throughout the book that Parkinson enjoys double and striped and unusual varieties, but as natural blessings.

Parkinson next lists alphabetically all the plants he knows and wishes to contribute to the gardens for pleasure and delight, for meat and for medicine. He has added many from the New World at a time when apparently any plants from the New World were at such a premium that someone said rather bitterly of a neglected flower, which he himself fancied, that one need only declare it to have come from America to make it immediately sought after.

Of interest is Parkinson's entry on the Jerusalem artichoke, if only to show how he maintained his judgment in the face of whatever rarities might be introduced. This is not a reference one can look up in a hurry, because Parkinson does not list either artichokes or hartichokes in his Index. He

Parkinson's suggestion for laying out garden beds.
From a reprint of the 1629 edition of his *Paradisus*.

‡ *Flos Solis Pyramidalis.*
Ierusalem Artichoke.

quite deliberately places Jerusalem artichokes under potatoes, of which he recognizes three kinds: Spanish, Virginian and Canadian. Parkinson says of those potatoes brought from Canada that the English have called them artichokes of Jerusalem only because the root when boiled tastes something like "the bottome of an Artichoke head," which explains one part of the name, anyway. Parkinson gives the uses of all three sorts of potatoes and says of the third: "The Potatoes of Canada are by reason of their great increasing growne to be so common here with us at London that even the most vulgar begin to despise them, whereas when they were first received among us they were dainties for a Queene. Being put into water they are soon boiled tender, after which they bee peeled, sliced and stewed with butter and a little wine, was a dish for a Queene being as pleasant as the bottome of an Artichoke, but the too frequent use, especially being so plentiful and cheape, hath rather bred a loathing than a liking of them."

Parkinson's *Theatrum Botanicum*, published in 1640, is engagingly dedicated to the King, as it is a "manlike worke of Herbs and Plants," whereas the 1619 "Feminine work of Flowers" was dedicated to the Queen. Originally intended, says Parkinson, as a "Physical Garden of Simples," time has changed it to a "Theater of Plants." It is organized into "Tribes" which are arranged in seventeen sections as follows:

1. Sweete smelling Plants. 2. Purging Plants. 3. Venomous, Sleepy, and Hurtful Plants and their Counterpoysons. 4. Saxifrages or Breakstone Plants. 5. Vulnerary or wound Herbs. 6. Cooling and Succory-like Herbes. 7. Hot and Sharpe-biting Plants. 8. Umbelliferous Plants. 9. Thistles and Thorny Plants. 10. Fearnes and Capillary Herbes. 11. Pulses. 12. Cornes. 13. Grasses, Rushes and Reeds. 14. Marsh, Water and Sea Plants, Mosses and Mushroomes. 15. The Unordered Tribe. 16. Trees and Shrubbes. 17. Strange and Out-landish Plants."

It is a very large and exhaustive work, obviously one on which Cotton Mather depended for reference for recommended remedies in his *Angel of Bethesda*. It is indeed, as Parkinson subtitled it, "*An Herbal of Large Extent.*" As, like all others, it leans in its turn upon Dioscorides and "the Arabian Physicians," to whom Parkinson is willing to pay tribute for extolling virtues "which the Greeks have not remembered," it becomes part of the great mass of general herbal lore of the seventeenth century. Individually the *Theatrum* forms the strong and definitive background of medical knowledge which appears more engagingly in the *Paradisus* as references to what people really did with various plants rather than purely what they were advised to do.

Nicholas Culpeper, the third of our triumvirate, was the first of the cure-yourself cult which of necessity became one of the longest lasting and perhaps still most doubtful blessings of the American countryside and character. He was born in Ockley, Surrey, October 18, 1616. His father was a well-connected and comfortably-off clergyman who died before Nicholas was born. His mother returned to live with her father, also a clergyman, from whom, presumably Nicholas acquired his inclination toward Puritanism, thought to have cost him his inheritance from his father's family who were all Royalists. Nicholas was sent well prepared to Cambridge and there acquired sufficient knowledge to be able, later, to translate the *London*

Pharmacopoeia into English and to have it published with his own interpretations. He did not stay long at Cambridge, however. A girl with whom he was planning to elope was struck by lightning. In a state of shock, he fled to his mother's home and refused ever to return to the university. Instead, he apprenticed himself to an apothecary in St. Helen's Bishopsgate and there studied herbal medicine, derived from Galen, and astrology, whose implications in medicine Galen had indicated. Culpeper became convinced that our lives are totally influenced by celestial phenomena, which is understandable in a young man whose love was killed by lightning. To everything he wrote he gave a strongly astrological interpretation which became a source of controversy for him all his life. Galen, to put it so simply as to risk the wrath of his spirit and Culpeper's, is reported to have said that no man can reasonably deny that the natural ground of medicine and disease depends much upon astral influence and elementary impression. This last phrase recalls Aristotle's theory that the elements were impressed into four categories — air, water, earth and fire — and within this framework Galen attaches to the last-named element, fire, all the "fiery signs" or constellations. Nicholas Culpeper became a dedicated, ardent and extremely biased propounder of the theory of our microcosmic dependence upon the macrocosm, and worked out a relationship of every part of the human body and every herb to some particular heavenly body.

The College of Physicians took great exception to his strong astrological bent, but part of their irritation with him may have begun, as he thought it did, with his rendering into English for the use of every man, the secrets of the *Pharmacopoeia*. Translated for us by Culpeper, they do not seem to have been worthy of great secrecy and high fees, but this may be in part due to their having become, even for us, as Culpeper intended, more or less common knowledge.

Although Nicholas Culpeper, upon whom so much of the home practice of medicine by the early settlers depended, began to influence his readers to become their own physicians only after the middle of the seventeenth century, much of the knowledge which he insisted upon sharing with anyone who could read was not new and had been in practice long before that. In translating the *London Pharmacopoeia* into English in mid-century, his feat lay in taking it from the preserves of the physicians and surgeons of seventeenth-century England and putting it into a form available to any of the poor and ailing who could read and could not afford the fees of the practi-

Portrait of Nicholas Culpeper from an edition of his
book *English Physician*, 1652.

tioners. The knowledge itself was mostly over a thousand years old, and
much of it should have been forgotten, such as various ancient historic
remedies having to do with worms and toads. In any case, the battle between medical professionals and amateurs was on and lasted for over a
hundred years. Culpeper died before much of his work was published, but
he saw much of it pirated within his lifetime. His influence was strong and
reached into every home, both in England and in the New World. Whether

or not the people who depended upon him believed in astrology — and most of them seem not to have — they believed in his nostrums, especially those they could grow in their own gardens and concoct in their own kitchens and stillrooms.

Culpeper seems from his writings to have been a man of considerable charm and originality, although it is hard to get back to the original Culpeper, he has been constantly amended and amplified. Apart from early pirated editions, got out in his own time and, as he claimed, full of errors for which no one could blame him, there were dozens of editions with additions or interpretations by others. Like *The Bay Psalm Book* and *The New England Primer*, all the early copies of what might be called the real Culpeper seem to have been worn completely away with use. The handsome editions with colored plates belong to the next century. So it is with the very greatest pleasure that we can refer here to the little edition of the *London Dispensatory*, Culpeper's English translation of the *Pharmacopoeia Londonensis*, published in London in 1683, of which sixteen copies were imported by a Boston bookseller.

In this little volume, in an "Appeal to the Reader," we hear the real Culpeper addressing us, pleading his wish to help the poor and to "do good to my Countrymen, yea them that are yet unborn." It is for their health, he says, that he has lost his own, writing "Seventeen Books of Physick (besides those already published) which will discover to you the whole Method of Physick." He offers an *Astrolog-Physical Discourse* as a foundation for the whole "Fabrick," and then launches upon proving the importance of the realization that God made "an Unity in all his Works, and a dependency between them, and not that God made the Creation to hang together like Ropes of Sand." Culpeper sees our inferior world as governed by a second superior one and influenced by it, and a third world ruling over all. The Elemental World in which we live, is governed by the Celestial, and that by the Intellectual. Anyone who denies this, he says, may as well deny the whole world was made for men. The celestial world of the stars which governs man's body is governed by the intellectual world of the angels. The elemental world of man is dependent upon the influence of the celestial in bodily matters, although man's mind is influenced by the highest world. The stars do not reach the rational part of man because that belongs to the intellectual, but to Culpeper it is obvious that there is a close relation between men's physical bodies and the celestial bodies. "But," he says, "because there is some dispute about it (I should have said Cavilling) by such as would fain have their Knaveries hidden, and therefore they would fain have the Stars made to stop bottles, or else for the Angels to play at Bowls with, when they had nothing else to do, but not rule the Elementary World, no, by no means." So he proceeds to prove the Stars rule over the Elementary World, first by Scripture, secondly by Reason, which is as far as we need to follow him. Bottle-stoppers and bowls are not Culpeper's idea of the use of the stars as we shall see in his descriptions of plants and their uses, taken from a later edition of his *Complete Herbal*.

This extreme theory, so fiercely held, estranged him from the leading physicians of his time, who were able to take the planets less seriously and directly. There were, however, other things that his medical contemporaries found it impossible to forgive. By his apparently easy access to printers he was able to dash off various pamphlets in which he did nothing to further the esteem in which the practicing physicians of mid-seventeenth-century London were held. In Culpeper's sincere zeal to be of the greatest, widest and most inexpensively come-by aid to the ailing, he derided the avarice of the lavishly overdressed practitioners who cared more for their fees than for their patients and who depended upon art rather than nature for their knowledge. "Poor ignorant men" is his gentlest name for them. In contrast to their greedy ignorance he offers his own services which are based upon the practices of the ancients from the days when, he is sure, the world was a healthier place than it had become in his day. Since Culpeper told people what they really wanted to hear, that health was within the reach of every man who could become his own "domestic physician," it is small wonder that his books ran into innumerable editions and that the College of Physicians fulminated against him in vain.

In 1640, when he was twenty-four, he took a bride of fifteen, Alice Field, who had a fortune. She must also have had wits and loyalty, for it was she who furthered his publications. They had seven children of whom only one lived to grow up. In 1642 Culpeper was wounded fighting on Cromwell's side at the Battle of Edgehill. He never recovered his health and died twelve years later, purportedly of tuberculosis as a result of his wounds, although his wife is recorded as saying it was due to "the destructive Tobacco he too excessively took."

Culpeper's translation of the *Pharmacopoeia Londonensis* appears as the *London Dispensatory*, in the mid-seventeenth century, and from then on it

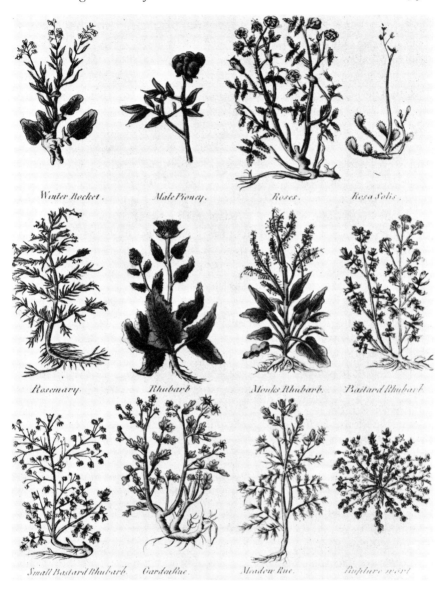

A page of Culpeper's medicinal plants from a 1798 edition of his *English Physician* which is too late for our purposes and yet shows nicely the confusions of what plant was really meant by "rhubarb."

becomes the springboard from which take off the innumerable editions of Culpeper's efforts to make every man his own physician. Dealing with the "natural appropriation of herbs, and the cure of all diseases," it begins as a small handbook with clearly designated divisions, defining all the medical terms, apportioning the commonest herbs to their appropriate planets and listing all the ordinary ills of the flesh with the herbs which will be most useful. He gives the temperature of each herb and then lists the various forms in which each may be taken, from waters and syrups through conserves and powders and electuaries to pills and troches. The remarkable characteristic of his work, in the edition of 1683, is its simplicity and clarity. There is great contrast between this book for home medical practice and Cotton Mather's rather swashbuckling effort to dazzle before he cures in the *Angel of Bethesda* which makes Nicholas Culpeper's sincerity so obvious that one can well believe that his continuing popularity is deserved.

In no time the *London Dispensatory* and the *English Physician* were joined to a *Complete Herbal* to form a "complete family dispensatory." Many editions were published, with emendations by many and corrections and objections by none, so that the physicians and surgeons were left far behind.

The only sharp criticism of Culpeper which has lasted as long as his reputation for do-it-yourself medicine is contained in William Cole's *Art of Simpling, An Introduction to the Knowledge and Gathering of Plants*, published in 1657. We have seen Cole defining simpling as "an Art which teacheth the knowledge of all Drugs and Physical Ingredients, but especially Plants, their Divisions, Definations, Differences, Descriptions, Places, Names, Times, Vertues, Uses, Temperatures, and Signatures." For one wholly receptive to the doctrine of signatures, William Cole is singularly fierce about astrology. In the edition of the *Art of Simpling* which was printed shortly after Culpeper's death, Cole says of astrological botanists, "Amongst which Master Culpeper (a man now dead and therefore I shall speak of him as modestly as I can, for were he alive I should be more plain with him) was a great Stickler; and he, forsooth, judgeth all men unfit to be Physicians who are not Artists in Astrology, as if he and some other Figure-flingers his companions had been the only Physicians in England, whereas for ought I can gather, either by his books, or learne from the report of others, he was a man very ignorant in the forme of Simples." Cole's most telling thrust against the belief that plants are dependent upon planets is made by pointing out that plants were created on the third day, when planets and stars had not yet been set in their courses. Cole enlists God on his side who did "even at first confute the folly of those Astrologers who goe about to maintain that all vegetables in their growth are enslaved to a necessary and unavoidable dependence on the influence of the Starres; whereas Plants were even when Planets were not."

As Culpeper's works were not illustrated originally, we cannot know exactly what various plants looked like to him.

Illustrations for Gerard and Parkinson are as different each from the other as the authors and their purposes. The illustrations for the original edition of Gerard are almost entirely credited to Bergzabern, whom Gerard refers to as Tabernaemontanus, from the blocks used in his *Elcones Plantarum*. This increases the fervor of Gerard's detractors, although the blocks had in themselves been copied or lifted from earlier works by Fuchs and others. To Gerard, however, is conceded by every critic the honor of first publishing an illustration of the potato.

When Johnson took over his one-year Herculean task of restoring order to the occasionally confused and misplaced illustrations, and of correcting and adding to the text, he introduced new illustrations from Plantin's collection.

It seems curious to us today that the borrowing of one man is the theft of another, when to our remoter view the whole of available botanical knowledge and illustration should have been free to all. The criteria for botanical illustration were fairly evenly divided between recognizability and artistic excellence. Whether or not to so arrange the specimen as to fill the entire block in an overall design, or to show it as architecturally impressive as a church spire, to focus on the roots and bend the blossoms downward to make room for them on the block although they naturally grew erect, or to concentrate upon the blossoms and foreshorten the stems — these were decisions for the engravers who could achieve lasting distinction for their work.

It is one of the appealing features of Parkinson's *Paradisus* that the illustrations are all of one kind, artistically undistinguished and botanically oversimplified and yet all easily recognizable by gardeners. Illustrations for the eighteenth-century editions of Culpeper have no artistic merit, but from them even the hastiest home physician can tell one plant from another, which was, after all, the main point.

9

"To Keep in Fashion"

Of one thing we can be sure in reconstructing the designs of seventeenth-century gardens in New England. It is most unlikely that they were intended to look in the least like Hampton Court or anything suitable for a French king to observe from his second-storey windows. To draw up such a design for a seventeenth-century Puritan garden in the New World is to miss the point entirely, however arresting the result. The early settlers were well aware of all the pretty tricks to do with sixteenth- and seventeenth-century gardens in England: embroidery patterns worked out in close-clipped green-and-gray-leaved herbs upon a groundwork of colored gravels; eye-teasing traceries of knots and mazes executed in low box hedges continually shaved; conceits of lively animals rendered in sheared green shrubbery. But it is doubtful they longed for anything like this in New England enough to strive to reproduce it. Not even nostalgia would have induced them to imitate such useless nonsense. And we can see from their letters that nostalgia was neither fashionable nor a sign of breeding and prestige, as it was, perhaps, to become in the next century and farther south. The wave of Puritan migrants who settled New England were refugees for ideological reasons primarily, not because of economic desperation or exploitation. When their thoughts turned homeward, it was not with a wish to effect such elaborations as they had taken pains to leave behind. There was always the possibility, of course, that when they had set up the perfect godly state in the New World, and the Old World had so profited by the example as to have changed for the better, they might go "home" again. The last thing they needed, however, as they set themselves to be examples to all, was gardens requiring meticulous grooming and yielding nothing to be readily harvested "for meate and medicine."

It is also extremely unlikely that they would take pains to reproduce any small, unrelated feature of the Old World garden, such as a complicated knot, a pool, a mound, a bit of topiary, or a *pâte d'oie*, although references to garden "conceits" are not lacking. The language of gardening and garden designs is always useful, as we have seen in Governor Winthrop's love letter to his future wife. Cotton Mather, also, is obviously using a word which will strike home to his New World listeners when he speaks of becoming "briered" in an argument. Those engaged in pushing back the wilderness understand him at once. Searching for hitherto unconsidered trifles to aid us in recreating garden design, we may rejoice to come upon a reference like that to a "mount" in the introduction to William Wood's *New Englands Prospect*, but we cannot fairly lift it out and reconstruct it.

Later American gardeners were charmed by mounts, and Jefferson is said to have adopted the idea in the two mounts flanking the White House. But the early settlers of New England were not troubled by too great a flatness in their landscapes, and water, as we have seen from all the early chronicles, was in such astonishing profusion that any householder with "a spring beside his door and the ocean before it" who wished to create an ornamental pond and a mount from which to take the prospect of his property would have had to be mad with homesickness.

Similarly, a garden conceit which rivaled the mount and was used as the climax of even grander garden designs, the so-called *pâte-d'oie*, was also probably not a solution for the New World. It is as unlikely that the settlers were attracted by this idea for a pattern for their gardens as that they grew grapes on rows of pollarded elms or floated marble plates across marble pools like Pliny. They knew about all these things, but elaborations of landscape design were not suited to their character, and no answer to their needs.

The requirements of the Puritan housewife had to be the deciding factors in the layout and materials of the garden near the house. What the men did with field crops far from the houses was their concern, but the garden close to the dwelling, neatly fenced, and bright with a "variety of flowers" was as much her domain as her kitchen and her stillroom. Throughout the seventeenth century she must be able to extract from it all she would need for flavorings and seasonings and garnishings, for insect repellents and deodorants, for changing the air in rooms and keeping out moths and rodents and snakes, for dyeing and fulling and "teasing," for concocting syrups and cordials and waters, for making plasters and salves and coated pills, for treating wounds and aiding in childbirths and in laying out the dead. Finally, she must find there her favorite plants, remedies or no, like pansies and pinks and violets with their own familiar country names which only to hear was to be sustained and comforted. And, of course, all of these plants, useful and dull though they may sound, were capable of bursting into fragrant bloom to make gardens gay and pleasant spots.

It was hard work. The housewife had many responsibilities, but gardening had been one of her chief duties since well before the seventeenth century, so that she must have been in a familiar groove by that time. In the cheerful verses of Thomas Tusser, described as one of England's first didactic poets, we see the housewife's lot described. Coming from East Anglia in the reign of Henry VIII, educated at Eton and Cambridge, Tusser "devised" his book, *Five Hundred Points of Good Husbandry*, published first in 1557. Most of his subject matter must have been common knowledge. When we quote his gardening instructions we can be sure he is stating as a model to be followed what was an already accepted practice.

In the chapter on "September's Husbandry" which begins his order of months, Tusser in mid-poem addresses the housewife:

> Wife, into thy garden, and set me a plot,
> With strawberry roots, of the best to be got;
> Such growing abroad, among thorns in the wood,
> Well chosen and picked, prove excellent good.

> The barberry, respies and gooseberry too,
> Look now to be planted as other things do.
> The gooseberry, respies and roses, all three,
> With strawberries under them, trimly agree.

An attractive idea for conserving garden space and labor.
In December, he has her out again:

> Hide strawberries, wife,
> To save their life,
> Knot, border and all,
> Now cover ye shall.

and amplifies it a little in:

> If frost do continue, take this for a law
> The strawberries look to be covered with straw,

Laid overly trim upon crochets and bows,
And after uncovered as weather allows
The gilliflowers also, the skillful do know
Both look to be covered, in frost and in snow.
The knot and the border, and rosemary gay,
Do crave the like succour, for dying away.

The settlers' gardens probably looked very like English cottage gardens of today, which Geoffrey Grigson claims to be direct descendants of the Elizabethan gardens with which early Puritans were familiar. What seems to us today complete confusion in the Puritans' thinking — their ability to accept seemingly most unrelated doctrines — must have been vividly reflected in their gardens, where it was both convenient and necessary to grow as many things as possible all together at the same time.

And yet there had to be order, to make upkeep possible, and enough of a plan to allow walking about easily, preferably in pairs. Gardens were reputedly healthy spots, convenient for thinking or for discussing matters with friends. Men of affairs like Samuel Sewall liked to walk in their gardens. In his old age John Winthrop, Junior, liked to "step out into" his, which places it as close to the house as even Parkinson could wish. The wills of Essex County in the seventeenth century abound in mentions of "the garden before the house with its fencing" and, nearby, "the little orchard" or "the great orchard."

There was obviously a difference between the gardens and the orchards, although both were as near as possible to the houses. One of the books in John Winthrop, Junior's library is entitled *A Short Method of Physick*, said to be "from the practice" of one C. B. Gent, London, 1651. It is loaded with garden-grown remedies for the cure of "Fourty-five Severall Diseases." Frequently a dozen plants appear in one cure with a jammed-in air that betrays the author's hope that at least two or three will do some good. The title page announces the book as "very necessary for young Practitioners or Chyrurgions that goe to Sea, or for Housekeepers in the Country who are remote from a Physician," which last certainly defines the New England housewife.

However, another book in the junior Winthrop's library, on the planting of fruit trees, is called *Designe for Plentie* and is by Samuel Hartlib, one of the younger Winthrop's valued scientific friends and correspondents. Published in London, also in 1651, it contains no suggestion that the housewife study fruit-growing. Nor is she required to study another Winthrop book, Sir Hugh Platt's *New and Admirable Arte of Setting of Corne* (London, 1600), with its handsome frontispiece of a sheaf of wheat. This is the same Sir Hugh Platt who published in 1609, *Delightes for Ladies*, a handbook on preserving, distilling, cookery, perfume-making and other arts which could be expected of the able housewife, including the making of "Irish Aquaevitae." Apparently, no book and no one required the housewife to be clever in the orchard or the fields, except possibly in an emergency, as when the elder Mistress Winthrop is requested by letter to get in the turnips. We remember the Beverly widow whose husband left her, besides the garden, all the beans from the field crops for which she might have any need. There is nothing about her growing them. While we have seen both Winthrops hoeing away to make new gardens, labor in the fields belonged to male hands working off their passage money and to those striving to maintain ownership of "improved land" which would revert to common ownership if they could not keep it productive. Nor were women expected to work in the fields, or, indeed, to work with livestock beyond chickens, goats and occasional cows, which are usually named "Cherry," and appear in wills as part of the household's necessary appurtenances. Although men like Governor Winthrop and Governor Endecott seem to have been ever ready to give or exchange bits and pieces from their gardens, their orchards were obviously their prime concern. And their livestock was always managed by lesser folk, frequently unreliable. Also among the prime responsibilities of the male were hedges. One wonders about the success of the recommended hedge which we have a record of Samuel Sewall planting — eglantine roses and juniper bushes in a ratio of one to two, a formula which was supposed to make an impenetrable barrier within a few years.

For a garden large enough to walk in, yet where no space would be wasted, where every effort would count and where every plant would flourish, draining and soil would be the first considerations. As in England, the simplest solution would have been to make raised beds edged with boards neatly arranged with stakes at each corner to keep them upright, and filled with the best soil available. We have a description of a raised bed especially for tobacco seedlings, by John Josselyn. These are the original, real flower "beds," raised above the level of the ground like beds for people. They are far easier to take care of than anything on the level where the gardener stands. Long and narrow, not so wide that one must step onto them to work in them, they were bordered with plants of a character easily controlled, to help keep their form, and the soil from falling over the edge — "to keep them in fashion" was the expression then. "Fashion" then means shape today.

It is not likely that the Puritan housewife in arranging her garden would group her plants by their uses, such as culinary, medicinal, cosmetic and household aids, although that would be the easiest way for us today to realize and exhibit her responsibilities. She was advised to arrange them chiefly with regard to those which she could expect to keep year after year in the same place, and those which would need resowing and resetting annually. Garden practice of the time allowed a sort of broadcasting across one whole bed of several sorts of seeds which would come up in different sizes at different times — for instance, the onions and lettuces and radishes Sir Peyton Skipwith sowed together in the next century in Virginia. And it would have been sensible to keep the big-rooted things away from the delicate plants with shallow roots. Beyond this she must have liked to have things so arranged that she could put her hand on what she needed instantly — something to staunch a wound, something to freshen a drink, something to flavor a meat pasty, something to lend a neighbor who was having hysterics or a baby, something, indeed, for a sick cow or goat or an ailing horse. She might also wish to reach for a nosegay, or some "strewing herbs" to change the air in the rooms.

The list of plants we know were commonly grown and used runs to well over one hundred names. Of some of these the housewife would have needed many, like clove gillyflowers. Of some she would grow, perhaps, only one or two, like horseradish and angelica. Of some she would wish to have enough plants to use handfuls at a time and leave the plants alive, like burnet for flavoring drinks. Others would probably have filled a substantial little plot, like elecampane for lung troubles and bugloss for almost everything as well as comforting the heart. Large plantings would be needed of feverfew against all fevers, of everlasting (for smoking or against moths) and of tansy which had a multitude of virtues — from flavoring to serving as a tonic to curing worms in humans and for laying out the dead. And of course there were the special emergency plants like balm for stings, and snakeroot to be kept always in the pocket and a few poisonous plants to be taken in moderation to dull pain. To say nothing of all the popular salad herbs.

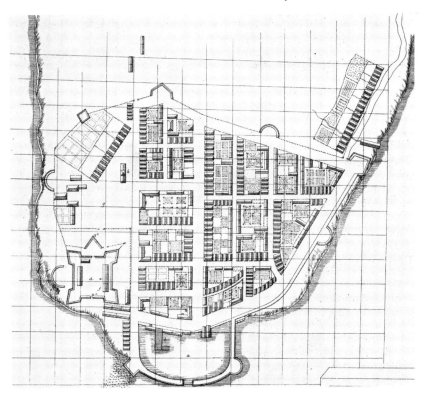

Plan of New York in the late seventeenth century, showing the houses and gardens, after thirty years of British rule, as done by a French hydrographer.

It is not surprising to discover that descriptions of copies of garden designs are few as compared with the richness of references to plants and seeds. The "dressing" of a garden was second nature to most experienced gardeners. Recording the plan of the garden mattered to them least of all. Champlain's sketch of the garden plots of the little settlement founded by him and Sieur de Monte on the Island of St. Croix, on the St. Croix River, is pure luck for us. The island is now in Maine, being slightly more on the New England side of the river than on the Canadian.

The little map of the settlement may be completely accurate. We trust the recorders more when we read that the garden plots did not indeed prove as fruitful as had been hoped, due to the very hot and burningly dry summer. The gardens on the mainland fared better, but of those there is no formal plan. Champlain's sketches of the Indian villages, showing fenced circles around little huts, more solid and rounded than Western wigwams, make us long to believe in the elaborate sketches of the layout of the first garden plots on New England soil, so like in design to those of sixteenth-century Europe. With geometrical small units in simple but attractive variations of circles and squares and diagonals, with paths between the larger beds, all situated close by the houses, Champlain's design seems a reasonable plan for the time.

Another interesting illustration of the planning and placing of gardens and houses in concentrated areas is seen in the map of Manhattan houses and gardens at the end of the century, by which time Manhattan, originally Dutch, had been English for thirty years. Here, too, we see the formal squaring away, the paths, the geometrical planting. The general idea is clear — great concentration in well-defined spaces, but it is possible that the engraver may have been doing a little landscape gardening on his own.

A case of the engraver's taking liberties appears in the deBry engravings of the White drawings of Virginia Indians. These illustrate a splendid book dedicated to Sir Walter Raleigh. The engravings show an Indian village laid out on the principles of the very best Italian and French garden styles, with even a hint of future efforts at space filling as an object in itself in the shape of a symmetrically placed round flower bed full of corn in the left front of one picture. On the right of the grand *allée* are three hedged and bordered square beds of corn or tobacco with a strip of melons most elegantly separating them from the main thoroughfare, and the places for a fire and dancing are equally formal in their layouts. Another picture shows an old Indian in winter clothing backed by a beautiful expanse of well-tended wheat, again laid out in rich squares about the center circle of the well-fenced village. Here it is fairly obvious that someone to whom it was inconceivable that even simple savages could lack a formal approach to garden design must have straightened up a few lines here and there. But when one compares these engravings with the original watercolors, one sees what a dream of formal Virginia has indeed been perpetrated by the German engraver. The whole surroundings of the ancient Indians are pure Frankfurt fiction, as are all the border plants at the feet of the savages, as are indeed, most of the elaborations of the Indian village layout which, under the brush and careful lettering of the exact White, show only three plots of corn in different

Plan of the village section of Secoton, as painted by John White in 1585.

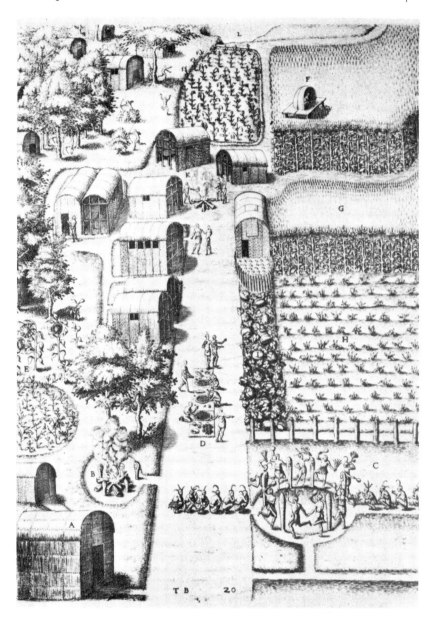

The village of Secoton, as engraved by Theodore de Bry, from the painting by
John White, 1590.

stages of ripening — an interesting point for the practical gardener, but not much to establish a sense of design. So, when we look at the St. Croix sketch, the original of which has been lost, and at the late seventeenth-century plan of Manhattan island, we may wonder if indicated arabesques may not be the engraver's way of suggesting garden plots rather than the way the French and Dutch and English sowed their cabbages.

But, for the design of New England gardens we have more reliable references than engravers. Of the books we know the settlers had in their hands, two especially stand out as solid foundations for what they knew and grew: Parkinson's *Paradisus*, and Leonard Meager's *English Gardener*.

Parkinson, last of the great herbalists and the first of the great gardeners, considers gardens as in part for pleasure. The principles of design which we are able to extract from the *Paradisus* are few and practical. He is concerned more with the contents of the garden than with its form, although he is very insistent that, while "many men must be content with any plot of ground . . . . gentlemen of the better sort and quality will wish the frame and forms to be laid out for their Garden . . . in such convenient manner, as may be fit and answerable to the degree they hold." This is a modest sight to set, and Parkinson's plea for propriety would not have upset the early Puritans.

Let us see what Parkinson calls "convenient." "To prescribe one forme for every man to follow would be folly," he asserts, "for every man will please his own fancie." Parkinson disposes of the "orbicular or round forme" as "not accepted anywhere . . . but for the generall Garden to the University of Padua." So Parkinson has, indeed, *his* feet upon the ground. The triangular form, also, he feels is "seldome chosen by any that may make another choice." He announces, "The four square forme is the most usually accepted with all, and doth best agree to any man's dwelling." This can be worked out, if always in equal squares, for any area, though preferably one close to the house, and with walks wide enough to allow "less Harme" to the "herbes and flowers . . . that grow next unto the allies sides."

Allowance can be made within this design of equal squares for all possible variations of edges and hedges, inner squares, knots, "trayles," "a maze or wildernesse, a rock or mount, and there can be even a fountaine in the midst thereof to convey Water to every part of the Garden. . . . Arbours also being graceful and necessary" may be placed in the corners or elsewhere "to serve both for shadow and rest after walking." In fact, "Let every man therefore, if hee like of these, take what may please his mind. . . ."

After having rid himself of this obligation in one simple paragraph, and appending one page of six squares which he has "caused to be drawne," he hurries on to list the "many sorts of herbes and other things wherewith the beds and parts of knots are bordered to set out the forme of them." This, we must remember, is for his "Garden of Pleasure," but he lists some plants for edging which he will later include in his kitchen garden also. In order, they are: thrift, hyssop, germander, lavender cotton, marjoram, savory, thyme, juniper and yew. Considered above all these, whose several "incommodities" he lists also, is "French and Dutch Boxe." There are also, he says, "dead materials" to "set or border up any knot": lead, which can be bent but is too hot or cold for Parkinson's liking, and boards, which must serve "for long upright beds, or such knots as have no rounds, halfe rounds or compassing in them." Some have chosen sheeps' shank bones, whitened by exposure, to edge up raised beds, or tiles, but these are subject to being easily broken. He prefers above all pretty pebble stones of "reasonable proportion." He feels he has now mentioned and explained every means of keeping the form of beds and knots except that "used in the Low Countries, and other places beyond the Seas, being too grosse and base . . . Jawbones," presumably of cattle. He mentions finally, the outer hedges for the garden. These might be of privet alone or of sweetbrier and whitethorn laced together, with roses here and there; or of a shrub "called in Latin *Pyracantha* . . . fit to be brought into the forms of an hedge though it have no Physicall use." Once again we realize how total was the idea that everything was or should be good for something, besides just growing.

Discussing the "Ordering of the Kitchen Garden," Parkinson feels that whereas the "garden for pleasant flowers" should go directly beneath the main windows of the house, the "Herb garden" belongs at one side, since cabbages and onions are to be included and their scent is not what one wishes to have coming in through the windows. Also, he says, by its very nature the herb garden must often present "breaches" from harvesting and cutting. However, he realizes that "private mens houses . . . [or those] who must like their habitations as they fall into them . . . must make a virtue of necessity . . . making one place serve for all uses." He knows that although the Garden of Pleasure is "wholly formable in every part with squares, trayles, and knots and to be still maintained in their due form and beautie," the usual form of the kitchen garden "is for the most part to bee still out of forme and order" because of the "continuall taking up of the herbes and rootes." Even so, resigned, he has suggested a rather magnificent if simple design made by planting artichokes or cucumbers or pumpkins or melons by themselves and saving space by edging them with cabbages. He advocates sowing radishes, lettuce and onions together as they can all come up together, though parsnips and carrots must be put where they can be left longer. And he then proceeds to a careful description of the cultivation and harvesting of all these, of "herbes for the pot, for meate, and for the table," "herbes and rootes for Sallets," and of "divers Physicall herbes fit to be planted in Gardens, to serve for the especiall uses of a familie." Then comes the listing and discussion of the individual herbs, except for "some sweet herbes" fit for the "hand or bosom" which he has described in his "Garden of Pleasant Flowers."

Parkinson also had a great deal to say about improving the soils. We can imagine how the early settlers appreciated the Indian idea of using fish as fertilizer, since they must have lacked what Parkinson most approved, quantities of well-rotted horse manure.

Our other chief reference book comes from a far humbler source than Parkinson. Leonard Meager's *English Gardener*, so popular it ran into six editions in England before the end of the century, is tied into the life of the colony by a little copy in the Winthrop library firmly inscribed "John Winthrop III," and by another dog-eared copy now in Salem, bearing the repeated signature in a seventeenth-century child's hand of "Abigail" — and an equally repetitive rendering of "Boston" in the rounded Gothic script.

Meager's thorough little handbook of instructions is the epitome of all the handy books on gardening which were becoming plentiful in a time when books of instruction from those purporting to be experts were greatly in vogue.

Meager's book is entirely practical, and makes a good instruction book for gardeners even today. Its title page, in toto, reads:

The English Gardener; or a Sure guide to young Planters and Gardeners in Three Parts. The first, Showing the way and order of Planting and Raising all sorts of Stocks, Fruit-trees and Shrubs with the divers ways and manners of Ingrafting and inoculating them in their several Seasons, Ordering and preservation. The Second, How to order the Kitchin-Garden, for all sorts of Herbs, Roots, and Sallads. The Third, The ordering of the Garden of Pleasure, with varietie of Knots, and Wildernesse-work after the best fashion, all cut in Copper Plates; also the choicest and most approved ways for the raising all sorts of flowers and their seasons, with directions concerning Arbors, and hedges in Gardens, likewise several other very useful things to be known of all that delight in Orchards and Gardens. Fitted for the Use of all such as delight in Gardening, whereby the meanest capacity need not doubt of success (observing the rules herein directed) in their undertakings. By Leonard Meager above thirty years a Practitioner in the Art of Gardening. London, Printed for T. Peirrepoint, and sold by the Booksellers of London, 1682.

The dedication is to "The Worshipful Philip Hollman" who has been "rather as an Indulgent Father, than a Master" to the author, who hopes to

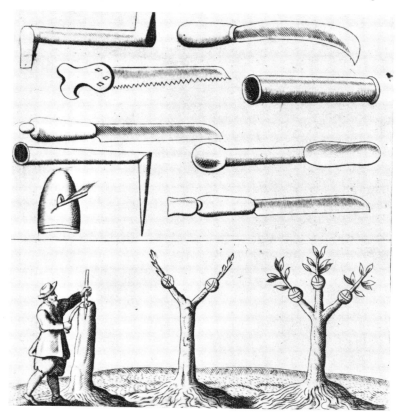

The tools and methods recommended by Leonard Meager for improving the sorts and production of fruit trees by grafting. It may have been methods such as this for converting an entire tree which led William Wood to have hopes that the choke cherries of New England could be turned into something less wild than the Indians.

have written something "Worthy to be practised, although by those that are of the lowest Orb." In the letter "To the Reader" which follows, he very briskly states that he has "set down very plainly without any deceitful dress and unnecessary flourishes, whereby it may become very useful for all sorts of practitioners, yes though of very weak capacities." There is no pretense about Leonard Meager. No flattery of anyone.

Leonard Meager gives careful directions as to how to "level and bring a Garden into some order and form." He begins by assuming a "wall or outside Fence," where one will have a border all around, and, inside that, a "Walk round your Plot or utmost Walk," and inside *that*, either a "next border or quarter," or a series of the same, depending upon one's plans for beds or knots. His is a far simpler conception than Parkinson's. He allows for some "plats" being simply grass well kept, which we see in the next century in the garden "plats" of Charleston, South Carolina. His list of "several herbs, etc. fit to set Knots with, or to edge borders to keep them in fashion, etc." begins with "Dutch or French Box" as the "handsomest, durable, and cheapest to keep." "Hysop is handsome" if kept well cut, as are also two or three sorts of thyme. "Germander was much used many years ago." "Thrift is well liked by some," but it is "apt to gape and be unhandsome." "Gilded Marjoram, or Pot-Marjoram with good keeping will be handsome." Also, "you may edge borders with divers things; as Pinks, they will be very handsome by cutting twice a year. Violets double or single, they will thicken and be handsome if oft cut." And he continues his list: "Grass cut oft. Periwinkle cut oft . . . Lavender-Cotton . . . Herbagrace . . . Rosemary . . . Lavender . . . Sage . . . Primroses and double Daisies." Box is still, however, "the most durable of any kind of herb wherewith knots are made," and he tells how to cut the roots regularly with a spade to prevent them from taking too much of the bed. Speaking of knots and hedges and borders, he cautions care "to fit your work to your ground, that it may be pleasant and sutable. . . ." with "convenient room for what you shall plant. . . . neither that you make your work so spacious but that you may have it pleasantly in your eye at a view. . . ." with "walks neither too little, neither too big, like a small City with over-large gates."

He proceeds to give "easie and plain directions. . . ." In the first place, there must be "dung or good earth" except where the walks are. There you will scrape off the good soil and, when the rest of the garden is finished, spread and roll fine gravel with coarser on top, or "Chalk or the like," graded to drain well but not so much as to be "Uneasie for such as wear high-heeled shooes." Starting with the wall, using a line and "handsome straight stakes about four or five foot long, being sharpened at one end," you may proceed to stake out all your borders "by your Wall or outside Fence," then, measuring the breadth you intend for your walk "round your Plot or utmost Walk," you place stakes to mark the level of the entire garden, wherever you intend to have your "next border or quarter" or your "quarters square or equal." He then tells how to have the whole garden slightly pitched if necessary for drainage or watering.

Proceeding now "to the digging and orderly finishing of your ground, beginning first with your borders." To make them "lye fast and handsome" set the edges "close and handsome" inside the line, with "Pinks, Violets, or any other thing you think fit that keeps always green." Or you can edge your borders with turf, or sow the edges of your borders with Pink seed, first making a trail or gutter straight and even." After you have finished your "utmost borders," you are to proceed to the finishing of your other borders or quarters, "perfecting of your intended work." After you have finished your borders, knots or quarters, "as occasion serves," you are to make your walks.

While there are plentiful rather elaborate copperplate designs included by Meager, it is in the asides that we get the best idea of how to set as simple a garden as we need. Under "Of the manner of Sowing small Seeds," discussing "such sorts of Herbs that are for Physick uses or to Still," he says to "tread out your beds handsome and straight by a line, it will be the pleasanter to look on," and "if you will, you may sow your seeds in rows or trails, either round about the edges of your beds to keep them in fashion, and Plant either Herbs or Flowers in the body of your beds." No fountains. Instead, after a discussion of how to water — "in a close or gloomy day, is better and more effectual than two in hot Sunshine weather" and "Evening watering is more effectual" — Meager recommends a tub full of water with sheeps' dung or other dung "to stand in the sun until it be in better case to use as aforesaid."

"Another thing worth the practicing is," he continues, "that you be careful to cut or top your herbs often, for it is not only handsome but causeth your herbs to last longer . . . by your often topping your sweet herbs, you may, if you will, make use of them to dry and make them into powder to use all winter. . . . Sage and Rosemary beds, are likewise to be cut smooth

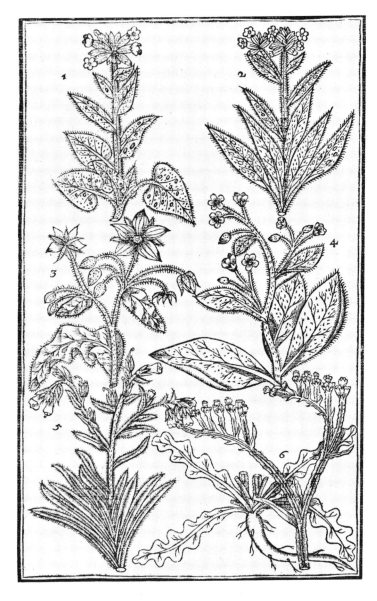

Parkinson's illustrations of members of the borage family, which he
says he included in his garden illustrations chiefly because ladies like
to show them in their needlework. From a reprint of the 1629 edition
of his *Paradisus*.

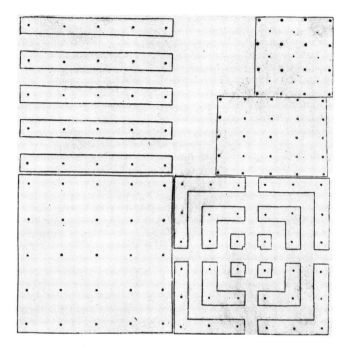

Leonard Meager's suggestions for planting fruit trees: in long beds, a broken square, squares and a quincunx.

and handsome, which being often done, a small matter doth it; and besides they will be useful as an hedge to lay small cloaths upon to white and dry, besides a handsome ornament in a Garden. Another convenient practice tending to handsomeness and good order, is that you sow or set together in one quarter, or Beds by themselves, all such herbs as are durable, and not to be renewed again every year, by which means that part of your ground will be always in handsome order; you may easily guess what the contrary practice will be. After this manner you may bring it to pass, sow Thyme, Winter-Savoury, Hysop, Pot Marjoram and Winter sweet Marjoram . . . increased only by slips, let such be near together; likewise Balm, Costmary, Mints and the like; in beds together, also Bugloss, Sorrel, Succory and the like; and for such as are, as I said, to be renewed every year, as sweet marjoram, summer savory, and sweet basel, etc. let these be near together; also all ordinary Pot-herbs that are yearly to be renewed by themselves. . . . I shall need to say no more as to this."

It is worth noting that when Leonard Meager speaks of a "Wildernesse" he does not mean what the early settlers were actually encountering, but a maze in which one may become happily and amusingly bewildered. With so much genuine wilderness all about them it is unlikely that a formal pattern to create a sort of festive confusion would have had any appeal whatever.

The other designs appended to this little book are all what an average person, given a ruler, a compass and perhaps a wine glass for the smaller circles, could concoct without too much effort. Simple designs made up of oblong beds in various dispositions within the squares are obviously what Meager is writing about, but he also includes some complicated ideas involving corners given over to "broderie," and even one or two really grand designs of arabesques and acanthus leaves to be rendered in living materials and seen against gravel or even, perhaps, colored sands.

Here would seem to be the place to include Leonard Meager's list of plants divided into what he deemed their proper categories. It helps us to judge what the early settlers may have brought and grown their plants for, and in what sorts of groupings they themselves considered them. These lists are given with Meager's spellings, but omit his asides on their culture and care. And I have not attempted to define them except when, later, they appear in the settlers' own lists.

First Meager gives us:

The names of divers sorts of herbs commonly called Sweet Herbs

Balm . . . Basil . . . Burnet . . . Coast-mary . . . Camomile . . . Callamint . . . Hyssop . . . Lavender . . . musked Crainsbill (or Muskomy) . . . Mints (spear, red, water, basel, pide) . . . Marjoram (sweet, wintersweet, yellow, pide, pot or wild) . . . Maudlin or Sweet Maudlin . . . Penroyal . . . Sage . . . Savoury (winter and summer) . . . Thyme (English or hard, French, pide, limon, musk, mastick) . . . Herb Mastick . . . Tansie. . . .

And then:

The names of divers ordinary Physick Herbs, usually Planted in Gardens

Angelica . . . Asarabacka . . . Bears-foot (Setterwort) . . . Carduus . . . Dragons . . . Dittander . . . Elecampane . . . Fetherfew . . . Goats Rue . . . Germander . . . Garlick

. . . Harts tongue . . . Horse radish . . . Liverwort . . . Lavender Cotton . . . Liquorish . . . Master-wort . . . Marshmallows . . . Mother-wort . . . Pelletary of the Wall . . . Pionies . . . Rubarb . . . Rue . . . Solomon's Seal . . . Scordium . . . Scorsonera . . . Scurvy-grass . . . Southern-wood . . . Smalage . . . Sneese-wort . . . Tansie . . . Greek Valerian . . . Great Valerian (or Setwell) . . . Winter Cherries . . . Wormwood (both English and Roman).

And for good measure and perhaps to encourage us, he tosses in here:

A sort of sallet commonly gathered in the Spring,

consisting of divers young buds and sprouts both of trees and herbs, which being gathered discreetly, with nothing but what is very young and tender, and so that no one thing do too much exceed another, but there be a fine agreement in their relish, if so it will be acceptable to many. Violets with some young leaves, Primroses with some young leaves, small sprouts of Burnet, also of Mints, Sorrel, and divers other of the like, also small buds of Gooseberries, Roses, Barberries, etc. Also when they are to be had, the Flowers of Bugloss, Cowpagles, Archandel, with divers others.

Then he begins to give us his lists, roughly alphabetized under their proper categories:

The names of divers ordinary Pot herbs, called also chopping herbs

Arach, red and white	Marygolds
Blood-wort	Nep
Burage	Orach
Bugloss	Parsley
Beetes	Sives
Carrots	Strawberries
Clary	Succory
Endive	Violets
Langebeff	Worts or Brocketts
Leeks	

And:

The names of divers Sallet herbs and roots, and other herbage for the Kitchin Uses

Alisanders	Onions
Beans (French and ordinary	Parsley
Garden beans)	Parsnip

A geometrical plan for a garden of "broken squares" from
The Countrie Farme.

Beets	Potatoes (of Virginia, and Canada
Cabbages	or Jerusalem Hartichoaks)
Carraway	Purslain
Carrots	Rocket
Corn-sallet	Rampions
Colworts	Ramsons
Colliflowers	Radish
Cucumbers	Horse-radish
Cress	Shelot
Dill	Scorsonera
Endive	Skerrets
Fennel	Sparragus
Hartichoaks	Sorrel (French and English)
Lettice	Spinage
Muskmillions	Tarragon
Mustard seed	Turnips

Peas he lists separately:

Pease, of divers sorts

Hot-spurs-pease
Redding-pease
Sandwich-pease
Sugar-pease
Tufted or Rose pease

Gray Windsor — pease
Great Maple pease
Great Bowlins pease
Great Blew pease

Meager includes:

These things following which are . . . by divers cooks and others pickled for Sallets to use in Winter

Cucumber
Purslain
Tarragon
Summer-Savoury
Broom-buds

Elder-buds
Onions
Leeke
Hartichoaks

Also some pickle-up Turnips, Beetroots. . . . Also some make a very acceptable boiled sallet of the young and tender stalks of both Turnips and of Cabbages when they first run up in the spring, they boil them and peel them and put Butter and Vinegar and Pepper to them.

These are put with White-wine Vinegar and sugar for Winter Sallets

Clove-gelly-flowers
Cowslip flowers
Bugloss flowers

Burage flowers
Arch-angel flowers

So much for Meager's "Kitchin Garden." Next he helps us with the "ordering" of the "Garden of Pleasure," and, after many pages devoted to "ordering" he gives us his flower lists. Some are slightly repetitive, but he groups them as he finds most convenient.

A Catalogue of flowers . . . of divers kinds, and because many are very much taken and affected with furnishing of their flower pots for the adorning of some room in their houses etc. . . . as also of such as are only for ornament in their places where they grow, or for Nose-gays. . . .

And first of:

Those called Annuals. . . .

Adonis-flower
African Marygolds
Scarlet Beans
Coventry Bell-flowers
Great blew bind-weed, or Connuvolus
    major
Small bindweed
Candy-tufts
Catch-fly
Fennelflower (or Nigella)
Fox-gloves, white and red
Small white flax
French Honey-suckles
Honesty, or white Sattin
Hollihocks, double and single
Larks-heels or Spurs

Lupins
Melancholy Gentleman
Mothmullins
Marygolds, double French
French Marygolds
Princes Feather, or Amorantus
Pearl-grass
Tree Primrose
Double Poppies
Indian or Musk Scabious
Sianus or Bottles
Snap dragons
Spanish Saffron
Stock Gillyflowers
Venus Looking-glass

And then, of those which can be:

increased by slipping, parting their roots, and laying . . . and are also *fit to furnish a flower pot.*

Champions
Canterbury Bells
Columbines
Cranes bill
Carnations or Gillyflower
Everlasting pease
Fraxanella
Goats Rue
Whitson Gillyflowers
Wall Gillyflowers
Stock Gillyflowers

Hollihocks
Sweet John
White Marygold
None-such, or flower of Bristol
Pinks
Snapdragons
Spanish tufts
Throat-wort, a sort of Bell-flower
Valerian, red and the greek, both
    white and blue

And for good measure, though with warnings about some:

Auster-aticus, or Italian starwort,
    apt to run in a Garden.
Batchellor Buttons, or double Campions

Ladies Smocke-double
Live ever or life everlasting
Blew Marygolds

Peach-leaved Bell-flowers, blue
and white
Double Crowfoot, besides the more
choice sorts known by the
latine name *Ranunculo's*.
Crimson Cardinals-flower
Double-Featherfew
Hungarian Dead-Nettle
Spanish knap-weed
Lichnes, or double none-such

Periwinkle
Double Rockets, or Whitson
July-flowers
Double Sope-wort, a busie runner
in a Garden
Double Sweet-Williams
Double Wall-flowers
Willow-flower, a troublesome guess
[guest] in a Garden though pretty
for a flower pot

. . . and also both the bulbous and tuberous rooted:

Bulbo's and Tubero's rooted fit to furnish a garden and adorn a flowerpot

Anemonies
Crown Imperial
Corn-flags
Fritillarias
Flowerdeluces
Hyacinths
Indian Juca
Kings-spear
Lillies

Molies
Martagons
Munks-hoods
Star-flowers
Tulips, abundant in varieties
Persian Lilly
Pionyes, six or seven varieties
Bulbo's Violets

. . . and *"divers other pretty flowers"*

Bulbo's violet (again)
Crocus and Saffron flowers
Colchecons
Grape-flowers
Hollow-root Flower, or ground hony-
suckle or a fumetary

Marracock, or Passion-flower
Ranunculo's
Sow-bread
Spider-wort
Winter-wolfe-bane

And, finally:

Other sorts fit to furnish a Garden . . .

Barren-wort
Bears ear, or French cowslips and
Bears-ear sanickle
Hepaticas
Jerusalem cowslips

Lily of the Valley
Marvel of the World
Mandrake-Golden Mouse-ear
Navel Worts
Pances or Hearts-ease

Crismass flower

Daisies

Indian Cresses, or Nastersian Indicum

Primroses and Polianters

Sultans flower or Turkey Cornflower

Violets

So, thank you, Mr. Meager.

10

"Of Such Plants . . ."

Originally, in searching for the names of plants grown in seventeenth-century New England, it seemed sufficient to concentrate on contemporary references, in writing and in print, in letters, accounts, wills, deeds, diaries, inventories, poems, receipts and/or recipes, prescriptions, and so on. New England had fostered our nation's origin and offered a still rich field for exploration into the gardens of our founders. The New England passion for never throwing anything away served us well. Ancient attics yielded clues to household uses of garden plants. Dog-eared herbals explained the purposes for which each plant had seemed ordained by an all-wise Providence. Seed lists included the most staple vegetables "for the pot" and for salads. All these put together seemed to form a brilliant, nourishing and fragrant garden, suitable for the "meate and medicine" of any distinguished family of settlers. And yet, missing were many references to plants that we know they had, plants they felt they could not do without.

So we began again, outside walls and books, and searched the countryside where they had lived to see if by any chance any of their plants were still left. And it suddenly became apparent to us that the gardens of our forebears are brightly scattered about the New England landscape; for better for worse, for richer for poorer, they and it are firmly wedded, to the enjoyment of us all throughout the New England summer.

By checking in their own reference books we know the "vertues" of these plants, and why the settlers brought them and we can take them back into the gardens again. Some of the plants grew wild in England, though handy for the country housewife, and have "escaped" to grow in fields again in the New World. Some were garden plants in Old England, strayed beyond the carefully impaled areas where they were first tenderly set out after their long journey. They mingle now with our own native wildflowers and look as if they had always been here. Carefully called "escaped," "ad-

ventive," "introduced," "naturalized" or whatever by those who write
about our native plants, they have become so familiar to us throughout the
years and seem so much a part of the native scene that it is a shock to discover they are as newly arrived on this continent as we ourselves. Sadly
enough, one of the best ways to check their identities is in books on
"weeds" written for embattled farmers. When we see their profusion today
we think of that great gardening truth: if it is given to you, it spreads. We
know how the upland pastures and lowland meadows of remote New England farms can be full of invaluable remedies for human ills and those of
cattle also.

Some of these once loved garden plants have wandered far. Anyone who
has trout-fished in New England woods will have found the river banks
matted with our wild lilies and the imported creeping "moneywort," in
close affinity. Yarrow and cudweed, goutweed, and horseheal, tansy and
feverfew, eyebright and viper's bugloss, chicory, St. Johnswort and a host
of other old friends to which we have become so accustomed as scarcely to
notice their abundance, suddenly take on a new importance, crowding in
with the native plants from which, to most of us, they are almost indis-
tinguishable. Cellar holes yield lilacs, damask and eglantine roses, lily-
of-the-valley, Solomon's-seal, vinca, mints, chamomile, fleabane, gill-over-
the-ground, boneset, Bouncing Bet, butterfly-weed, teasel, early red peonies
and violets. Coltsfoot and mullein and plantain bloom on old banks.
Good-King-Henry takes over old barnyards. Scarlet pimpernel is everywhere. Stonecrop, too. And among them all are the native plants for which
so much was hoped, plants that went to the Jardin du Roi in Paris and to
the Chelsea Physic Garden in London to see what might be made of them
there, and may even have come back here as seeds and roots to be planted
again on their native continent. There are also a few pretty flowers which
obviously came for their looks alone, like the ragged-robin which stains the
Newbury meadows with wine-pink, and, perhaps, the yellow flags which
had long before become the *fleur-de-lis* of France, according to legend. The
settlers may have brought these last for strewing their floors to keep their
feet warm, as Josselyn said, and then found there were plenty of iris here
which could be used to the same end. Our seemingly most American wildflowers, the daisy and buttercup and the so-called Indian paintbrush may,
indeed, have come in the cattle-feed, as is often said, but they may also have
been brought in with a purpose. The little yellow broom, *Genista tinctoria*,
which no good farmer loves today, was called "witches blood" in Salem

because it bloomed on Gallows Hill; but it was brought originally as a dye plant for "sad" or sober colors. And so the lists grow. Amid this welter of color and fragrance and aromatic odors, we are left to refer to our check lists in Gothic handwriting and little printed books which capitalize most nouns and spell every word as differently as anyone pleased, even in the same line. We begin again with the written and printed words, but newly encouraged by all the living witnesses awaiting only the dignity of inclusion, as their names may be called.

Our chief sources of information on what the early gardens contained, are: first, the seed bill which was sent to John Winthrop, Junior, as he sailed to join his father in the new land; and, second, the account by John Josselyn of the plants he found in New England. The Winthrop list is printed here just as it appears in the Winthrop papers. John Josselyn's account is taken from a reprint of his *New-Englands Rarities* with added plants from his *Two Voyages to New England*, also a reprint. Both these invaluable books are out of print, even as reprints, so we have quoted them freely, as it would be a pity for readers themselves not to see John Josselyn at work.

These are the main bulwarks. Additional plant names come from an assortment of lesser sources, some very humble and fragmentary.

In the Appendix all the plant names will be listed, with the information on each one available to the early settlers, as it was recorded by their recognized authorities in the colonists' reference books: John Gerard as edited by Thomas Johnson, John Parkinson, Nicholas Culpeper and John Evelyn, expert on salads.

I

First comes the John Winthrop, Junior, seed list, with the spelling of the seed merchant and the quantities and prices as ordered by Winthrop.

This list appears in the collection of the Winthrop Papers by the Massachusetts Historical Society, Volume III, among invoices of goods shipped on the *Lyon* or *Lion* whose Captain was Mr. Pierce in the summer of 1631. Among the list of quantities of hogsheads of oil and vinegar, "bookes and cloth," gunpowder and butter, bundles of leather and rope and ironware, the two lists which most interest us are the "bill of glasses" endorsed by John Winthrop, Junior, containing all his alchemical equipment, and the "bill of garden seeds," also endorsed by him.

ROBERT HILL'S BILL TO JOHN WINTHROP, JR.

Bought of Robert Hill gr(ocer) dwelling at the three Angells in lumber streete 26th July 1631.

	s	d	
1 oz Alisander seeds at 2d		2	
1 oz Angelica seeds at 4d		4	
1 oz Bassill seeds at 3d		3	
1 oz Buglos at 2d		2	
1 oz Burradg seeds 4d		4	
1 oz Burnett 3d		3	
1 oz Beets at 2d		2	
1 oz Bludwort seed 2d		2	
1 oz Carduus benidictus 6d		6	
1 oz Colewort seeds 3d		3	
1 oz Cullumbine seeds 3d		3	
1 oz Cresses seed 3d		3	
8 oz Cabedg seed 2s per li	1		
1 li Carrett seed 12d per li	1		
1/2 oz Charnill seed 3d per oz	0	1	1/2
1/2 oz Cicory seeds at			
3d per oz	0	1	1/2
1/2 oz Clary seeds at 3d per oz	0	1	1/2
1/2 oz Corn sallet at 2d	0	2	
2 oz Culiflower seeds			
2s 6d per oz	5		
1/2 oz dill seed 3d per oz	0	1	1/2
1/2 oz endiue seed 3d per oz	0	1	1/2
1 oz fennell seed 1d	0	1	
1/2 oz sweet fennell 1d	0	1	
1 oz hysopp seed at 2d	0	2	
1/2 oz hollihocks seeds at 2d	0	2	
1 oz louadg seeds 2d	0	2	
3 oz lettice seeds 2d per oz	6		
1/2 oz Lang de beefe at 1d	1		
1 oz leekes seeds at 3d	0	3	
1/2 oz Walflower seed at			
4d per oz	0	2	
for orradg seeds 1d	0	1	
flower of the sonne	0	2	

	s	d	
1/2 oz mallow seed at 1d	0	1	
1/2 oz marigold at 2d	0	2	
1 oz sweet marjoram at 8d	0	8	
1 oz pott maioran at 4d	0	4	
1/2 oz munkhoods seeds at			
3d per oz	0	1	1/2
1/2 oz maudlin seed 2d	0	2	
1 oz nipp seed 2d	0	2	
1 li new onyon seed at			
2s 8d	2	8	
4 oz parsley seed at			
16d per li	0	4	
1 oz pursland seed at 4d	0	4	
1/2 oz popey seed at 2d	0	2	
8 oz pompion seed at			
2s 8d per li	1	4	
1 li new parsnipp seed			
at 20d	1	8	
8 oz Radish seed at			
12d per li	0	6	
1/2 oz Rockett seed at			
4d per oz	0	2	
1 oz Rosemary seed at			
8d per oz	0	8	
1 oz Sorrell seed at 2d	0	2	
1 oz winter sauory at 6d	0	6	
1/2 oz summer sauory	0	2	
1 oz spynadg at 2d	0	2	
1/2 oz stockjelliflower 3d	0	3	
3 oz skerwort seed			
3d per oz	0	9	
1 oz Thyme seed 6d	0	6	
Tansy seeds	0	2	
violett seeds	0	2	
1/2 oz walflower	0	2	
1/2 oz hartichockes 2s per oz	1	0	

                                            £1 6 0

(Endorsed by John Winthrop, Jr.:) "bill of garden seeds."

This list of seeds could have belonged to any distinguished Pompeian householder, except for a few additions of hardier plants culled from the English countryside and brought into garden cultivation before the colonizers of William the Conqueror arrived with a few reliable herbs of their own. There is no concrete example of the many thrilling new discoveries which suddenly burst upon gardeners and willing experimenters in the art of physic from the Spanish conquests in South America—even to tobacco. On the whole Winthrop's seed list is a very old and reliable one indeed, well tried for more than a thousand years.

II

With John Josselyn's list we step onto a wilder shore. On the identification of many of these plants, I have no intention of locking horns with the experts, whose horns are already interlocked. When it comes to what was growing here from the beginning and what was brought in by the settlers, we can see that what Josselyn thought is not entirely borne out later by one of his editors, Professor Edward Tuckerman in 1865, nor by one of our earliest medical botanists, Dr. Jacob Bigelow, who appears to have written his *Flora Bostoniensis*, published in 1814, while riding about making calls. Nor does Asa Gray's *Manual of Botany*, eighth edition, 1950, as expanded by M. L. Fernald, always agree with any of them. Even the most modern of wildflower identification handbooks do not agree on indigenous plants and those "introduced" or "escaped" or "adventive." Gray often uses the more comprehensive term "naturalized." Who are we to say and how are we to know exactly what plants the early settlers intended by "snakeroot" and "eyebright"? When even the great Gerard, corrected by Johnson, presents his eyebright or euphrasia first among the forget-me-nots and then among a handful of buglosses, who are we to decide the juice of which plant is to be put into white wine and dropped into our eyes? Even when *Euphrasia officinalis* grows today upon the Maine coast and may well appear to solve the dilemma? As for snakeroot—that invaluable remedy against both snakes and their bite—we have several candidates, but who is to know which one John Winthrop carried in his pocket? It behooves us to stay within seventeenth-century bounds. The "curious" reader may seek out present-day identifications and evaluations from modern studies of herbal medicine. It is profitable — and astounding — to see how much of what they knew is still deemed sound and how often they ardently courted disaster.

John Josselyn's serious effort in his *New-Englands Rarities* lists plants as he saw them, and sorts them into several categories: those common also in England; those he believes to belong "properly" to the new country; those, like the former, to which he thinks no really correct names have been attached; those which he thinks have started to grow since the English planted their gardens and pastured cattle; and, finally, those he has observed growing in gardens and how well, or otherwise, they grow.

Throughout, he includes asides on what he believes about these plants, or has done with them, or what Indians and others have done. His other book, *Two Voyages to New England*, naturally includes many of these comments, but in a less consciously organized form, being generally found in directions for cooking (as in his eel "receipt") and planting (as in using eglantine combined with juniper for a hedge which Samuel Sewall records doing in his diary). These have been quoted in their contexts in the chapters on "Meate" and "Medicine," and we have not repeated them in full here. But to be quite fair to his industry we have appended his own additional list in his second book, which he gave, fearing he had been too brief in the first account. Since Josselyn considered as "God's trees" all those not planted by the hand of man, we have left them out of this account as not pertaining to gardens, but we have indicated where Josselyn mentioned them.

For convenience of reference I have listed in the left-hand margin what Professor Edward Tuckerman gave as the botanical names of these plants in 1865. One or two which he passed over as beyond identification and which seemed to me obvious from my weeding, I have dared to name in parentheses. In cases where the species name is now different I have quoted only the name of the genus as Tuckerman gave it.

However, in determining which plants Josselyn meant when he began his list of the plants common to both England and New England, we are vastly helped and entertained by finding that he follows the order given by Gerard, as edited by Johnson, Josselyn's reference book to which he most respectfully refers. He appears to have gone through it page by page, so that where we may feel some doubt about "matweed" we need be in none as to what it looked like to Josselyn, as its picture follows a few pages after Gerard's (and Josselyn's) "hedghog grass."

So we begin Josselyn's listing with Tuckerman's identifications included only to help those of the "curious" today who may wish to bring these identifications further up to date.

From John Josselyn's *New-Englands Rarities*, 1672, as reprinted in 1865 by William Veazie, with notes by Edward Tuckerman.

1. *Of such Plants as are common with us in England.*

[*Tuckerman's Names*]	[*Josselyn's List*]
Carex flava	*Hedghog-grass.*
Calamagrostis areneria	*Mattweed.*
Typha latifolia	*Cats-tail.*
Stellaria graminea	*Stichwort,* commonly taken by ignorant People, for Eyebright; it blows in June.
Iris	*Blew Flower-de-luce;* the roots are not knobby, but long and straight, and very white, with a multitude of strings. It is excellent for to provoke Vomiting and for Bruises on the Feet or Face. They Flower in June, and grow upon dry and sandy Hills as well as in low wet Grounds.
Erythronium	*Yellow Bastard Daffodill;* it flowereth in May, the green leaves are spotted with black spots.
Orchis	*Dogstones,* a kind of Satyrion, whereof there are several kinds groweth in our Salt Marshes. I once took notice of a wanton Womans compounding the solid Roots of this Plant with Wine, for an Amorous Cup; which wrought the desired effect.
Nasturtium officinale	*Watercresses.*

Lilium philadelphicum   **Red Lillies** grow all over the Country
                            innumerably amongst the small Bushes,
                            and flower in June.
Rumex acetosella   **Wild Sorrel.**
Ophioglossum vulgatum   **Adders Tongue** comes not up till June;
                            I have found it upon dry hilly grounds,
                            in places where the water hath stood all
                            Winter, in August, and did then make
                            Oyntment of the herb new gathered; the

Humming Bird Tree, from John Josselyn's
New-Englands Rarities, 1672.

fairest leaves grow amongst short Hawthorn Bushes, that are plentifully growing in such hollow places.

*Smilacina bifolia**   **One Blade.**

*Clintonia borealis***   **Lilly Convallie,** with the yellow flowers grows upon rocky banks by the Sea.

Alisma plantago   **Water Plantane,** here called Water suck-leaves. It is much used for Burns and Scalds, and to draw water out of swelled Legs. Bears feed much upon this Plant, so do the Moose Deer.

Plantago maritima   **Sea Plantane,** three kinds.

Sagittaria sagittifolia   **Small-water Archer.**

Gentiana saponaria   **Autumn Bell Flower.**

Veratrum viride   **White Hellibore,** which is the first Plant that springs up in this Country, and the first that withers; it grows in deep black Mould and wet, in such abundance, that you may in a small compass gather whole Cart-loads of it. The Indians Cure their Wounds with it, annointing the Wound first with Racoons greese, or Wild-cats greese, and strewing upon it the powder of the Roots; and for Aches they scarifie the grieved part, and annoint it with one of the foresaid Oyls, then strew upon it the powder. The powder of the Root put into a hollow Tooth, is good for the Toothach: The Root sliced thin and boyled in Vineagar, is very good against *Herpes Milliaris.*

Polygonum lapathifolium   **Arsmart,** both kinds.

Polygonum persicaria   **Spurge Time,** it grows upon sandy Sea

(Euphorbia serpyllifolia)   Banks, and is very like to Rupter-wort, it is full of Milk.

*From the picture in Gerard it seems to be *Maianthemum canadense.*
**Gerard's picture is of *Convalleria majalis.*

(Herniaria glabra)	Rupter-wort, with the white flower.
Saxifraga virginiensis	Jagged Rose-penny-wort.
Salicornia herbacea	Soda bariglia, or massacote, the Ashes of Soda, of which they make Glasses.
Salicornia virginica	Glass-wort, here called **Berrelia**, it grows abundantly in Salt Marshes.
Hypericum perforatum	St. John's-Wort.
Hypericum quadrangulum	St. Peter's-Wort.
(Ascyrum stans)	
Veronica arvensis	Speedwell Chick-weed.
Veronica officinalis	Male fluellin, or Speed-well.
Hedeoma pulegioides	Upright Peniroyal.
Mentha aquatica	Wild-Mint.
Nepeta cataria	Cat-Mint.
Agrimonia eupatoria	Egrimony.
Xanthium strumarium	The lesser Clot-Bur.
Nuphar advena	**Water Lilly,** with yellow Flowers, the Indians Eat the Roots, which are long a boiling, they last like the Liver of a Sheep, the Moose Deer feed much upon them, at which time the Indians kill them, when their heads are under water.
Arum	**Dragons,** their leaves differ from all the kinds with us, they come up in June.
Viola	**Violets of three kinds,** the White Violet
Viola blanda	which is sweet, but not so strong as our Blew Violets; Blew Violets without sent, and a Reddish Violet without sent; they do not blow till June.
(Parthenocissus quinquefolia)	**Woodbine,** good for hot swellings of the Legs, fomenting with the decoction, and applying the Feces in the form of a Cataplasme.
(Polygonatum multiflorum)	**Salomons-Seal,** of which there is three kinds, the first common in England, the
Polygonatum virginianum	second, Virginia Salomons-Seal, and
Smilacina stellata	the third, differing from both, is called

	Treacle Berries, having the perfect tast of
Smilacina racemosa	Treacle when they are ripe; and will keep
	good a long while; certainly a very
	wholesome berry and medicinal.
Geranium carolinianum	*Doves-Foot.*
Geranium robertianum	*Herb Robert.*
Geranium maculatum	*Knobby Cranes Bill.*
Geranium maculatum	*Ravens-Claw,* which flowers in May,
	and is admirable for agues.
Potentilla canadensis	*Cinkfoil.*
Potentilla canadensis	*Tormentile.*
Geum strictum	*Avens,* with the leaf of Mountane-Avens,
	the flower and root of English Avens.
Fragaria vesca	*Strawberries.*
Angelica atropurpurea	*Wild Angelica,* majoris and minoris.
Smyrnium aureum	*Alexanders,* which grow upon Rocks by
	the Sea shore.
Achillea millefolium	*Yarrow,* with the white flower.
Aquilegia canadensis	*Columbines,* of a flesh colour, growing
	upon Rocks.
Chenopodium botrys	*Oak of Hierusalem.*
Ambrosia eliator	*Oak of Cappadocia,* both much of a na-
	ture, but Oak of Hierusalem is stronger
	in operation; excellent for stuffing of the
	lungs upon Colds, shortness of wind,
	and the Ptisick; maladies that the na-
	tives are often troubled with; I helped
	several of the Indians with a Drink made
[The following are put	of two Gallons of Molosses wort, (for in
in as a convenience]	that part of the Country where I abode,
	we made our beer of Molosses, Water,
	Bran, chips of Sassafras Root, and a
(Sassafras officinalis)	little Wormwood, well boiled), into
(Artemisia absinthum)	which I put of Oak of Hierusalem, Cat-
(Chenopodium botrys)	mint, Sowthistle, of each one hand-
(Nepeta cataria)	ful, of Enula Campana Root one Ounce,
(Sonchus oleraceus)	Liquorice scrap'd brused and cut in

(Inula helenium) pieces, one ounce, Anny-seed and sweet
(Glycyrrhiza) Fennel-feed, of each one Spoonfull
(Pimpenella anisum) bruised, Sassafras Root cut into thin
(Foeniculum vulgare) chips, one Ounce; boil these in a close
Pot, upon a soft Fire to the consumption
of one Gallon, then take it off, and strain
it gently; you may if you will boil the
strained liquor with sugar to a Syrup,
then when it is Cold, put it up into
Glass Bottles, and take thereof three or
four spoonfuls at a time, letting it run
down your throat as leasurely as possibly
you can; do thus in the morning, in the
Afternoon, and at Night going to bed.

Galium aparine   *Goose-Grass, or Clivers.*
Aspidium   *Fearn.*
Pteris aquilina   *Brakes.*
Oxalis corniculata   *Wood Sorrel.*

[Here he lists three trees: elm, "line tree" and maple.]

Drosera   *Dew-Grass.*
Apios tuberosa   *Earth-Nut,* which are of divers kinds, one bearing very beautiful Flowers.
Fungi   *Fuss-Balls,* very large.
Fungi   *Mushrooms,* some long and no bigger than one finger, others jagged flat, round, none like our great Mushrooms in England, of these some are of a Scarlet colour, others a deep Yellow, &c.
Anagallis caerulea   *Blew flowered Pimpernel.*
Hepatica triloba   *Noble Liver-wort,* one sort with white flowers, the other with blew.
Rubus   *Black-Berry.*
Rubus   *Dew-Berry.*
Rubus   *Rasp-Berry,* here called Mul-berry.
Ribes hirtellum   *Goose-Berries,* of a deep red Colour.

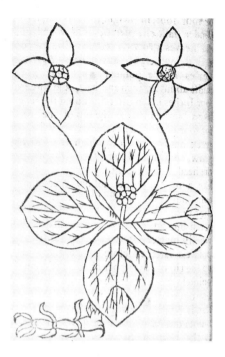

Herb True Love, from John Josselyn's
New-Englands Rarities, 1672.

[Here he mentions the hawthorne tree and the "Haws being as big as Services, and very good to eat, and not so astringent as the Haws in England."]

Linaria vulgaris	*Toad flax.*
Teucrium vulgaris	*Pellamount,* or Mountain time.
Antennaria plantaginifolia	*Mouse-ear Minor.*
Juniperus	*Juniper,* which Cardanus saith is Cedar in hot Countries, and Juniper in cold Countries; it is here very dwarfish and shrubby, growing for the most part by the Sea side.

[Here he mentions the willow tree.]

Kalmia angustifolia   **Spurge Lawrel,** called here Poyson
                         berry, it kills the English Cattle if they
                         chance to feed upon it, especially Calves.
Myrica gale   **Gaul, or noble Myrtle.**

[Here he lists several shrubs and trees and their uses: elder, dwarf elder,
alder, hazel, "filberd," walnut, chestnut, beech, ash, "quickbeam or wild
ash," birch, poplar. As we left in his reference to the cherry trees, we leave
in also the next entry, as having to do primarily with domestic matters.]

Prunus maritima   **Plumb Tree,** several kinds, bearing
Prunus americana   some long, round, white, yellow, red,
                         and black Plums; all differing in their
                         Fruit from those in England.
Portulaca oleracea   **Wild Purcelane.**
Genista tinctoria   **Wood-wax,** wherewith they dye many
                         pretty Colours.
Ribes   **Red and black Currans.**

2. *Of such Plants as are proper to the Country.*

Zea Mays   **Indian Wheat,** of which there is three
                         sorts, yellow, red, and blew; the blew
                         is commonly Ripe before the other a
                         Month; Five or Six Grains of Indian
                         Wheat hath produced in one year, 600.
                         It is hotter than our Wheat and clammy;
                         excellent in Cataplasma to ripen any
                         Swelling or impostume. The decoction
                         of the blew Corn, is good to wash sore
                         Mouths with: It is light of digestion,
                         and the English make a kind of Lob-
                         lolly of it to eat with milk which they
                         call Sampe. . . .

[Josselyn then gives the recipes for "The New England's Standing Dish"
which appear in Chapter VI.]

Acorus calamus	**Bastard Calamus Aromaticus,** agrees with the description, but is not barren; they flower in July, and grow in wet places, as about the brinks of Ponds.
	The English make use of the Leaves to keep their Feet warm. There is a little Beast called a Muskquash that liveth in small houses in the Ponds, like Mole Hills, that feed upon these Plants. Their Cods sent as sweet and as strong as Musk, and will last a long time handsomely wrap'd up in Cotton wool; they are very good to lay amongst the Cloaths. May is the best time to kill for then their Cods sent strongest.
Allium canadense	**Wild-Leekes,** which the Indians use much to eat with their fish.
Lepidium virginicum	**A Plant like Knavers-Mustard,** called New-England mustard.
Lilium superbum *Lilium canadense*	**Mountain-Lillies,** bearing many yellow Flowers, turning up their leaves like the Martigon, or Turks Cap, spotted with small spots as deep as Safforn, they Flower in July.
Cornus canadensis	**One Berry,** or Herb True Love. See the Figure.
Nicotiana tabacum	**Tobacco,** there is not much of it planted in New England. The Indians make use of a small kind with short round leaves called Pooke.

[Someone else will have to smoke this one out. In Gerard all "tobaccos" are familiarly called "henbane."]

With a strong decoction of Tobacco they Cure Burns and Scalds, boiling it in Water from a Quart to a Pint, then wash the Sore therewith, and strew on the powder of dryed Tobacco.

Sarracenia purpurea   ***Hollow Leaved Lavender,*** is a Plant that grows in salt Marshes over-grown with Moss, with one straight stalk about the bigness of an Oat straw, better than a Cubit high; upon the top standeth one fantastical Flower, the Leaves grow close from the root, in shape like a Tankard, hollow, tough, and always full of Water; the Root is made up of many small strings, growing only in the Moss, and not in the Earth, the whole Plant comes to its perfection in August, and then it has Leaves, Stalks, and Flowers as red as blood, excepting the Flower which hath some yellow admixt. I wonder where the knowledge of this Plant hath slept all this while, i.e. above Forty Years.

It is excellent for all manner of Fluxes.

Antennaria margaritacea   ***Live for ever,*** a kind of Cud-weed.

Oenothera biennis   ***Tree Primerose,*** taken by the ignorant for *Scabious.* A Solar Plant, as some will have it.

Adiantum pedatum   ***Maiden Hair,*** or ***Cappellus veneris verus,*** which ordinarily is half a yard in height. The Apothecaries for shame now will substitute Wall-Rue no more for Maiden Hair, since it grows in abundance in New England, from whence they may have good store.

Pyrola   ***Pirola,*** Two kinds. See the Figures, both of them excellent Wound Herbs.

Allium tricoccum   ***Homer's Molley.***

Epilobium angustifolium   ***Lysimachus or Loose Strife,*** it grows in the open Sun four feet high, Flowers from the middle of the Plant to the top, the Flowers purple, standing upon a small sheath or cod, which when it is ripe breaks and puts forth a white silken doun, the stalk is red, and as big as ones Finger.

Helianthus strumosus *Marygold of Peru,* of which there are two kinds, one bearing black seeds, the other black and white streak'd, this beareth the fairest flowers, commonly but one upon the very top of the stalk.

Smilacina racemosa *Treacle-Berries.* See before Salomons Seal.

Chenopodium botrys *Oak of Hierusalem.* See before

Ambrosia eliator *Oak of Cappadocea.* See before

Apios tuberosa *Earth-Nuts,* differing much from those in England, one sort of them bears a beautiful white Flower.

Cakile americana *Sea-Tears,* they grow upon the Sea banks in abundance, they are good for Scurvy and Dropsie, boiled and eaten as a Sallade, and the broth drunk with it.

Phaseolus vulgaris *Indian Beans,* falsly called French beans, are better for Physick and Chyrurgery than our Garden Beans. *Probatum est*:

Cucurbita *Squashes,* but more truly Squonter-squashes, a kind of Mellon, or rather Gourd, for they oftentimes degenerate into Gourds; some of these are green, some yellow, some longish like a Gourd, others round like an Apple, all of them pleasant food boyled and buttered, seasoned with Spice; but the yellow squash called an Apple Squash, because like an apple, and about the bigness of a Pomewater is the best kind; they are much eaten by the Indians and the English, yet they breed the small white Worms (which Physicians call *Ascarides*), in the long Gut that vex the Fundament with a perpetual itching, and a desire to go to stool.

Citrullus   *Water-Mellon,* it is a large Fruit, but nothing near so big as a Pompion, colour smoother, and of a sad Grass green rounder or more rightly Sap-green; with some yellowness admixt when ripe; the seeds are black, the flesh or pulpe ex-

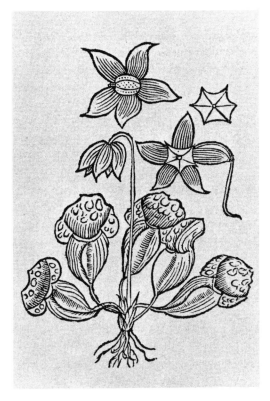

Pitcher Plant, from John Josselyn's *New-Englands Rarities*, 1672.

ceeding juicy. It is often given to those
sick of Feavers, and other hot Diseases
with good success.

Saxafraga virginiensis **New England Daysie, or Primrose,** is
the second kind of Navel Wort in John-
son upon Gerard; it flowers in May,
and grows amongst Moss upon hilly
Grounds and Rocks that are shady. It
is very good for Burns and Scalds.

[Here Josselyn interrupts himself to entitle his use of this plant . . .]

An Achariston, or Medicine deserving thanks. An Indian whose Thumb
was swell'd, and very much inflamed, and full of pain, increasing and creeping along to the wrist, with little black spots under the Thumb against the
Nail; I cured it with this **Umbellicus veneris** Root and all, the Yolk of an
Egg, and Wheat flower, f. Cataplasme.

Convolvulus sepium **Briony of Peru,** (we call it though it
grows here) or rather Scammony; some
take it for Mechoacan: The green juice
is absolute Poyson; yet the Root when
dry may safely be given to strong Bodies.

Ribes **Red and Black Currence.** See before

Rosa carolina **Wild Damask Roses,** single, but very
large and sweet, but stiptick.

Comptonia asplenifolia **Sweet Fern,** the Roots run one within
another like a Net, being very long and
spreading abroad under the upper crust
of the Earth, sweet in taste, but withal
astringent, much hunted after by our
Swine: The Scotchmen that are in New-
England have told me it grows in Scotland.

Aralia nudicaulis **Sarsaparilla,** a Plant not yet sufficiently
known by the English: Some say it is a

Aralia hispida kind of Bind Weed; we have, in New-

England two Plants, that go under the
name of Sarsaparilla: the one not a foot
in height without Thorns, the other having the same Leaf, but is a shrub as high
as a Goose Berry Bush, and full of sharp
Thorns; this I esteem as the right, by
the shape and favour of the Roots, but
rather by the effects answerable to that
we have from other parts of the world;
It groweth upon dry Sandy banks of
Rivers, so far as the Salt water flowes;
and within Land up in the Country, as
some have reported.

Vaccinium pennsylvanicum   **Bill Berries,** two kinds, Black and Sky
Vaccinium corymbosum   Coloured, which is more frequent. They
are very good to allay the burning heat
of Feavers, and hot Agues, either in
Syrup or Conserve. A most excellent
Summer Dish. They usually eat of them
put into a Bason, with Milk, and sweetened a little more with Sugar, and Spice,
or for cold Stomachs, in Back. The Indians dry them in the Sun, and sell them
to the English by the Bushell, who make
use of them instead of Currence, putting
of them into Puddens, both boyled and
baked, and into Water Gruel.

Rubus chamaemorus   **Knot Berry, or Clowde Berry,** seldom
ripe.

Rhus   **Sumach,** differing from all that I ever
did see in the Herb lists; our English
Cattle devour it most abominably, leaving neither Leaf nor Branch, yet it
sprouts again next Spring. The English
use to boyl it in Beer, and drink it for
Colds; and so do the Indians, from
whom the English had the Medicine.

Cerasus virginiana **Wild Cherry,** they grow in clusters like
Cerasus serotina Grapes, of the same bigness, blackish,
red when ripe, and of harsh taste. They
are also good for Fluxes. Transplanted
and manured, they grow exceeding fair.

[Here he mentions the pine tree, hemlock, larch, fir "or Pitch Tree," and spruce.]

Sassafras officinale *Sassafras, or **Ague Tree.*** The Chips of
the Root boyled in Beer is excellent to
allay the hot rage of Feavers, being
drunk. The Leaves of the same Tree are
very good made into an Oyntment for
Bruises and dry Blows. The Bark of the
Root we see instead of Cinamon; and it
is sold at the Barbadoes for two shillings
the Pound. And why may not this be
the Bark the Jesuits Powder was made
of, that was so Famous not long since in
England, for Ague?

Vaccinium macrocarpum **Cran Berry, or Bear Berry,** because
Bears use much to feed upon them, is
a small trayling Plant that grows in
Salt Marshes that are over-grown with
Moss; the tender Branches (which are
reddish) run out in great length, lying
flat on the ground, where at distances
they take root, overspreading sometimes
half a score Acres, sometimes in small
patches of about a Rood or the like; the
Leaves are like Box, but greener, thick
and glittering; the Blossoms are very
like the Flowers of our English Night
Shade, after which succeed the Berries,
hanging by long small foot stalks, no
bigger than a hair; at first they are of a

pale yellow Colour, afterwards red, and
as big as a Cherry; some perfectly round,
others Oval, all of them hollow, of a
sower astringent taste; they are ripe in
August and September. They are excellent against the Scurvy. They are also
good to allay the feaver of hot Diseases.
The Indians and English use them
much, boyling them with Sugar for

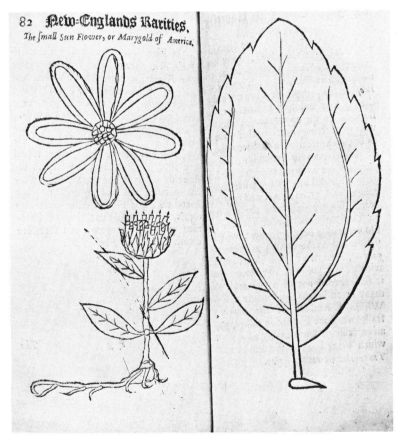

82   New-Englands Rarities,
The *small Sun Flower*, or *Marygold of America*,

American Sunflower, from John Josselyn's *New-Englands Rarities*, 1672.

Sauce to eat with their Meat; and it is
a delicate Sauce, especially for roasted
Mutton: Some make Tarts with them as
with Goose Berries.

Vitis labrusca   *Vine,* much differing in the Fruit, all of
Vitis aestivalis   them very fleshy, some reasonably pleasant; others have a taste of Gun Powder,
and these grow in Swamps, and low wet
Grounds.

[Josselyn has also mentioned in this category of native plants "One
Berry, Herba Paris, or True Love" which was listed in his first category and
will appear in the next.]

3. *Of Such Plants as are proper to the Country and have no Name.*

[This section need not detain us, long though it be, and showing Josselyn
at his best as an inquiring mind and a careful reporter. I have briefed it,
mercilessly.

He describes first what he believes to be a "kind of Pirola," sketching
only the leaves. Tuckerman identifies the plant as *Goodyera pubescens*. The
second plant is one of which Josselyn saw the leaves only, which he sketches;
it was brought to him by a neighbor who got lost hunting for his cattle.
Both Josselyn and Tuckerman are inclined to give up on this one and even
the great Plukenet describes Josselyn's sketch as "sufficiently unhappy" and
thinks it is a fern. Josselyn's next discovery is a "Clownes all heal of New
England" which he says is "another Wound Herb not Inferiour to ours, but
rather beyond it" and he describes what Tuckerman takes to be *Verbena
hastata*. Then, with a fine flourish of a drawing, he describes our skunk cabbage, *Symplocarpus foetidus*, according to Tuckerman, who says Josselyn has
placed the plant in the correct category as native to the country and name-
less. This is followed by a gay sketch of the plant loved by humming birds
called "Humming Bird tree" by Josselyn, but *Impatiens fulva* by Tuckerman, who says it is known in England, too, although Josselyn did not
know it and could never have recognized the poor picture in Gerard. In any
case, our hummingbirds were one of the wonders of the New World and

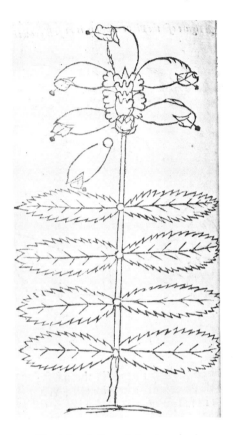

Turtlehead, from John Josselyn's
New-Englands Rarities, 1672.

could easily have made the plant seem unique. The next plant is indeed a
genus peculiar to America, as Tuckerman says: snake-weed, or *Nabalus
alba*, of which we are given a sketch. A graphic if somewhat stylized drawing of *Chelone glabra*, another American genus follows, and then we have a
fine picture of what Josselyn hopes is a "variegated Herb Paris." Tuckerman says Plukenet copied the picture and also Josselyn's poor guess, as the
plant is really *Cornus canadensis*. With a rather out-of-scale drawing of the
native American *Helianthus strumosus*, according to Tuckerman, or "a kind
of small Sun Flower" according to Josselyn, the list of his plants "proper to
the country" but without names is brought to a very creditable end.]

4. *Of such Plants as have sprung up since the English Planted and kept Cattle in New-England.*

Holcus mollis or *Triticum repens*	Couch Grass.
Capsella bursa pastoris	*Shepherds Purse.*
Taraxacum dens leonis	*Dandelion.*
Senecio vulgaris	*Groundsel.*
Sonchus oleraceus or *Chenopodium atriplex*	*Wild Arrach.*
Solanum nigrum	*Night Shade,* with the white Flower.
Urtica dioica	*Nettlestinging,* which was the first Plant taken notice of.
Malva sylvestris	*Mallowes.*
Plantago major	*Plantain,* which the Indians call English-Man's Foot, as though produced by their treading.
Hyoscyamus niger	*Black Henbane.*
Artemisia absynthium	*Wormwood.*
Rumex crispus	*Sharp pointed Dock.*
Rumex patientia	*Patience.*
Rumex sanguineus	*Bloodwort.*
Ophioglossum	And I suspect *Adders Tongue.*
Polygonum aviculare	*Knot Grass.*
Stellaria media	*Cheek weed.*
Symphytum officinale	*Compherie,* with the white Flower.
Maruta cotula *(Anthemis cotula)*	*May weed,* excellent for the Mother; some of our English housewives call it Iron Wort, and make a good Unguent for old Sores.
Lappa major	*The great Clot Bur.*
Verbascum lychnitis	*Mullin,* with the white Flower.

Q. What became of the influence of those Planets that produce and govern these Plants before this time?

I have now done with such Plants as grow wild in the Country in great plenty, (although I have not mentioned all). I shall now in the Fifth place give you to understand what English Herbs we have growing in our Gardens that prosper there as well as in their proper Soil, and of such as do not, and also of such as will not grow there at all.

[And here we come to what most concerns us. As Tuckerman thought it unnecessary to identify more than a very few, we give Josselyn's list with our identifications as from L. H. Bailey, guided sometimes by illustrations from Gerard and help from Geoffrey Grigson's *Englishman's Flora*.]

5. *Of such Garden Herbs (amongst us) as do thrive there and of such as do not.*

Brassica oleracea	**Cabbidge** growes there exceeding well.
Latuca sativa	*Lettice.*
Rumex acetosa	*Sorrel.*
Petroselinum hortense	*Parsley.*
Calendula officinalis	*Marygold.*
Malva crispa (Gerard)	*French Mallowes.*
Anthriscus cerefolium	*Chervel.*
Poterium sanguisorba	*Burnet.*
Satureja montana	*Winter Savory.*
Satureja hortensis	*Summer Savory.*
Thymus vulgaris	*Time.*
Salvia officinalis	*Sage.*
Daucus carota	*Carrats.*
Pastinaca sativa	**Parsnips** of a prodigious size.
Beta vulgaris	*Red Beetes.*
Raphanus sativus	*Radishes.*
Brassica rapa	*Turnips.*
Portulaca oleracea	*Purslain.*
Triticum aestivum	*Wheat.*

Secale cereale	**Rye.**
Hordeum vulgare	**Barley,** which commonly degenerates into Oats.
Avena	**Oats.**
Pisum sativum	**Pease** of all sorts, and the best in the World; I never heard of, nor did see in eight years time, one Worm eaten Pea.
Vicia faba	**Garden Beans.**
Phaseolus vulgaris	
Avena nuda	**Naked Oats,** there called Silpee, an excellent grain used instead of Oat Meal, they dry it in an Oven, or in a Pan upon the fire, then beat it small in a Morter.
Mentha viridis	**Spear Mint.**
Ruta graveolens	**Rew,** will hardly grow.
Chrysanthemum parthenium	**Fetherfew** prospereth exceedingly.
Artemisia abrotanum	**Southern Wood,** is no Plant for this Country. Nor
Rosmarinum officinalis	**Rosemary.** Nor
Laurus	**Bayes.**
Lunaria rediviva	**White Satten** groweth pretty well, so doth
Santolina chamaecy parissus	**Lavender Cotton.** But
Lavendula vera	**Lavender** is not for the climate.
Mentha pulegium	**Penny Royal.**
Petrosilinum hortense	**Smalledge.** (Parsley)
Glechoma hederacaea	**Ground Ivy,** or Ale Hoof.
Dianthus caryophyllus	**Gilly Flowers** will continue two years.
Foeniculum vulgare	**Fennel,** must be taken up, and kept in a warm cellar all Winter.
Sempervivum tectorum	**Houseleek** prospereth notably.
Althea rosea	**Holly hocks.**
Inula helenium	**Enula Campana,** in two Years time the Roots rot.
Symphytum officinale	**Comferie,** with white Flowers.
Coriandrum sativum	**Coriander,** and
Anethum graveolens	**Dill,** and

Pimpinella anisum	**Annis** thrive exceedingly, but Annis Seed, as also the Seed of Fennel seldom come to maturity; the Seed of Annis is commonly eaten with a fly.
Salvia sclarea	**Clary,** never lasts but one Summer, the Roots rot with the Frost.
Asparagus officinalis	**Sparagus** thrives exceedingly, so does
Rumex acetosa	**Garden Sorrel,** and
Rosa eglanteria	**Sweet Bryer, or Eglantine.**
Rumex sanguineus	**Bloodwort** but sorrily, but
Rumex patientia	**Patience,** and
Rosa	**English Roses,** very pleasantly.
Chelidonium majus	**Celandine,** by the West Country men called Kenning Wort, grows but slowly.
Moschata	**Muschata,** as well as in England.
Lepidium sativum	**Dittander, or Pepper Wort,** flourisheth notably, and so doth
Tanacetum vulgare	**Tansie.**
Cucumis	**Cucumbers.**
Cucurbita	**Pompions,** there be several kinds, some proper to the Country, they are dryer than our English Pompions, and better tasted; you may eat them green.

[And now we present Josselyn's additional listing from his *Two Voyages to New England*, to which we have added a few further identifications.]

I have given you an Account of such plants as prosper there and such as do not; but so brief that I consider it necessary to afford you some more of them.

Plantago major	**Plantain.** I told you sprang up in the Country after the English came, but it is but one sort, and that is broad-leaved plantain.
Dianthus caryophyllus	**Gilliflowers** thrive exceedingly there and are very large, the Collibuy or humming bird is much pleased with them. Our

English dames make syrup of them without fire, they steep the Wine till it be of a deep colour, and then they put to it spirit of Vitriol, it will keep as long as the other.

Rosa eglanteria   **Eglantine or sweet Bryer** is best sown with Juniper berries, two or three to one Eglantine-berry put into a hole made with a stick, the next year separate and remove them to your banks, in three years time they will make a hedge so high as a man, which you may keep thick and handsome with cutting.

Trifolium pratense   **Our English Clover-grass** thrives very well.

Raphanus   **Radishes.** I have seen them as big as a man's arm.

Linum usitatissimum   **Flax and Hemp** flourish galantly.
Cannabis sativa

Avens sativa   **Our Wheat,** i.e., summer wheat many times changeth into Rye. And is subject to be blasted, some say with a vapour breaking forth out of the earth, others, with a wind North-east or North-west, at such times as it flowereth, others say again it is with lightning. I have observed that when a land of Wheat hath been smitten with a blast on one corner, it hath infected the rest in a weeks time, it begins at the stem (Which will be spotted and goes upwards to the ear making it fruitless) in 1669 the pond that lyeth between Watertown and Cambridge cast its fish dead upon the shore, forc'd by a mineral vapour as was conjectured.

Our fruit trees prosper abundantly. Apple trees, Quince trees, Cherry trees, Plum trees, and Barberry-trees. I have observed with admiration that the Kernels sown or the succors planted produce as fair and good Fruit without grafting, as the Tree from whence they were taken: the Countrey is replenished with fair and large Orchards. It was affirmed by one Mr. Woolcut (a magistrate in Connecticut Colony) at the Captains Messe (of which I was) aboard the ship I came home in, that he made Five hundred Hogsheads of Syder out of his own Orchard in one year. Syder is very plentiful in the Country, ordinarily sold for Ten shillings a Hogshead. At the Taphouses in Boston I have had an Alequart spiced and sweetened with sugar for a groat, but I shall insert a more delicate mixture of it. Take of Maligo-Raisons, stamp them and put milk to them and put them in an Hippocras bag and let it drain of itself, put a quantity of this with a spoonful or two of Syrup of Clove-Gilliflowers into every bottle, when you bottle your Syder and your Planter will have a liquor that exceeds passada, the Nectar of the Countrey.

The Quinces, Cherries, Damsons, set the Dames at work, Marmalad and preserved Damsons is to be met with in every house. It was not long before I left the Countrey that I made Cherry wine, and so many others, for there are good store of them both red and black.

We have seen Josselyn's little panegyric to the Plants of New England in his description of the country, but we may remind ourselves here that:

"The plants of New England for the variety, number, beauty, and vertues, may stand in Competition with the plants of any Countrey in Europe. Johnson hath added to Gerard's Herbal 300, and Parkinson mentioned many more. . . ."

III

Finally we come to lesser lists though not, by any means, by lesser folk. The earliest sources of medicinal plant names with which the settlers were familiar were contained in two small books on health written in mid-sixteenth-century England by Andrew Boorde and preserved in Newbury, Massachusetts, for nearly three hundred years. They are worn with use, coverless, dog-eared and stained. And yet the words, some in the old Gothic black letter and some in the newer Roman type, still spring to our attention. The style is quick, original and direct, and the text is full of good moral as well as medical advice. There is a tender understanding of human nature, as witness this remedy for love-sickness: "I do advertise every person not to set to the hart that another doth set at the hele. Let no man set his love so far out but that he may withdraw it betime. And muse not but use mirth and mery company, and be wise and not foolish." He is a staunch advocate of moderation in eating and drinking. "Use a good dyet and sit not up too late and use some labour or manuel occupation to sweat at the browse. . . ." His chapters on the soul and the mind of man show as much practical concern as other chapters on other parts of man and their afflictions. Some of his names for the diseases take on a sort of pathetic charm, as when he calls the affliction of not being able to move one's joints "Elephany" because, he says, elephants have no joints at all.

The plants recommended by Dr. Andrew Boorde in his *Breviarie of Health* and its sequel *Extravagantes* are not many. The same ones are used over and over again. And they all turn up later in accounts of the settlers. "Enula campana," St. Johnswort and plantain are his standbys. Lilies and water-lilies "stamped" make useful plasters. Roses and "mirtilles" and house-leeks (or "syngreen") purslain, calamint, tansy, "Materwort," "calamus aromaticus,'' betony, violets, fumitory, ''doves' dung'' (presumably *ornithogallum* which was supposed to be the doves' dung of the Bible), daisies, small-age, hoarhound, hyssop, chamomile and sumach are about all he requires from the garden. It is interesting that John Winthrop, Junior, and Cotton Mather share his respect for elecampane and Cotton Mather for plantain. Boorde especially advises women to use tansy, which confirms what an elderly lady on Cape Cod said she had been told as a child — that every woman should curtsey on passing tansy, it has done so much for the female sex. This is a pleasanter association than one of my childhood when an elderly woman in rural Maine bore out what Samuel Sewall recounted of tansy by saying she did not like the smell, as it reminded her of dead people. (Samuel Sewall viewed the exhumed body of a long dead and very distinguished friend with interest because it was so well preserved by the tansy in which it had been packed.)

Still, Dr. Andrew Boorde's list of plants gives us an idea of what the settlers knew long before they came. The situation of plants other than medicinal when they arrived, can be sampled from a letter to the settlers from the Governor and Company of the Massachusetts Bay in New England in 1629, containing a list of plants ''to provide and send for New England.'' This list includes tame turkeys and rabbits, but in the main it consists of ''. . . wheat, rye, barley, oats . . . beans, pease, stones of all sorts of fruits as peaches, plums, filberts, cherries, pears, apple, quince kernels, pomegranates, woadseed, saffron heads, liquorice seed (roots sent and madder roots), potatoes, hop roots, hemp seed, flax seed, currant plants. . . .''

Potatoes are interesting here, as they were fairly recent then in English cultivation, Gerard seeing fit to have his portrait painted for his first edition of 1597 with a potato leaf in his hand. The pomegranate seeds must have proved just another disappointment to those settlers hoping the New England climate would enable them to grow things which did well in Spain or around the Mediterranean. Along with grapes for wine, many things must have been abandoned with regret.

We have seen Endecott's letter to Winthrop with his portmanteau list of remedies. While this was a message of help in toto, we have other instances of cures almost offhandedly included in long letters about other subjects. A typical indication of this is in a letter from Samuel Torrey, received by Thomas Hinckley, sixth and last governor of New Plymouth before it was absorbed into the Massachusetts Bay Colony in 1692. The letter, among other subjects, refers to a baby who has just died and volunteers "what to do in this sickness." Inevitably the treatment begins with a vomit and cathartic, followed by treacle water colored with saffron to "drive out malignity by sweat." If the patient is "distempered" in the head, a "glister" is administered, that is, an enema, and "blister-plasters" on the ankles for delirium. But now come the healing herbs. "A good quantity" of anise seed, liquorice, sliced figs, stoned raisins, maidenhair and pimpernel make a good drink boiled in barley water. And a posset drink in which featherfew has been boiled is advised. "What we do," indeed. In this instance, it is interesting to note that at the end of the next century in Philadelphia, remedies had not changed much and yellow fever was being treated in much the same way.

Another, more cheerful, letter to Hinckley comes from friends who moved from Boston to Dorchester in Carolina at the end of the century. As they went down the coast by boat, it is uncertain if they mean England or Boston in New England when they say "over." The sea voyage down the coast must have seemed a long way then, too. The letter comes, from Joseph Lord, whose wife still remembers the roughness of the trip.

"Angelica," he says, "that we brought over with us is grown out five and six inches long already. So are some damson trees, the roots whereof we brought o(ver). Currant and gooseberry bushes that we brought with us are green and flourishing. Apple seeds sown by us since we came, came up in January and are about three inches high, some of them. All other herbs that we brought over seem to take very well and grow here."

A slight reference from an inventory gives us an idea of the stillroom practices. In Captain John Whipple's Inventory of his Ipswich house in 1683, we have listed as three separate items:

Item. A great Bible, sixteen shillings. . . . Books five pounds, eight shillings and nine pence . . . five bottles of syrup of clove gilly flowers.
Item. Three bottles of rose water six shillings, two bottles of mint water three shillings.
Item. A glass bottle of Port wine two shillings, Angelica water, syrup of gilliflowers, strawberry water three bottles four shillings.

Perhaps the first five bottles of syrup of clove gillyflowers, or clove pinks, were included with the lot of books — and how we would like to know their titles!

To give us a fair view of what householders knew they needed and must provide from their gardens more or less as a matter of course, we may turn to the "Physical Receipts" from the journal of Captain Lawrence Hammond of Charlestown. His journal dates from 1677 to 1694. Captain Hammond would seem to have been one of those prepossessing citizens to whom others turn for various sorts of advice, including medical. He was made a free-man of the colony in 1666, and for several years was chosen to be Deputy of the General Court from Charlestown. In 1686 and for several years after, he was Recorder of Middlesex County, which was like being Register of Deeds and Probate today. He was lieutenant and captain of a foot company in Charlestown, and died in Boston in 1699. His "Physical Receipts" are neatly listed.

AN EXCELLENT WATER FOR YE SIGHT

Take Sage, Fennel, Vervain, Bettony, Eyebright, Celandine, Cinquefoyle, Herb of Grass, pimpernel. Steep them in white wine one night, distil them altogether, and use the water to wash the Eyes.
The juice of Eyebright is Excellent for ye sight.

ANOTHER

Take good White wine, Infuse Eyebright in it 3 dayes, then Seeth it with a little Rosemary in it, drink it often, it is most excellent to restore and strengthen the sight. Also Eate of the powder of Eyebright in a new laid egg rare roasted every morning.

ANOTHER

Take Fennel, Anniseed and Elicompaine, dry and powder them, mix it with good Nants-brandy, and dry it againe; Every morning and evening eate a pretty quantity it is excellent for the sight.

A Medicine to recover ye colour and complexion when lost by sickness.

Take two quarts of Rosewater red, take five pounds of clean White Wheat, put it into ye Rosewater. Let it lie till the Wheate hath soaked up all ye liquor, then take the Wheat and beat it in a mortar all to mash.
Nettle Seeds bruised and drank in White Wine is Excellent for the Gravel.

Physical Receipts for Comforting the Head and Braine.

Take Rosemary and Sage of both sorts of both, with flowers of Rosemary if to be had, and Borage with ye flowers. Infuse in Muscadine or in good Canary 3 dayes, drink it often.

The fat of a Hedge-hog roasted drop it into the Eare, is an Excellent remedy against deafness.

Also a Clove of Garlick, make holes in it, dip it in Honey and put it into the Eare at night going to bed, first on one side, then on the other for 8 or 9 days together, keeping in ye Eares black wooll.

For Hoarseness. Take 3 or 4 figs, cleave them in two put a pretty quantity of Ginger in powder, roast them and eat them often.

For the Palsey

Take a pint of good mustard, dry it in ye oven till it be as thick as a pudding, then dry it over a chafing dish of coales till it may be beaten to powder mix with it a handful of powder of Bettony leaves, put some sugar to it and Eate it every morning.

For the Megrum

Mugwort and Sage a handfull of each, Camomel and Gentian a good quantity, boyle it in Honey, and apply it behind and on both sides ye Head very warm, and in 3 or 4 times it will take it quite away.

So much for the helpful Captain Hammond.

From a shopkeeper's account book comes this rather lengthy set of directions for taking care of a measles case. Abraham How kept his book in his shop on the border between Ipswich and Rowley. While this advice for the treatment of measles is based upon the Reverend Thacher's broadside on the subject, which in turn was taken from the medical writings of the English doctor Sydenham, it has a certain verve of its own and deserves quoting in full.

ABRAHAM HOW, HIS BOOK, 1682

A letter about a good management under the distemper of the measels here published for the benefit of the poor & such as may want the help of Able physicians. Directions not to kill with kindness by over dosing or overheating, giving things to force Nature out of its own ordinary way. When the symptoms of the sickness is upon you, dont throw away your life by not being sick soon enough. When the red spots are upon you, take as much brimstone firmly powdered as will ly upon a sixpence twice or thrice a day, ordinarily sweetening is good. Feed spareingly on very easey food. A convenient warmth acording as ye season may requior but over heating will spoyle all as by hot Beer or Rum. When full of the measels &

thirsty, what shall he doe if he will ventur to Drink anything cold — it may be he will find death in the pot, Let him Drink a tea of Balm and Scabious. If he be faint or sick, add a little Saffron; if he be loose let the tea be of Sage or Rosemary; if he be gript it may have pennyroyall in it. But what is to be done if a cough annoy him: then a parcel of maiden hair, anniseed, liquirice, raisions, figs. Liquorice alone in a tea may doe, or, if that cant be had, a syrup of maiden hair of hyssop or colts foot. Ye cough must not be kept. You may allay it with sugar canded and with buttered pils, or hot honey will do very well alone. In hot constitution espechally a proper drink will be water with ripe apples in it. Some will add a little ginger to it. Hot cyder if not hard, is allowable drink. Toward the heighth of the distemper hyssop is preferable to Scabious. Let the Diet throw the whole course be thin enough, gruels, caudles, watter pottage & ye like given offen enough. About the third or fourth day will come burning: and this will be offten atended with frightfull circumstances: greevious oppresions as fainting, vomiting, purging, and vapours. In one word its all that is terriable but all will be presently and safely relieved if heaven afford a blessing to the proper managements. Sometimes the sudden disaperance of the measels gives a needless fright: they have performed their course but the frighted patient flys to hot things, to exspelers which endangers the raising of putrefaction heat, & that by emmoderat use of hot things: for apprsion of them take wine and oile, or, if ye patient be loose, let him have Rum and oile. Give it pretty hot, two or three spoonfulls every houer as there may be occasion or take almost any time that will cause a vomit, which at this time has reprieved many a Life that was just expiring. When a vomit wont relieve the oppressed the wine and oile, with a decoction of the cooling seeds, has presently don it: For vomiting or purging take scalding hot water. Do it nine or ten times if there be need and as he can bare it; then give a little hot wine with a bit of softened bread. For a considerable while after this the patient must be mighty careful of cold, nor eat too soon of any flesh meat. Let him not be well too soon & throw himself into a fever and so throw away his life, as many have presumtiously don. Let him take time. If a cough continues let him fly to the usual remedy, take a spoonfull of castle soap shaveings in a glass of wine or Beer for a few nights following. If a flux follow, wether common or Bloody, a tea made of Rubbub & sweetned with a little syrup made of malows giuen dayly so much as to cause one or two stools, is a way to carey it safely off. The same tea will carey off the worms that so offen follow the measels, espechally in children. A purge is needful to all that will not have the venom of the measels remain in them and followed with many euil consequences. In the letter is preposed all to direct in ye plainest maner without any term of art, and so that any nurs may understand how to administer what is to be done in the singular case of ye measels, which in itself is not so dangerous yet upon ye least error in treating of it proues as deadly as most that ye children of Death fall before.

The only other seventeenth-century manuscript list to be considered is that compiled by the painstaking Edward Taylor in his little *Dispensatory*, but as that runs to nearly five hundred plants and does, indeed, seem to be condensed Culpeper, we are noting it only when he has something original to add to this list of most-used, most-noticed, most-loved and most-depended-upon plants in the seventeenth-century European world.

Which is really about as far as we need go. In the descriptions of plants which follow, the adventives or "naturalized weeds" will be noted in their places. But we have come to the end of harvesting our plant names and finished raking the stubble for any last bits.

It would be charming here to toss in a nosegay, a little flowery tribute to the seventeenth-century gardeners. Surely, somewhere from their own time there must be a pretty compliment paid to them and their flowers? Gardening in New England cannot, after all, have been so serious as not to evoke some lovely comment upon its joys? The answer is that it was. Anne Bradstreet's modest request for a thyme and parsley wreath is the gayest token we can find to lay at the feet of the gardening housewife. Not even in the poems of Edward Taylor can we find references to gardens and flowers unencumbered with mysticism and not combining the plants of the Bible and those of Connecticut into one whole. How odd to come to the end of a book and find that it is as true as one can make it.

Still, we must have something festive to end with. Here is one of Edward Taylor's most horticulturally embroidered "Meditations." In "Meditation 62" we catch a glimpse of his garden. Taylor has been bidden to a heavenly feast:

> I'll surely come, Lord fit me for this feast:
> . . . .
> . . . .
> . . . .
> Give me my Sage and Savory; me dub
> With Golden Rod, and with Saints Johns wort good.
>
> Root up my Henbain, Fawnbain, Divells bit,
> My Dragons, Chokewort, Grasswort, Ragwort, vice,
> And set my knot with Honeysuckles, stick
> Rich Herb-a-Grace and Grains of Paradise
> Angelica, yea Sharons Rose the best
> And Herba Trinitatis in my breast.

Another Meditation sums up the general scheme of things somewhat translated from everyday life:

These all as meate and medicine, emblems choise
Of Spirituall Food and Physicke are which sport
Up in Christs Garden

Perhaps we are now ready to hear a clarion call from Cotton Mather in which he turns us into a plant, an operation we may feel by this time is easily accomplished.

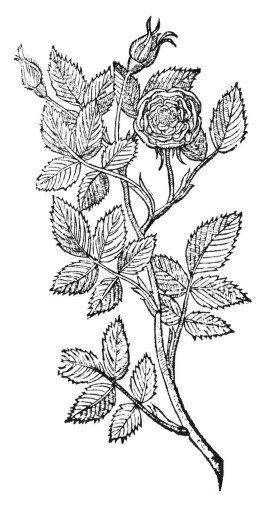

Rose without thorns, from 1633 edition of Gerard's *Herball*.

Cotton Mather addresses us in his *Angel of Bethesda:*

The Vertues of every Plant call for thy praises to the glorious God who has made the Plant and taught us the Vertues of it. And if thou are a Plant of Righteousness thou wilt study to be one, upon other Accounts, of greater Vertues than any that are to be found, from the Cedar that is in Lebanon even unto the Hyssop that springs out of the Wall.

On the whole, George Herbert's

> What is fairer than a rose?
> What is sweeter? Yet it purgeth.

may have to suffice as a final, howbeit medicinal, posy .

Part II

An Appendix of the Plants Mentioned by Explorers, Settlers, Underwriters and Visitors as Growing in Seventeenth-Century New England, Either found Growing or Planted by Those who came here to Live

Following is an Index of the plants commonly referred to by those living in seventeenth-century New England as those they used "for meate or medicine." To these have been added some of our best known "wild-flowers," called "escaped" or "adventive" or "naturalized," which we now recognize as having been brought to this country with a purpose. Some of them seemed so obviously necessary to the economy of living as to have missed particular mention, such as those used for dyeing or fulling. A few plants have also been included as "Indian plants," those adopted upon advice from the Indians.

From illustrations and descriptions in Gerard's *Herball*, edited by Thomas Johnson in 1633, one can have a fairly good idea of what was meant by most of the common names. It has been a fascinating experience to translate these names into modern botanical terms, checking through the English reference books of Geoffrey Grigson's *Englishman's Flora* and the *Modern Herbal* of Grieve and Lyel, to compare the names with those given in L. H. Bailey's *Standard Encyclopaedia of Horticulture* and Asa Gray's *Manual of Botany*, eighth edition. I have done as well as I can. If I am found to have made mistakes, I can only say what a pleasure it will be to anyone who may follow me through those fascinating labyrinths and come out at a different place. Every step will have been rewarding.

The greater part of the list of plants for replanting the gardens of the founders of our country in New England, has depended upon the seed list of John Winthrop, Junior, and the accounts of plants seen growing by John

Josselyn. I have included also those mentioned by Cotton Mather, Samuel Sewall, William Endecott, William Bradford, the senior John Winthrop, storekeeper How and Captain Hammond. Other plants have come from inventories, receipts, letters, and prescriptions with the owners' and users' names attached where possible.

Each plant is listed under the name they called it, described as the settlers themselves knew it from information available to them in Gerard's *Herball* of 1633, used as reference book by Josselyn; in the charming *Paradisi in Sole Paradisus Terrestris* of 1629 of John Parkinson, which was owned by President Hoar of Harvard and later by the Mathers; in the works of Nicholas Culpeper; and in the *Acetaria* of the omniscient John Evelyn. I am fortunate to have a copy of Culpeper's translation of the *Pharmacopoeia Londonensis*, the *London Dispensatory* of 1683, an edition ordered in some quantity by Boston booksellers, and to have found a copy of his *English Physician* of 1652. These two works were apparently in constant demand on the Boston book market. In addition, since there was more Culpeper lore than these two volumes contain in the hands and minds of the early settlers, I have used two later editions of his *English Physician and Complete Herbal*, of 1788 and 1798, checking these with each other and with a modern reprint chiefly depended upon by Culpeper's modern admirers, who must also find clear, large print a blessed relief.

I have let Parkinson be represented almost entirely by his *Paradisi in Sole Paradisus Terrestris* of 1629 rather than his more pretentious *Theatrum Botanicum* of 1640, obviously one of Cotton Mather's sources for his *Angel of Bethesda*. I have described the plan of the *Theatrum*, but I felt that for our purposes his friendly style in the *Paradisus* would serve us better in our attempt to give a general and rounded view of the information then readily available, much of it probably common knowledge, about the plants our early settlers knew and grew.

I made use of the little reprint of John Evelyn's *Acetaria, a Discourse of Sallets*, as he was another of Cotton Mather's authorities. Seeing what he considered edible salads helps us define what Captain John Smith and others, like Champlain, meant when they planted their gardens with "sallets" especially in mind.

And a word about the modern botanical names for these — many, very old — plants. I have depended chiefly upon Asa Gray's *Manual of Botany*, as expanded by Merritt Lyndon Fernald in the eighth edition, 1950, for all the plants I could trace to this masterful collection of the flowering plants of the northeastern United States and adjacent Canada. This was, after all, the settlers' region and these are their plants. I have included the various versions of their introduction and naturalization as given in this book, without trying to inquire into the differences between the terms. To me, even the "adventives" and "escapes" seem all very much of the same shipload.

Vegetables and fruits I checked in the latest edition of the *Standard Cyclopedia of Horticulture*, L. H. Bailey, 1928–1929, feeling that here again it would be better to stay on this side of the Atlantic Ocean with the names as well as the plants. In every case, however, I checked back to the names and pictures in Gerard and Parkinson. It is amazing and rather a relief to discover how such plants as the radish, for instance, have been improved. When we read of a seventeenth-century success with a radish "as big as a man's arm," we cannot but be incredulous until we realize this was the kind of radish the Romans used to boil, and no wonder.

The genus names for the plants mentioned by the settlers are fairly obvious: *Brassica*, *Thymus*, and so on. It is the descriptive names which identify the species which are harder to select. However, if one remembers that *officinalis* means "of the shop," or the kind the apothecaries sold, the choice for predominantly medicinal plants becomes obvious. And when one realizes that *oleraceus* means the plant is a potherb or used for cooking, and *sativus* means cultivated, these are heartening indications that one is not out in the cold. For more complicated translations, may I refer you to my husband's book, *A Gardener's Book of Plant Names* without which I would never have had the courage to pick the John Winthrop, Junior, seed list to fascinating bits and pieces.

As for the spelling of the names — I left the early settlers on their own. The botanical names I willfully refused to capitalize except for the names of each genus.

The plants are listed alphabetically in the order of their familiar names of that time, with their modern botanical names as given by L. H. Bailey and Asa Gray, an indication of who mentioned them and used them, and a description of each plant and its "vertues" as understood at the time by Gerard and Johnson, Parkinson, Culpeper and Evelyn. The terms "adventive," "naturalized" and "introduced" are quoted as Asa Gray uses them. "Escapes," noted in wild flower identification books, is a graphic term but not very helpful. The differences between the three nomenclatures I have used, I leave it to Asa Gray to determine. "Introduced" has an aura of being wanted. "Adventive" signifies coming in unaided. "Naturalized" can include both. But the reader will be able to judge by the context how often a plant considered "adventive" may have been brought in as a "sovereign" remedy.

The initials to designate the users are:

C. M.   for Cotton Mather
J. W.   for the first Governor Winthrop
J. W. Jr. for his son, whose seed list and whose cases have proved most rewarding
J. J.   for John Josselyn
R. W.   for Roger Williams
W. W.   for William Wood

and the others' names are spelled out in full as they are found in the text.

ૐ

ADDERS TONGUE *Ophioglossum vulgatum*   J.J.

Listed by Josselyn as one of the plants common to both England and New England, this is recommended by Gerard for green wounds, when the leaves have been stamped and boiled in olive oil. As a "balsam for greene wounds" it is comparable to "oyle of S. Johns wort" and may even surpass it.

Josselyn found this plant and made an ointment of "the herb new gathered."

Culpeper says it is under the Moon and Cancer and so cures by sympathy all diseases in parts of the body governed by the Moon and by antipathy those in parts governed by Saturn. The juice of the leaves drunk with distilled water of horsetail (*Equisetum arvense*) is good for all wounds or bleeding.

This is also one of the plants which George Herbert's exemplary parson's wife was to use in a salve to help the parson heal his parishioners.

AGRIMONY *Agrimonia supatoria*   J.J.
EGRIMONIE   "Cases" – J.W.Jr.
                                                        (Adventive)

As usual, Gerard's description of the plant is as good as a picture. "The leaves of Agrimonie are long and hairie, green above, and somewhat grayish underneath, parted into divers other small leaves snipt round the edges, almost like the leaves of hempe; the stalk is two foot and a halfe long, rough and hairy, whereupon grow many small yellow floures, one above another upward. . . ."

Named for Eupator, its discoverer, it has, according to Pliny, an appropriately "royal and princely authority." Hot and moderately dry according to Galen, it "cutteth and scoureth" and removes stopping and obstructions of the liver. A decoction of the leaves is good for those with "naughty livers." The seed, taken in wine, helps "the bloody flixe," and people bitten by serpents. The leaves stamped in "old swines grease" close up old ulcers. Boiled in wine and drunk, agrimony helps "inveterate hepaticke fluxes in old people."

Parkinson does not grow agrimony in either his kitchen or his flower garden.

Culpeper agrees with all that Gerard reports and places agrimony under Jupiter at the sign of Cancer. He adds that, outwardly applied, it is good for the gout, will take out thorns and nails from the flesh, strengthen "members that be out of joint," and is admirable for those whose livers are either hot or cold. The liver, he says, forms blood, blood nourishes the body, and agrimony strengthens the liver.

Those who have seen agrimony covering the banks of a New England river will now understand there is more behind "escaped" wild flowers than merely outlasting old gardens.

ALE HOOF *Glechoma hederacea*   C.M.
GROUND IVY   (Naturalized)

Gerard calls it "Ground Ivy" first, then "Ale Hoofe," and says it is beneficial for those who are hard of hearing, or have a "ringing sound and humming noise" in their ears. With equal quantities of celandine and daisies, stamped and strained with a little rose water and dropped with a feather into the eyes, "it is proved to be the best medicine in the world." This same concoction mixed with ale and honey, strained and squirted into the eye of a horse or cow will take away "the pinne and web." Here, however, Gerard stops, fearing to become "over eloquent among Gentlewomen, to whom especially my Workes are most necessarie." Why ladies are given hard facts about the human frame and then spared details of animals' ills is an interesting social insight. Back on safer ground, Gerard quotes Dioscorides as teaching that ground ivy is good against sciatica, jaundice, to stay the terms and to help weak backs. Women in Cheshire and Wales "tunne the Ale-hoofe into their Ale," so it will purge the head "from rhumaticke humours flowing from the braine." In ointments it is good for burns from fire and gunpowder. Bound in a bundle or chopped like herbs for the pot, it stays the flux in women.

Parkinson does not give it garden room, perhaps foreseeing its ability to cover the ground as it has in the New World.

Culpeper says it is hot and dry, agrees with all above, and adds it is "excellent for wounded people."

ALEXANDERS *Smyrnium olusatrum: Angelica atropurpurea*
ALISANDERS   Early explorers
                                                        Morton
                              "1 oz Alisander seeds at 2d" – J.W.Jr.

Gerard says the "leaves of Alexander are cut into many parcells like those of Smallage, but they be much greater and broader, smooth also, and of a deep greene colour; the stalke is thick, oftimes a cubit high; the floures be white, and grow upon spokie tufts. . . ." Its root is like a radish-root, good to eat, with a juice tasting like Myrrhe (Sweet Cicely, *Myrrhis odorata*). It is called Great Parsley or Horse Parsley in England, in other parts *Olus atrum* or "the blacke pot-herbe," or *Petroselinum Alexandrinum*, since the best was supposed to come from Alexandria. (*Petroselinum* is the modern botanical name for parsley.)

The seed and root, says Gerard, like those of garden parsley, are hot and dry, so they "cleanse and make thin." Dioscorides commended it as a salad herb. The leaves and stalks boiled, served alone or with fish, are good for the stomach. "In our age," says Gerard, "it is served raw in salads." Medicinally, the seeds "bring downe the floures, expell the secondine, break and consume winde, provoke urine, and are good against the stranguary." A decoction of the root made with wine will also do all this.

Parkinson calls it *Hipposelinum sive Olus atrum*, to get it all in, and, after agreeing with all the above attributes, says that the stalk and roots are used in Lent especially to make a broth of an "aromaticall or spicie taste, warming and comforting the stomach and helping to digest . . . waterish and flegmaticke meats . . . in those times much eaten." The roots, raw or boiled, are eaten with oil and vinegar.

Culpeper agrees with others, adds a touch about its being good against snakebite, and ends, "And you know what alexander pottage is good for, that you may no longer eat it out of ignorance, but out of knowledge."

Evelyn understands that Alexanders cleanse and comfort the stomach like parsley, approves them raw or balanced for salads, and says they make "an excellent vernal Pottage."

A N G E L I C A   *Angelica archangelica: Archangelica officinalis*
Early explorers
Whipple   Hinckley
"1 oz Angelica seeds at 4d" – J.W.Jr.

Gerard counts three sorts of angelica, that of the garden, that of the water, and a wild angelica growing upon the land. Of the garden variety he says "it hath great broad leaves, divided againe into other leaves, which are indented or snipt about . . . among which leaves spring up the stalkes, very great, thicke, and hollow, sixe or seven foot high, joynted or kneed; from which joynts proceed other armes or branches, at the top whereof grow tufts of whitish floures like Fennell or Dill; the root is thicke, great, and oilous, out of which issueth, if it be cut or broken, an oylie liquor; the whole plant, as well as leaves, stalkes, as roots, are of a reasonable pleasant savour. . . ." Gerard says that as it is hot and dry in the third degree. "It attenuateth or maketh thin, digesteth, and procureth sweat." Its virtues are many. The root is a "singular remedy against poison, and against the plague, and all infections taken by evill and corrupt aire; if you do but take a piece of the root and hold it in your mouth, or chew the same between your teeth, it doth most certainely drive away the pestilentiall aire, yea, although that corrupt aire hath possessed the hart, yet it driveth it out again by urine and sweat. . . ." A dram of the powder taken in thin wine will cure "pestilent diseases." "If the fever be vehement" it can be given with "the distilled water of *Carduus benedictus* or *Tormentill* . . . or with Treacle of Vipers added." Vipers here would be the vipers bugloss, or *Echium vulgare* elsewhere in Gerard, that sovereign remedy for snakebite which still infests the New England countryside. Treacle is often used to signify a remedy of unfailing efficacy and comfort.

After a little practice we should realize that what is hot and dry in the third degree "openeth the liver and spleen, draweth down the termes, drieth out or expelleth the secondine." "The decoction of the root made in wine is good against the cold shivering of agues. . . . It attenuateth and maketh thin, grosse and tough flegme; the root being used greene and while it is full of juice, helpeth them that be asthmaticke. . . ." Gerard somewhat disassociates himself from the following statement by saying only; "It is reported that the root is available against witchcraft and enchantments, if a man carry the same. . . ." But he returns to stating facts as he sees them when he announces further that it aids digestion, benefits the heart, and "cureth the bitings of mad dogges, and all other venomous beasts."

Parkinson puts "Garden angelica" into his kitchen garden foremost among the "herbes . . . for the profit and use of Country Gentlewomen and others . . . of the most especiall use for those shall need them, to be planted at hand in their Gardens, to spend as occasion shall serve, and first of Angelica."

"The distilled water of Angelica," says Parkinson, "eyther simple or compound . . . is of especiall use . . . in swounings, . . . tremblings and passions of the heart, to expell any windy or noysome vapours from it. The green stalkes or the young rootes being preserved or candied are very effectual to comfort and warme a colde and weake stomacke; and in the time of infection is of excellent good use to preserve the spirits and heart from infection." He adds a new quality in saying he has had related to him upon credit that the root dried and powdered and taken in wine "or other drinke will abate the rage of lust in young persons." And he recommends, for expelling phlegm from the lungs and "to procure a sweet breath," making a syrup by pouring white sugar into the stalk as it stands and grows, leaving it there for three days, and then draining off the syrup to keep it for "a most delicate medicine."

Culpeper considers angelica "of admirable use." It is an herb of the Sun in Leo. It resists poisons by defending and comforting the heart, blood and spirits, and is good against all epidemic diseases caused by Saturn. He enumerates all the medicinal uses noted by Gerard and adds a few of his own, such as dropping the juice into the eyes and ears for blindness and deafness. The root made into powder and applied with a little pitch "doth wonderfully help" in cases of venomous bites. It is beneficial for toothache, gout and sciatica. The wild angelica has similar powers. No wonder the early voyagers rejoiced to see it growing beside the Maine rivers.

A N I S E *Pimpinella anisum*		J.J.	C.M.
A N N I S		Hammond	Endecott
			How

Even considering that anise is "one of the four great hot seeds" and is allowed to be diuretic, carminative, aphrodisiac, and a corrector of bad breath, it is surprising to see how staunchly our forebears depended upon it in their remedies. According to Gerard: "The stalke of Annise is round and hollow, divided into divers small branches, set with leaves next the ground somewhat broad and round; those that grow higher are more jagged, like those of young Parsley, but whiter; on the top of the stalkes do stand spokie rundles or tufts of white floures, and afterward seed, which hath a pleasant taste as everie one doth know."

Gerard disagrees with Galen about the degrees of hotness and dryness, but agrees that it "ingendreth milke." And Gerard announces that "the seed wasteth and consumeth winde, and is good against belchings, and up-braidings of the stomacke, allayeth gripings of the belly, provoketh urine gently, maketh aboundance of milke, and stirreth up bodily lust. . . . Being chewed it maketh the breath sweet, and is good for them that are short-winded, and quencheth thirst, and therefore it is fit for such as have the dropsie: it helpeth the yeoxing or hicket, both when it is drunken or eaten dry; the smel thereof doth also prevaile very much. The same being dried by the fire and taken with honey clenseth the brest very much from flegmaticke superfluities; and if it be eaten with bitter almonds it doth helpe the old cough. It is to be given to young children and infants to eate which are like to have the falling sickness or to such as have it by patrimonie or succession. It taketh away the Squinancie or Quincie (that is a swelling in the throat) being gargled with honey, vinegar, and a little Hyssop gently boiled together."

Parkinson does not plant it in his gardens. Culpeper also does not seem to give it garden room.

APPLE *Malus*   Everyone

Gerard says the Latin name *Malus* "reacheth far among the old writers" and declares he will "briefely first intreate" of *Mali*, "properly called Apple trees, whose stocke or kindred is so infinite." His illustrations, Johnson declares, were taken from the figures of Tabernaemontanus, and are of the "Queening, or Queen of Apples," the "Sommer Pearmaine," the "Winter Permaine," the "Pome Water," and the "Bakers ditch Apple tree." Johnson has allowed only the last two to be shown. Gerard treats of apples generally as sweet and sour, or "harsh or austere, being unripe." He declares sweet apples yield more nourishment and keep their shape better when cooked than others; sour apples run to juice when cooked and are less nourishing. However, they "speedily passe through the belly" and so are especially beneficial taken before meat. Unripe apples bring colic. Roasted apples are better than raw. Outwardly applied, apples are good for hot swellings. Apple juice is good for melancholy tempers. An ointment made with apple pulp and called *Pomatum* in shops is good for the complexion. The pulp of roasted apples mixed to a froth in water and drunk by the quart, twice running, has benefited those with gonorrhea. Raw apples distilled with camphor and buttermilk take away smallpox scars, if the patient at the same time is given a drink of milk and saffron.

Parkinson likes the Latin *Poma* for his apple name and announces that the number of apples is infinite. He lists sixty kinds. His uses begin with the serving of apples as a last course at dinner "in most mens houses of account, where, if there grow any rare or excellent fruit, it is then set forth to be seene and tasted." Apples also may be baked for pies; stewed with rose water and sugar, cinnamon or ginger; or, roasted in the wintertime and

3 *Malum regale.*
The King of Apples.

dropped into a warm cup of wine, ale or beer. Some are "fittest to scald for Codlins" (half-grown apples) to cool the stomach and please the taste with rose water and sugar added. Some, not good for anything else, are best for cider, of which the West Country people make "hogsheads and tunnesfull" to be carried to sea in long voyages and drunk with water. The juice of pippins and pearmaines particularly is good for the melancholy, as is the distilled water of the same apples. He recommends the "fine sweet oyntment . . . called *Pomatum*," for chapped skin, to make it smooth and supple.

In his sixty-odd "kindes or sorts" of apples he includes many of apparently small repute and named for their appearance, like "cowsnout" and "cats head," but his many recommended varieties of pippins and pearmaine, russets and greenings, are also recommended by Leonard Meager and are, indeed, still going strong today. And of old ones mentioned singly like "sops in wine," "so named both of the pleasantness of the fruite and beautie of the apple," many of us have nostalgic memories. Today's Baker's Sweet, an old variety formerly much grown in parts of New England, according to the Department of Agriculture, may be Gerard's Baker. Blackstone's "sweeting," an apple unfamiliar to early English settlers of

Boston when the first, solitary settler of the Boston area gave it to them, may well have been a variety of this apple. A Pommewater still exists, as well as a Marigold (another name for the Orange Pippin, still one of the favorite "dessert apples" in England), and the Newtown Pippin or the Rhode Island Greening. This last was Benjamin Franklin's favorite apple. He took it to England, and it is sold in our supermarkets today for pies. Kentish, Queen, Gillyflower and Seek-no-farther are apple names still to be conjured with.

Mr. Blackstone, the dissident Church of England clergyman who had seen the light about not being able to continue living and preaching in England, and had acted upon it some time before the Puritans made their concerted move, had planted the first orchards on the Boston peninsula. Living there alone, with no intention of creating anything more in the New World than a comfortable and secure life for himself among his horticultural successes, he was generous enough to welcome Winthrop from relatively waterless Watertown to the bubbling springs of the hills soon to be settled as Boston. Blackstone's subsequent move to the area of the future Providence, where he again planted orchards and rode into the settlement on a bull trained to the saddle, seems to have been made without rancor or regret. He was a scholar with a sufficient library and a gentleman with sufficient mental resources. That he was the first orchardist in New England is probably all the memorial he would have wished.

To those who think of Johnny Appleseed, and the men who saved seeds of a good apple to be planted later, with pity for their lack of knowledge of fruit propagation, and who doubt that the original sorts of apples brought by the settlers could possibly be still around and edible, Parkinson has a message. We must remember that the settlers brought scions and exchanged them and that, while there was a good deal of dependence upon seeds, grafting was common practice. Parkinson speaks of the "Paradise or dwarfe apple tree" which is no good of itself but has a quality which makes up for its "former faults." Being a dwarf tree, "whatsoever fruit shal be grated on it, will keep the graft low like unto itself, and yet bear fruit reasonably well. And this is a pretty way to have Pippins, Pomewaters, or any other sort." He likes them in a hedge row in an orchard "all along by a walke side," a pretty idea and within the grasp of most gardeners, then and now.

Culpeper does not concern himself with the fruit, not even to keep the doctor away.

A R A C H   *Atriplex hortensis*
O R A C H
O R R A D G   "for orradg seeds 1d." – J.W.Jr.

Gerard is brief: the Garden Orach has "an high and upright stalke, with broad, sharp-pointed leaves . . . floures are small and yellow, growing in clusters . . . the leaves and stalkes at the first are of a glittering grey colour and sprinkled as it were with a meale or floure." In Latin, he says, it is *Atriplex*. He quotes Galen and Dioscorides as agreeing it is both moist and cold. Dioscorides says that, boiled and eaten like other salad herbs, "it softeneth and looseneth the belly." Raw or "sodden," it "consumeth away the swellings of the throat." The seed, drunk with mead or honied water, is a remedy against the "yellow jaundice." Galen thinks it may open stoppings of the liver.

Parkinson calls it *Arrach* or *Atriplex sive Olus Aureum*, and says it is cold and moist and of a slippery quality whereby it "passeth quickly through the belly, and maketh it soluble, and is of many used for that purpose, being boyled and buttered, or put among other herbes into the pot to make pottage." And "when they are young, being almost without flavour themselves, they are the more convertible into what relish any one will make them with. Sugar, Spice, etc."

Evelyn recognizes Orach, *Atriplex* "as cooling and allaying the Pituit Humour." The tender leaves can be mingled in a cold salad, but it is better in a pottage, and, like lettuce, the leaves can be boiled in their own moisture.

Culpeper says of Garden Arrach, "called also Orach and Arage," all that has been said above, but qualifies the moist and cold by saying this is due to its being under the Moon, and like her. He agrees with the above virtues and then launches himself into "Arrach, wild and stinking," which, since it is a notable "escape" we include here. In the invaluable *Manual of Weeds* it is called "Spreading Orache" or *Atriplex patula*, and is related to the so-called "Pig-weeds," sharing the honors with Good-King-Henry, or *Chenopodium Bonus Henricus*, as one of the once favored potherbs now relegated to the barnyard. Culpeper loves it. He begins: "Called also Vulveris, from that part of the body upon which the operation is most; also Dogs Arrach, Goats Arrach, and Stinking Motherwort," it is used "as a remedy to help women pained and almost strangled with the mother, by smelling to it; but inwardly taken there is no better remedy under the moon for that disease. I would be large in commendation of this herb were I but eloquent . . .

under the dominion of Venus and under the sign Scorpio it is common almost upon every dunghill. The works of God are given freely to man, his medicines are common and cheap, and easy to be found (Tis the medicines of the College of Physicians that are so dear and scarce to find.) I commend it for an universal medicine for the womb, and such a medicine as will easily, safely, and speedily cure any disease thereof . . . you can desire no good to your womb but this herb will effect it; therefore, if you love your children, if you love health, if you love ease, keep a syrup always by you . . . and let such as be rich keep it for their poor neighbours; and bestow it as freely as I bestow my studies upon them. . . ."

ARCHER, SMALL WATER *Saggitaria latifolia*   J.J. ARROWHEAD

Josselyn lists this as among those plants common also in England. He knows of no recorded "vertues" but assumes they are "cold and drie in qualitie, and are like Plantaine in facultie and temperament."

ARSEMART *Polygonum hydropiper*   J.J. WATERPEPPER SMARTWEED

Gerard finds it remarkable that this plant with the purple flowers "after which cometh forth little seeds . . . of an hot and biting taste, as is all the rest of the plant . . ." has "no smell at all" as a plant. According to Galen, the plant is hot and dry, like pepper and the leaves laid onto swellings will "dissolve and scatter congealed blood." Gerard obviously considers the following its chief value. "The leaves rubbed upon a tyred jades backe and a good handful or two laid under the saddle and the same set on againe, wonderfully refresheth the wearied horse and causeth him to travell much the better."

Culpeper says "Arsemart" is good for toothache, stone in the bladder, and swollen joints. He also knows about its value placed under the saddle of a tired horse. He adds that there is "scarce anything more effectual to drive away flies" if cattle are "annointed with the juice of arsemart." Strewed in a chamber in the hottest time of summer, it will soon kill all the fleas and drive away the flies.

ARTICHOKES *Cynara scolymus*
"1/2 oz hartichokes 2s per oz" – J.W.Jr.

Gerard says there are three sorts of Artichokes, two tame and one wild, which the Italians esteem as the best to be eaten raw. He says it is called in Latin *Cinara*, from "*Cinis*, Ashes, wherewith it loveth to be dunged." In low Dutch it is Artichoken. Considering its elaborate cultivation "in fat and fruitful soile" with "water and moist ground," it has fairly unusual virtues. It is best to eat it boiled, although many eat it raw. It "is good to procure bodily lust" raw or boiled. "It stayeth the involuntary course of the natural seed either in man or woman. . . ." The buds steeped in wine and eaten are especially good "to stir up the lust of the body." But the final observation is the most arresting. "I finde moreover that the root is good against the ranke smell of the arm-holes, if when the pitch is taken away the same root be boyled in wine and drunke; for it sendeth forth plenty of stinking urine, whereby the rank and rammish savour of the whole body is much amended."

Parkinson calls them "*Cinera*, Artichokes" and tells of various kinds and their cultivation in France and Holland, although "by daily experience . . . the English red Artochoke is . . . the most delicate meate of any." As for the use: "The manner of preparing them for the Table is well knowne to the youngest Housewife . . . to be boyled in faire water, and a little salt, until they be tender, and afterwards a little vineger and pepper, put to the butter, poured upon them for the sawce, and so are served to the Table. They use likewise to take the boyled bottomes to make Pyes, which is a delicate kinde of baked meate."

Culpeper in the *London Dispensatory* mentions artichokes solely as deodorant.

One may also consider here what Evelyn says of them in '*Acetaria*,' *A Discourse of Sallets* which may well afford us a clue as to the uses for some of these now rather baffling vegetables. "Artichaux" he defines as *Cinera* or *Carduus Sativus* and says the heads slit in quarters may be eaten raw with oil, vinegar, salt and pepper, and "gratefully recommend a Glass of wine Dr. Muffet says, at the end of Meals. They are likewise while tender and small, fried in fresh Butter crisp with Parsley." They become a "most delicate and excellent restorative" when full grown and "boiled in the common way." "And the Bottoms are also baked in Pies with Marrow, Dates and other rich ingredients." Evelyn also describes parboiling the bottoms and preserving them in butter in a "small earthen glazed pot" or stringing the bottoms on a packthread in a dry place, or pickling them.

The thistle family was well received in the seventeenth century, and Evelyn gives special mention to "Carduus Mariae, our Lady's milky or dappled thistle" as fit for salad consideration. *"Carduus benedictus,"* however, brought by the younger Winthrop, would seem to have been wholly a "sovereign" medicine and perhaps a dye.

ARTICHOKE, JERUSALEM *Helianthus tuberosus*
Indian food

Johnson-upon-Gerard — who must in this entry by consideration of dates, be only Johnson — begins rather fiercely, "One may well by the English name of this plant perceive that those that vulgarly impose names upon plants have little either judgement or knowledge of them." He says it was actually brought from America, and that Cardinal Farnesius has *his* artichoke roots from the West Indies, where Fabius Columna, one of the plant's "first setters forth" saw it. Mr. Parkinson, says Johnson, has "exactly delivered the history of this plant" under the name of "Canada." The author, now certainly Johnson, has his own history of the plant from his friend Mr. Goodyer, who was first to have it in England. He gives a full description and then adds that in 1617 he received two roots no bigger than hen's eggs; the one he planted brought him a peck of roots, wherewith he "stored Hampshire." The virtues may be quoted in full and seem to be signed by John Goodyer himself. "These rootes are dressed divers waies, some boil them in water and after stew them withe sack and butter, adding a little ginger; others bake them in pies, putting marrow, dates, ginger, raisons of the sun, sack etc. . . . But in my judgment, which way soever they be dressed and eaten they cause a filthy loathsome stinking wind within the body . . . a meat more fit for swine than men. . . . 17 October, 1621. *John Goodyer."*

Parkinson is more moderate, but as aware of the situation. He deals with the so-called Jerusalem artichoke as one of three kinds of potatoes. These he enumerates as "Spanish," then, "of Virginia," and finally, "of Canada." This last he says was brought from Canada by Frenchmen and given its name of artichoke "by reason of their great increasing" so that even "the most vulgar begin to despise them," although when they first came and were "boiled tender, peeled, sliced, and stewed with butter and wine, they were esteemed a dish for a Queene."

AVENS *Geum urbanum* *Geum rivale*   J.J.
CHOCOLATE ROOT   (Adventive)

When John Josselyn claims to have cured the man who had "melted his grease," trying to outmow another man, he administered a decoction of "Avens-roots and leaves in water and wine, sweetening it with Syrup of Clove-Gilliflowers." Avens belongs to the list of plants to which titles of sanctity are attached. In this case, Herb Bennet is Josselyn's other name for it. Gerard entitles his chapter "Avens, or Herbe Bennet," and the aura lingers in the French Canadian name for it, *Benoite*.

It is, however, impossible to tell which of the large family Josselyn used. The roots of some of the geums are said to have the same curative properties as Cinchona, or Peruvian bark, or what Cotton Mather and Josselyn called "Jesuits bark," which everyone was hoping to equal with some plant growing in New England. (Josselyn wondered if our hemlock might not be it.) Today *Geum rivale* is known as "Indian chocolate" but it is likely that Josselyn used the avens he found growing and resembling the "herb-bennet" he knew in England, since he takes pains to illustrate the "more masculine quality in the plants growing in New England" with his story of the cure described above.

Gerard recommends the roots and leaves in wine against internal injuries of those who have fallen from high places, stitches in the sides, and venomous bites. Culpeper concurs. Gerard adds that the dried root gives clothing a good odor and protects it from moths.

BALM *Melissa officinalis*   How
BALSAM   "Cases" – J.W.Jr.
  (Introduced)

Balsam, "bawme" and balm, in seventeenth-century accounts seem to mean the same plant. Gerard tells of the many varieties of "Bawme," but his first

is "called also Melissa" and has "a pleasant smell, drawing neere in smell
and savour unto a Citron." Hot and dry in the second degree, its virtues are
many. Drunk in wine, it is good against the bitings of venomous beasts,
comforts the heart, and drives away melancholy. The juice "glueth together" green wounds. The herb rubbed on hives keeps bees together and
induces others to join them. The Arabs particularly affirmed balm to be
good for the heart and to make it merry. Dioscorides advised applying the
leaves outwardly to the stings of venomous beasts and bites of mad dogs,
and recommended it against toothache and for those who cannot breathe
"unless they hold their necks upright." Pliny goes so far as to say that the
herb tied to a sword which has inflicted a wound will staunch the blood.
[So Sir Kenelm Digby was in distinguished company with his "powder of
sympathy."]

Parkinson speaks of "Melissa, Baulme," which, he says, is of common
use among "other hot and sweete herbes" for baths in summer to comfort
the sinews and, steeped in ale, as a balm water for use instead of *Aqua vitae*
against "suddaine qualmes or passions of the heart." He says this last virtue
should be apparent to anyone who so much as smells of the herb. In salves,
he says, it is good for green wounds; rubbed inside the hive, it attracts bees;
and he adds that Pliny says it is good to prevent stings. Culpeper agrees
with all of the above and adds that it will break boils and expel the afterbirth.

Evelyn records it as "cordial and exhilarating."

B A S I L   *Ocimum basilicum*

"1 oz Bassill seeds at 3d" – J. W. Jr.

Gerard recognizes several kinds of basil and says, although it is called
Indicum, Hispanicum is correct. He says it is sown in gardens, and in earthen
pots — which rings a bell. Parkinson gives it no room in a garden or a pot.
Gerard is careful, as was Dioscorides, to warn that if it be much eaten it
dulls the sight. Who can say that this was not Isabella's brothers' first
worry, as shown in the *Decameron* and by Keats, when they saw their sister
tenderly nursing a pot of basil, and not the subsequent discovery that the
pot contained also a bit of her lover? In any case, one must be careful of
basil. "There be that shunne Basill and will not eat thereof, because that if

it be chewed and laid in the Sun, it ingendereth wormes." On the other hand, "They of Africke" affirm that if those who have been stung by a scorpion eat basil, they feel no pain. And Dioscorides follows his first warning by adding that, "it mollifieth the belly, breedeth winde, provoketh urine, drieth up milke, and is of hard digestion." However, later writers, says Gerard, teach that the smell of basil is good for the heart and for the head, and that the seed takes away sorrowfulness.

Understandably Culpeper is somewhat cautious, since he finds Galen and Dioscorides and Crysippus against taking basil inwardly, although Pliny and the Arabian physicians defend it. He concludes that, as it is a herb of Mars and under the Scorpion and as "every like draws its like," no wonder it draws poisons. But he feels "something is the matter" since rue will not grow near it and rue is "as great an enemy to poison as any that grows."

"Basil, *Ocimum* (as Basil)," says Evelyn, "imparts a grateful flavour, if not too strong, somewhat offensive to the Eyes, and therefore the tender Tops to be sparingly used in our Sallet."

One last word from Culpeper: "To conclude," he says, and we feel he has reached his own conclusion, "it expelleth both birth and afterbirth, and as it helps deficiency of Venus in one kind, so it spoils her actions in another."

BEANS *Phaseolus*   Indian food

It is interesting to note that John Winthrop, Junior, ordered no beans for planting in gardens. On the other hand, beans seem to have been a staple food and medicine. "Indean beans" said Josselyn, are "better for physick use than other beans," and "Indean beans, falsely called French beans, are better for Physic and Chyrurgery than our garden beans. *Probatum est.*"

Gerard calls "the great garden Beane" *Faba major hortensis* and begins by quoting Galen who says that the "Beane is windie meate, though it be never so much sodden and dressed anyway." Parched, they lose their windiness but are harder to digest. The meal of beans will cleanse the skin, and so also the belly. Outwardly applied the pulp will cool and soothe. "The decoction of them serveth to die wollen cloth withall."

On the "Kidney Beane," which he calls "*Phaseolus* or *garden Smilax*," Gerard becomes eloquent and lists and illustrates a variety, differentiated mainly by color and place of origin. Gerard says the kidney beans are also called French, and "sperage" beans. Boiled in their pods before they are ripe, and buttered, they do not engender wind, but make a delicate meat. They loose the belly, provoke urine, and "ingender good bloud." They must, however, only be taken when young and tender, green and boiled, strung, put in a stone pipkin with butter and stewed gently, when they become "wholesome, nourishing and of a pleasant taste."

Parkinson does not have them in his garden under any name I can find, even "sperage," his second name for asparagus.

But Culpeper has so much to say on each bean — garden, field, and French or kidney — that we are persuaded that beans were considered as much medicine as meat in well-to-do and well-ordered households. Culpeper says they all belong to Venus, which is why the distilled flowers of the garden bean are good for cleansing the complexion. He mentions other uses also mentioned by Gerard but not noted above, such as drying up milk in swollen breasts and reducing bruises caused by "dry" blows. He refers often to them as "Windy meat" and recommends the "Dutch fashion" of cooking them halfway, husking them and then putting them back to stew again to make them "wholesomer food." Of the French or kidney bean, of which he admires particularly the "scarlet-flowered," he says they strengthen the kidneys and are of easy digestion. They also incite to venery. The scarlet-colored beans, set near a quickset hedge, will climb it to the "admiration of the beholder at a distance," but they will "go near to kill the quickset."

Still, one wonders about the Winthrops' listing no beans among their seeds.

BEDSTRAW   *Galium verum*   (Naturalized)
LADY'S BEDSTRAW

I am sorry to be unable to find again a manuscript reference to lady's bedstraw saying of the New England women "they stuff their pillows with it." However, this is not the sole reason this pretty plant grows in delicate yellow drifts by dry roadsides. It has the property of curdling milk, and is called "Cheese Rennet." It will color cheese yellow and it will also dye stuffs red, like its related madder, *Rubia tinctorum*. While this particular bedstraw was brought from Europe, there are many indigenous sorts which the settlers must have been glad to recognize. Sweet woodruff (*Asperula*), a somewhat similar plant, shares the quality of being fragrant when dried and is also an "escape."

Gerard says that "there be divers of the herbes called Ladies Bedstraw, or Cheese-renning" and describes it as having "small round even stalkes, weak and tender, creeping hither and thither upon the ground . . . very fine leaves . . . set at certaine spaces as those of Woodroofe, among which come forth floures of a yellow colour, in clusters or bunches thicke thrust together, of a strong sweet smel. . . ." Dioscorides recommended an ointment made from this plant for burns, and said it staunched blood. Steeped in the sun in olive oil, it is good to anoint the weary traveler. Gerard says that in Cheshire, where the best cheese is made, the herb is highly esteemed.

Parkinson does not entertain it in his garden.

Culpeper recommends it against bleeding and to heal inward wounds. It is a herb of Venus. It is also good for the scab or itch in children.

Whether it is a singular lady's bedstraw with implications of a holy tradition or plural ladies all busily stuffing pillows with it, does not matter today when we have it spelled either way.

BEETS *Beta*   "1 oz Beets at 2d" –J.W.Jr.

Gerard says that the red beet, being eaten when boiled, quickly descends and loosens the belly, especially when taken with the broth "wherein it is sodden." He says it "nourisheth little or nothing, and is not so wholesome as Lettuce." The juice "conveyed up into the nostrils will purge the head." Then he cheers up. "The great and beautiful Beet . . . may be used in winter for a sallad herbe, with vinegar, oyle and salt, and is not only pleasant to the taste but also delightfull to the eye. . . . the greater red Beet, or Roman Beet, boyled and eaten with oyle, vinegar and pepper, is a most excellent and delicate sallad; but what might be made of the red and beautiful root (which is to be preferred before the leaves, as well in beauty as in goodness) I refer to the curious and cunning cooke, who no doubt when he hath had the view thereof and is assured that it is both good and wholesome, will make thereof many and divers dishes, both faire and good."

Parkinson agrees with all the above, quotes Gerard directly, and says that eaten hot or cold the roots make a good salad. He adds a little grace note in telling us that of late the root of the red beet is of much use among cooks to trim up their dishes of meat, being cut into "divers formes and fashions." Parkinson notes that the great red beet which Mr. Lete gave to Gerard and which grew very well with him in 1596, as recorded by Gerard in his *Herball*, had leaves which were as good as those of cabbages, when boiled.

Culpeper says the red beet is good to stay the bloody flux and helps yellow jaundice. He says little of it as a dish except that the ancients noted its insipid taste, and he recommends snuffing it up the nose to evacuate phlegmatic humors from the brain, which remedy he says is considered by some to be a great secret.

BERRIES

It will be easier to get a sense of the berries and their plenty and variety by reading about them in the chapter dealing with the "forerunning signals of fertilitie." Their uses are described in the sixth chapter on "meate." There are also numerous references in lists compiled by the redoubtable John Josselyn. However, to help those wishing to check in a hurry, here is the lot.

a. BARBERRY *Berberis*
Gerard refers to the "Barberry bush" as in "Here we go round . . ." being "planted in gardens in most places of England." Both leaves and fruit are eaten in sauces and are "good for the same things" medicinally, especially against fevers and to stop bleedings. It was one of the first fruits "sent over."

b. BAYBERRY *Myrica pensylvanica*
This is one of the truly North American plants the settlers found. The only reference to its use lies in its familiar name of "candleberry."

c. BEARBERRY *Arctostaphylos uva-ursi*
Here again is a plant the settlers found but failed to mention specifically, although we have references which sound like it. John Josselyn may have known of it as a corn remedy. His "bear berry," however, is another name for cranberry, which he says bears are said to enjoy.

Here would be the place to note that Gerard includes these and several berries later listed here all in one handsome chapter called "Of Worts or Wortleberries." Beginning with *Vaccinia nigra, rubra* and *alba,* as black, red and white "worts or wortleberries," he goes on to *Vitis Idaea,* or "Hungarie Wortleberries" and *Vaccinia Ursi sive Uva* Ursi, or "Beare Wortleberries."

d. BILBERRY *Vaccinium*
. . . "two kinds, Black and Sky Coloured, which is more frequent," says Josselyn, trying to fit his New World berries into his Gerard. "*Vaccinia nigra,* the black Wortle or Hurtle" may presage our huckleberry.

e. BLACKBERRIES *Rubus*
Following the eglantine rose in Gerard's *Herball* and followed by the "Holly Rose," the "Bramble or black Berry-bush" hardly seems to have caught Josselyn's attention although he must have become, in Cotton Mather's term, "briered" by the many varieties. He lists "Black-berry" and then "Dew-berry," its common English name today.

f. BLUEBERRIES *Vaccinium*
are Josselyn's sky-colored bilberries, cooling and a substitute for currants in "Puddens."

g. CLOUDBERRIES *Rubus chamaemorus*
The only variety, both in Gerard's *Herball* and modern botany. "Seldom ripe," according to Josselyn.

h. CRANBERRIES *Vaccinium oxycoceus*
Josselyn calls these "Cran Berries and Bear Berries," and says bears are reputed to like them. See his good description of them and their uses in his own list in Chapter X. Gerard calls them "Marish Worts or Fenne Berries" and says they "take away the heate of burning agues" and that the juice boiled "till it be thicke, with sugar added that it may be kept" is better than the berries.

i. CURRANTS *Ribes*
"Red and black," says Josselyn, and leaves it there. We have wild varieties and the garden varieties have "escaped." They were imported very early.

j. ELDERBERRIES *Sambucus canadensis*
Like the elder which was used in England, but without "a smell so strong." Gerard lists several sorts, all with "wholesome" qualities.

k. GOOSEBERRIES *Ribes*
Englishmen are always delighted to see gooseberries. Josselyn notes those "of a deep red colour." Gerard is moved to reveal his own fondness for them above the blueberries and blackberries and bilberries, all cool and binding, by finding them appetizing and adding that they are also called "Sea-berrybush" in Cheshire, "my native countrey."

l. RASPBERRY *Rubus*
"Rasp-berry, here called Mul-berry," says Josselyn, which may explain Captain John Smith's thinking there were mulberries on Plum Island.
Gerard speaks of the "Raspie bush or Hinde berry" as being like the "Bramble" in temperature and "vertues," such as making an excellent lotion for healing sores and fastening teeth and healing eyes that "hang out."

m. STRAWBERRY *Fragaraia*
The variety called *canadensis* or *virginianum* was sent to England by the early settlers in Virginia, and said to be the best thing that had come to England since the "hautbois" (were they speaking of the *fraise du bois?*) In any case, praise of our New World strawberries was excessive, compared to that for any other berry, for their quality, size, and quantity. Gerard places them between herb bennet and angelica among most beneficial plants, and recommends them highly as a washing water, to remove spots from the face, to fasten the teeth, allay inflammations, and make the heart merry.

BETONY *Stachys officinalis*   Boorde
Hammond
C.M.

Gerard says, "Betony groweth up with long leaves and broad, of a dark greene colour, slightly indented about the edges like a saw. The stalke is slender, four square, somewhat rough, a foote high more or lesse. It beareth eared floures, of a purplish colour, and sometimes reddish. . . ." Gerard found a white betony growing in Hampstead. Gerard quotes Galen, and says betony is hot and dry in the second degree, and "hath force to cut." He ascribes to the plant an assortment of virtues. It is good for those who have the falling sickness and those who have "Ill heads upon a cold cause." It cleans the lungs, chest, liver, and the mother. It makes a man have a good appetite and prevails against "sower belchings." It is good for the ague, for cramps, ruptures and convulsions. Inwardly or outwardly it is a remedy against mad dog bites and those of venomous serpents. It is "most singular against poyson." It kills worms and "looseth the belly very gently" when taken in a powder with meat. It has a great virtue to heal and, in a conserve made of the flowers and sugar, it is very good for a headache.

Parkinson does not give it garden room.

Culpeper says this herb is Jupiter's, and quotes a physician to the emperor Augustus Caesar as saying it preserves men from epidemical diseases and from witchcrafts. On his own, Culpeper finds the root very displeasing . . . "whereas the leaves and flowers, by their sweet and spicy taste, are comfortable both in meat and medicine . . . it is a very precious herb, that is certain, and very proper to be kept in a man's house, both in syrup, conserve, oil, ointment, and plaister. The flowers are usually conserved."

Cotton Mather quotes a story of a poor countryman with an intolerable toothache being cured by betony thrust up his nose. Hammond includes betony with eight other herbs in his "Excellent water for ye Eyes."

BLOODROOT *Sanguinaria canadensis* C.M.
(Native)

The Indians appear to have indicated this plant to the early settlers as a source for dye and quite possibly (considering their use for such things, as see hellebore) as a powerful emetic. Cotton Mather mentions it, but by his time its other, gentler properties as an expectorant may have proved valuable. Its use for those afflicted with pulmonary consumption may be due to the doctrine of signatures. In any case, its beauty must have gladdened many a first early spring in "this wildernesse."

BLOODWORT   *Rumex sanguineus*   How
BLUDWORT   Hammond
BLOUDWORT
                            "1 oz Bludwort seed at 2d" – J.W.Jr.

Gerard puts all the "Dockes" together in one splendid collection, setting
them into four classes like Dioscorides: wild or sharp-pointed; garden
dock; round-leaved dock; and sour dock, or sorrel. To these he adds others
from "later Herbarists." The two which most concern us are what he calls
Patience, or Monk's Rhubarb, *Rumex patientia* mentioned as doing well in
New England by Josselyn, and "bloudy Patience," or "Bloudwort," *Rumex
sanguineus*. It is worth noting here that another name given the latter is
Sanguis Draconis, which is often met with in medicinal concoctions, because
of "the bloudie colour wherewith the whole plant is posset." Gerard considers this plant, which he prefers to call "Bloudwort," as "of pot-herbes

‡ 5 *Lapathum sativum sanguineum.*
Bloudwoort.

the chiefe or principall, having the properties of the bastard Rubarbe; but of lesse force in his purging quality." Gerard says that it is sown for a pot-herb in most gardens and describes it as having "long thin leaves sometimes red in every part thereof, and often stripped here and there with lines and strakes of a dark red colour, among which rise up stiffe brittle stalkes of the same colour: on the top whereof come forth such floures and seed as the common wild docke hath. The root is likewise red, of a bloudie colour." This is the kind of dock "best known to all."

Parkinson says: "Among the sorts of pot-herbes Bloodworte hath alwayes been accounted a principall one, although I do not see any great reason therein, especially seeing as there is a greater efficacie of binding in this Docke, then in any of the other, but as common use hath received it, so I here set it downe. . . . The whole and only use of the herbe almost serveth for the pot, among other herbes . . . accounted a most special one for this purpose. The seed . . . commended for any fluxe in man or woman . . . and so no doubt the roots, being of a stipticke qualitie."

Culpeper confines his attention to the dock called Patience, or Monk's Rhubarb.

Meager includes bloodwort in his "chopping herbs." "Bloodwort" can also mean *Sanguinaria officinalis*, or burnet.

B O R A G E   *Borago officinalis*

"1 oz Buradg seed at 4d" – J.W.Jr.

Many members of the borage family can be treated as one, as they are so often grouped together by the seventeenth-century authorities, who also, somewhat confusingly, include "lang de beefe" with them because it also has long, roughly hairy leaves. [To save others' searching, "lang de beefe," the twenty-eighth seed name on the Winthrop list, is now found under the familiar name of ox-tongue, botanically *Picris echioides*, and appears under L.]

Gerard treats first "Of Borage," says it is called *Borago* "in shops," which is the true meaning of *officinalis*, and mentions that the Greeks and the "old writers" referred to it as what became in Latin *Lingua bubula*, referring to the "rough and hairie" leaves. Pliny, he says, called it *Euphrosinum* because it makes a man merry and joyful, as in the old verse which he also quotes and translates for us as "Borage brings always courage."

Gerard describes borage as having "broad leaves, rough, lying flat upon the ground, of a black or swart green colour . . . upon which riseth up a stalke two cubits high . . . whereupon do grow gallant blew floures, composed of five leaves apiece." There is also a white-flowered borage and one that Gerard calls *"semper virens"* because, unlike the others, it does not die back in the winter.

Gerard says borage is evidently moist, but between hot and cold, and he lists its virtues — all happy ones — after he mentions loosening the belly and curing hoarseness.

"Those of our time," he says, "do use the floures in sallads, to exhilarate and make the mind glad. There be also many things made of them, used everywhere for the comfort of the heart, for the driving away of sorrow, and encreasing the joy of the minde." He quotes Dioscorides and Pliny as saying the flowers and leaves put into wine make men and women glad and merry and drive away sadness, dullness and melancholy. So do the flowers made into a syrup, and — even more efficacious — made up with sugar. The leaves eaten raw are good for the blood of those who have been lately sick.

Parkinson treats of "Buglosse and Borage" together and announces that he puts them into his "Garden of pleasant Flowers" although they more properly belong in the Kitchen Garden, "in regard they are wholly in a manner spent for Physicall properties, or for the Pot, yet because anciently they have been entertained into Gardens of pleasure, their flowers having been in some respect; in that they have alwaies been interposed among the flowers of womens needle-worke, I am more willing to give them a place here, then thrust them into obscurity, and take such of their tribe with them as may fit for this place, either for beauty or rarity." He describes bugloss as having narrower leaves than borage, whose leaves are "broader, shorter, greener, and rougher" than those of bugloss. He then gets into as much confusion as anyone else. Borage has blue flowers, though sometimes reddish and sometimes white, and the ever-living borage [about which he finds Gerard in error as to its lasting qualities] is rather like Comfrey. He includes a "Sea Buglosse or Alkanet" which would seem to be our *Anchusa*, which will color things red, and tosses in a "marsh buglosse" whose Latin name is *Limonium*. He declares that "by consent of all the best moderne Writers," the ordinary garden borage he describes is the "true Buglossum of Dioscorides. . . . Our Buglosse was unknown to the ancients."

In any case "Borage and Buglosse are held to be both temperate herbes, being used both in the pot and in drinkes that are cordiall, especially the flowers, which by Gentlewomen are candied for comfits. The Alkanet is drying and held to be good for wounds, and if a piece of the roote be put into a little oyle of Peter or Petroleum, it giveth as deepe a colour to the oyle, as the Hyperion doth or can to his oyle, and accounted to be singular good for a cut or greene wound."

Culpeper, also, treats of "Borage and Bugloss" and adds a third sort, called "Langue-de-beuf," whereupon he, also, describes a plant with rough and hairy leaves but yellow flowers rather like a dandelion. They are all three, he says, herbs of Jupiter and under Leo, "all great cordials and strengtheners of nature." The uses are all as above, with rather more attention to sweating, loosening, healing, gargling, and washing the eyes.

Evelyn states the base for borage (which he treats alone) very simply: "hot and kindly moist, purifying the Blood . . . an exhilarating Cordial, of a pleasant Flavour. The tender Leaves and Flowers especially, may be eaten in Composition; but above all, the Sprigs in Wine, like those of Baum, are of known Vertue to revive the Hypochondriac, and chear the hard Student." And then even he adds, "See Bugloss."

BOUNCING BET *Saponaria officinalis*
SOAPWORT   (Naturalized)

One wonders about this indispensable little plant, as one does about beans and teazel, why no one mentions having brought it over, and yet so many obviously did so. Called "the fullers' herb," it grows in great profusion around the sites of old cloth mills. Called soapwort, it is still used for delicate cleaning of valuable articles. Called Bouncing Bet still in this country, although, apparently no longer so-called in England, it is an ornamental garden plant in its double form and recalls the practice of lathering cloth in a decoction of its leaves boiled in water before the cloth was stamped.

Gerard calls it "Sope-wort or Bruse-wort," although "it is commonly called *Saponaria*," because of "the great scouring qualities that the leaves have; for they yeald out of themselves a certaine juyce when they are bruised, which scoureth almost as well as Sope." Gerard knows of "no use in physicke set down by any Author of credit." Johnson here becomes officious and sets down what *he* had heard, which comes from a Spanish authority who used a decoction "against the French Poxes . . . and men-

tions the singular effect of this herb against that filthy disease." And Johnson has heard of its being used for green wounds.

Parkinson groups "Sopewort" with the gentians and prefers the double-
flowered kind. He says wild soapwort is used in many places "to scoure the
countrey womans treen, and pewter vessels, and some physically make great
boast to performe admirable cures in Hydropicall diseases, because it is
diureticall, and in *Lue Veneria*, when other Mercuriall medicines have failed."

Geoffrey Grigson reports that the early settlers took it to New England
and there found it useful to treat the "horrible rash caused by Poison Ivy."

Culpeper says that Venus owns it. He neglects to mention any household
cleaning qualities, but says that country people use it bruised to cure cuts.
Some consider it diuretical and also "singular good to void hydropical
waters; and they have no less extol it to perform an absolute cure in the
French pox, more than either sarsaparilla, guiacum, or China can do; which,
how true it is, I leave others to judge."

BUGLE   *Ajuga reptans*   (Naturalized)

Gerard and Culpeper commend this herb against "inward burstings," and
for strengthening the liver, dissolving clotted blood and making all wounds
sound. Culpeper says it is under Venus and that one should always have it
by one, in a syrup to take inwardly and in an ointment and a plaster to use
outwardly.

BUGLOSS   *Anchusa officinalis*
                                          "1 oz Buglos at 2d" – J.W.Jr.
                                                              (Adventive)

It is splendid to find L. H. Bailey quite definite that what the seventeenth
century meant by Borage was *Borago officinalis* and by Bugloss, *Anchusa
officinalis*.

Under "Buglosse" Gerard gives us first three sorts, the "Garden Buglosse," our old friend "Lang de beefe," which he shows in all its dande-
lion-like flowers, and "a small wilde Buglosse." He next deals with "Alkanet or Wilde Buglosse," and produces a bugloss called *Anchusa* by the

Greeks, because it will color things red, and then includes a yellow plant with quite different flowers but undeniably hairy leaves, and a third plant he calls the "small alkanet." The chapter after this assortment he devotes to "Wall and Vipers Buglosse."

The roots of alkanet, he says, are cold and dry and biting and will clean away "cholericke humours." Medicinal uses include being taken for jaundice, the spleen, and agues, with the added virtue that, used in a pessary, the leaves will bring forth the dead birth. As a coloring agent, the roots will make sweet butter red, when it is then to be melted and drunk by those who have fallen from high places. Syrups also can be colored.

Gerard quotes John of Arden, whom we recognize as one of the earliest of English physicians, mentioned by Chaucer. John of Arden had a composition he called *Sanguis Veneris*, which Gerard says is "singular" as a remedy in deep punctures, or wounds made "with thrusts." "Singular" it is. "Take of oile olive a pint, the root of Alkanet two ounces, earth worms purged in number twenty, boile them together and keep it to the use aforesaid."

Gerard quotes its use in ointments for "women's paintings" and later clarifies the reference for us by ending with, "The Gentlewomen of France do paint their faces with these roots, it is said."

We have seen that Parkinson says alkanet is drying and good for wounds and that it will color oil "as well as Hypericon," which must be St. John's-wort, a plant he does not entertain in his gardens.

Culpeper refers separately to bugloss only to say its roots are effectual made into a licking electuary against coughs and "rheumatic distillations upon the lungs," so we are led to wonder if he is not thinking of the other plant called "bugloss" by country people in England and New England, or *Pulmonaria officinalis*, whose spotted leaves were thought to indicate its efficacy as a remedy for lung trouble. (This is added only to help others seeking to replant the gardens of the times and becoming dazed by the popularity of the borage and bugloss tribes and their hangers-on.)

Evelyn, who has closed his borage reference with, "See bugloss," says it is in nature "much like Borrage yet something more astringent. The Flowers of both, with the intire Plant, greatly restorative, being Conserv'd." He then also becomes confused about what bugloss *really* is, since he feels that "what we now call Bugloss was not that of the Ancients but rather Borrage, for the like Virture named Courage."

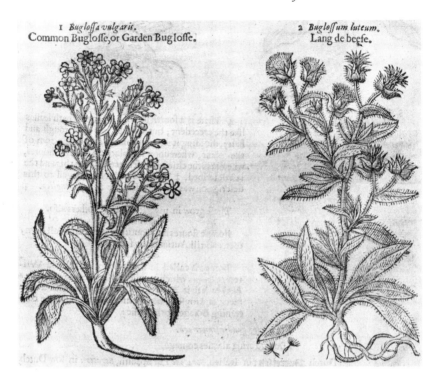

1 *Buglossa vulgaris.*
Common Bugloffe, or Garden Bugloffe.

2 *Bugloffum luteum.*
Lang de beefe.

How fortunate we are to have come *after* brave botanists who undertook to classify all the plants, regardless of their familiar names and "singular" uses!

BUGLOSS, VIPERS *Echium vulgare*   (Naturalized)

But we must add one more bugloss to the list, just as Gerard does, at the end. This is the little "wild buglosse," according to Gerard, which causes so much pain and trouble to farmers and pasture-keepers today. To Gerard it looks like the garden bugloss except that its flowers grow on one side of its stalk. The seed is small and black and "fashioned like the head of a snake or viper." It is cold and dry, and the root drunk with wine is good for those who have been bitten by serpents. The leaves and seeds operate in the same way. Those who have drunk before being bitten, or rather before meeting with a viper or serpent, will not be bitten. This same drink will cause plenty of milk in women's breasts, and the herb chewed and swallowed is a most singular remedy against poisons and poisonous bites. Chewed and laid on the bite, it is equally effective.

Fortunately, it grows wherever alkanet will grow, and today in many other places as well.

BURNET *Sanguisorba officinalis* (great burnet)
Poterium sanguisorba (salad burnet)
"1 oz Burnett at 3d" – J.W.Jr.

Gerard calls it garden burnet, or *Pimpinella hortensis*, and says that it differs from *Pimpinella*, which is also called *Saxifrage*, so it may be called fitly *Pimpinella sanguisorba hortensis Maxim.* He says, "the whole plant doth smell something like a Melon or Cucumber" and reports from the ancients that "it doth marvellously agree with wine, to which also this *Pimpinella* doth give a pleasant sent." "Burnet, besides the drying and binding facultie that it hath, doth meanly coole." Burnet, Gerard says, is also a "singular good herb for wounds."

The decoction of the powder of the dried leaves; the juice; or the herb and seed made into powder, drunk with wine or with water in which iron has been quenched . . ." will stop bleeding of all sorts. "The leaves are good on wounds. . . ."

"The leaves of burnet steeped in wine and drunken, comfort the heart, and make it merry, and are good against the trembling and shaking thereof."

Parkinson says, "the whole plant is of a stipticke or binding taste or quality but of a fine quicke sent, almost like Baulme." "The greatest use that burnet is commonly put unto," Parkinson says, is to put a few "leaves into a cup with claret wine, which is presently to be drunk, and giveth a pleasant quick taste thereunto, very delightful to the palate, and is accounted a help to make the heart merrie." Sometimes the leaves "when young" are put among "other sallet herbes to give a finer relish thereunto" and into "vulnerary drinks, to stay fluxes and bleedings," and are used in "contagious and pestilential agues."

Culpeper agrees with all the above. He says, "It is an herb the Sun challengeth dominion over, and is a most precious herb, little inferior to betony; the continual use of it preserves the body in health, and the spirits in vigour; for if the Sun be the preserver of life under God, his herbs are the best in the world to do it by."

Evelyn admits it to all salads, but says a fresh sprig in wine is "its most genuine element."

CABBAGE   *Brassica oleracea*
COLEWORT   "1 oz Cabedg seed 25 per li"
CULIFLOWER   "1 oz Colewort seeds 3d per li"
                                        "2 oz Culiflower seed 2s 6d per oz"
                                                                    J.W.Jr.

Despite the three different varieties of *Brassica* on the Winthrop seed list, where cauliflower takes popular and costly precedence, we shall treat them in a group here, since Gerard, Parkinson and Culpeper list coleworts and cabbages together, making only, like Winthrop, a nice difference with cauliflower. As even L. H. Bailey's *Cyclopaedia* admits that the *Brassica*'s are still much in their ancient state of confusion, we come with the greater sympathy to Gerard's struggles to differentiate between them. He begins by putting them all under Coleworts, of which he says Dioscorides recognized two sorts, tame and wild, and Theophrastus three, curled, smooth and wild. With which last, the Romans agreed. By Gerard's time, however, "Herbarists" had recognized many more kinds, and Gerard launches himself upon sixteen with illustrations of each, some in full bloom. All *Brassica* in Latin, they have a variety of names in English: Garden Colewort, Curled Garden Cole, Red Colewort (all straggling and loose), and then a closed head — White Cabbage Cole. Then Red Cabbage Cole and Open Cabbage Cole and "*Brassica florida*, Cole-florie." After this, some Savoy Cole, Parsley Cole and English Sea Coleworts, all before he gets to the wild kinds.

All the coleworts, Gerard says, are binding and drying. The white cabbage, he says, is best, though only next to the Cole-florie, called in Latin *Cauliflora*; so we can see Winthrop was knowledgeable about his cabbages.

Virtues of the cabbage family are many. Dioscorides found it good for those with dim eyesight, shaking palsy, trouble with the spleen, and the bites of venomous beasts. The leaves are good for hot swellings, stamped and laid upon them. The juice is good for those with gout. Sniffed into the nostrils, it purges the head. Dropped with wine into the ears, it cures deafness. Made up with barley meal into a pessary, it works wonders. The leaves laid on are good for the complexion. But a most arresting virtue is one Gerard records as "reported," a way he has when he dares not endorse an entry and yet cannot leave it out. Eaten raw before meat, coleworts preserve a man from drunkenness, the reason being that there is a natural enmity between coleworts and "the vine," of such an extent that if coleworts are planted near vines, the vines will die.

Parkinson says the "Cole flower" is the best of the coleworts with a pleasanter taste than either colewort or cabbage and is therefore "of the more regard and respect at good mens tables." Cabbages and coleworts are usually boiled in beef broth . . . ribs of deep green coleworts are boiled and served with oil and vinegar in Lent. . . . In the cold countries of Russia and Muscovia they are "powdered up" and serve "the poorer sort" for their most ordinary food in winter, although they "stinke most grievously." Parkinson quotes "Crisippus, an ancient physician," who wrote a whole volume on the virtues of the cabbage family applied to all parts of the body. Which need not seem remarkable since "the old Romans having expelled Physitians out of their Common-wealth, did for many hundreds of yeares maintaine their health by the use of Cabbages, taking them for every disease."

Culpeper includes all the foregoing, adding only that they are "extremely windy" and that the Moon challenges the dominion of the herb.

Evelyn knows all the foregoing but wonders at "the Veneration we read the Ancients had for them, calling them Divine and Swearing *per Brassicam.*"

C A L A M I N T   *Satureja calamintha*   Boorde
                                                       (Naturalized)

Called "Mountain Mint" by Gerard and Culpeper, and disregarded by Parkinson, the calamint recommended by Andrew Boorde to the settlers of Old Newbury is now rated as a variety of savory by Linnaeus, and allowed several varieties by Asa Gray, "all naturalized." It is indeed a very strong herb, as its name of "mountain" indicates, it being common knowledge that all mountain herbs are stronger than their lowland sorts, hence the superiority of the bezoar stones of mountain goats.

Gerard gives it almost unlimited virtues. Eaten with honey, it provokes sweat and cures shivering fits. Rubbed on in oil, it will do the same and help those with sciatica. The juice is good against worms and for jaundice, for those with leprosy, nosebleed, bruises and trouble with breathing. The decoction provokes urine and the terms, and will bring down the dead child, inwardly taken or outwardly applied. It makes a good gargle.

Culpeper says it is "an herb of Mercury, and a strong one, too, therefore excellent good in all afflictions of the brain." He lists all the above but adds that it "hindereth conception in women, being either burned or strewed in the chamber," which is really very strong indeed, we must admit, Mercury or no. But Culpeper adds this warning, "Let not women be too busy with it, for it works very violently upon the female subject."

CALAMUS AROMATICUS *Acorus calamus*   Boorde
BASTARD CALAMUS AROMATICUS   J.J.
                                        (Naturalized)

"Sweetflag" to us, "early introduced and naturalized," according to Asa Gray, would appear, from what Gerard says of it, to have been one of the ancient remedies of mankind. It is good for practically every organ of man or woman; smoked in a pipe it is good for coughs. It is used in baths for women, and in plasters "for the smells sake." It has powers as a counter-poison. Johnson adds to Gerard that the Turks preserve the root and take it in the morning against corrupt air and the Tartars will not drink water, "their usual drink," unless some of this root has been steeped in it.

The interesting thing about this plant is that, sterile in Europe, it is not so here. A fact noticed by the astute Josselyn. *"Bastard Calamus Aromaticus,"* he calls it, as it "agrees with the description [presumably in his copy of Gerard as corrected by Johnson] but is not barren." He says the English make use of the leaves to keep their feet warm. Obviously the habit of strewing floors with rushes had been brought to the New World as well as the rushes.

The beautiful yellow iris *Iris pseudacorus*, which has "escaped" to a lavish extent from early gardens, is so named because it resembles the "sweetflag," until it blooms, and grows in much the same places. See under Iris.

CAMOMILE   *Anthemis nobilis*   Boorde
CHAMOMILE   J.J.   Hammond
CHAMMOMEL   (Naturalized)

As both Gerard's "garden" and "wild" chamomile have "escaped" to an extent that marks them as valued herbs of the early settlers, we treat of them both here, listed first without the now approved *h*, since that is the way Gerard and Parkinson preferred. Culpeper in 1652 lists Chammomel. Gerard describes four kinds of "Cammomill," so called because it smells of apple. He allows other names: *Chamaemelum*, which implies the apple fragrance, *Anthemis*, and *Leucanthemum*, the preferred name for the double sort. All of which goes to show us how difficult it must have been for early

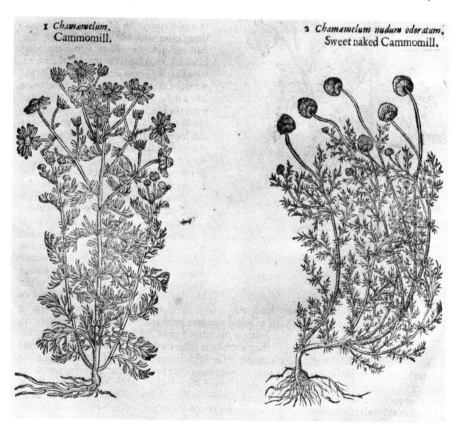

1 *Chamæmelum.*
Cammomill.

2 *Chamæmelum nudum odoratum.*
Sweet naked Cammomill.

herbalists to choose one name for anything from the varied selections offered them. Gerard described the common "Cammomill" with white petals around yellow "thrums"; then the "Sweet Naked Cammomill" which appears to have no petals and only the yellow "thrums"; the "Double flowered Cammomill"; and the "Romane Cammomill" which stands stiffer than the others. The first kind has a trailing habit and is the one we know was recommended for "lawns" in Elizabethan gardens.

Our own listing today would be: *Anthemis cotula* (Mayweed), *A. arvensis* (Field chamomile), *A. nobilis* (Garden chamomile) and *A. Tinctoria* (yellow or ox-eye chamomile). All are naturalized.

Chamomile is good, Gerard says, against colic and stones, aches and pains, coldness in the stomach, and the ague. It is good for the bladder and for women's diseases, and in a bath will produce sweat. Galen called it hot and dry in the third degree, so it digests, slackens and rarifies. It "mollifieth and suppleth" and is a special remedy for "wearisomeness." Gerard exclaims, "All these operations are in our vulgar Cammomill."

Parkinson describes chamomile as "well knowne to all, to have many trailing branches, set with very fine small leaves, bushing and spreading thick over the ground . . . white flowers with yellow thrummes in the middle, very like unto the Featherfew . . . and the whole herb of a very sweet sent." The "naked" sort is rarer in gardens but equally fragrant. The double, as with double Featherfew, will produce double-flowered plants from its seeds. The "vertues" are "divers and sundry . . . for pleasure and profit . . . for inward and outward diseases . . . for the sicke and the sound. . . . In baths it is beneficial to all. The flowers boiled in a posset drink provoke sweat and expell colds and aches. The syrup is used for jaundice and dropsy caused by the evil disposition of the spleene."

Culpeper agrees with all the above, but adds a rather arresting story of how good it is against the stone, telling that a stone recently removed from a man and wrapped in chamomile "will in a short time dissolve."

Captain Hammond had a good idea for treating the "Megrum" which is mild if messy. "For the Megrum, Mugwort and Sage, a handfull of each, Camomel and Gentian a good quantity, boyle it in honey and apply it behind and on both sides of ye head very warm, and in 3 or 4 times it will take it quite away."

CARAWAY   *Carum carvi*   (Naturalized)

Caraway would seem to be one of those plants that have accompanied man since the most ancient times. Gerard finds it comparable to anise, another of man's plant companions, and says Galen calls it hot and dry in the third degree, with a "moderate biting qualitie." "It consumeth wind," is "delightfull to the stomacke and taste, it helpeth concoction, provoketh urine, and is mixed with counter-poisons; the root may be sodden, and eaten as the Parsenep or Carrot is." He adds that the seeds are made into "comfits" with sugar.

Parkinson says first [because we are in the Kitchen Garden] that the roots boiled may be eaten as carrots and comfort the cold, weak stomach by their spicy taste. They help to "dissolve winde (whereas Carrots engender it) and is a very welcome and delightful dish to a great many. . . ." The seeds are used in baked fruit and bread, to give a relish and to help digestion. Made into comfits, they are taken against cold and wind, and are served at the table with fruit.

Culpeper says it is Mercurial and agrees with the above. He adds that caraway seed is good for eyesight and colds in the head. The roots are particularly good, sodden, for the elderly.

CARDUUS BENEDICTUS   *Cnicus benedictus*
                    "1 oz Carduus benedictus 6d" – J.W.Jr.
                                                (Adventive)

Gerard says that *Carduus Benedictus*, or Blessed Thistle, resembles Wild Bastard Saffron with its yellow flowers. Found growing wild on "champion grounds" in Lemnos, an island in "the Midland Sea," it is "diligently cherished in Gardens" in England. Taken in meat or drink, it is good for swimming and giddiness of the head, strengthens memory and is a "singular remedy against deafness." It can be used in many forms. Dried, green, powdered, "sodden," it is good for purgings, against poisons, stings and pestilence, for ulcers, and is "excellent good against the French disease."

Parkinson says although the leaves are hairy and rough they are without sharp thorns or prickles and the tenderest hand may touch them. He is brief about the uses. The distilled water is to be drunk against all agues, "pestilential or humoral." It "helpeth to expell worms because of the bitterness" and is in many ways "a friend to the stomack."

Culpeper spares us a description, as "almost everyone may describe them," and proceeds to analyze the herb in a "rational conception." A herb of Mars and under the sign of Aries, by sympathy it helps vertigo, because Aries is in the house of Mars. It is a remedy against jaundice because Mars governs choler. It strengthens the attractive faculty in man, clarifies the blood, and cures red faces, tetters and ringworms because Mars causes them, as is the case with mad dog bites, boils and the itch. By antipathy to other planets, in this case, Venus, it cures venereal disease. By antipathy to Saturn, it cures deafness and strengthens the memory. In summing up, Culpeper declares, "By this we may in part understand with how great virtue God hath indued, and I may say, blessed, this herb." Which is a splendid example of the Puritans' total acceptance of seemingly very unrelated theories.

Meager includes "carduus" among "the names of divers ordinary Physick Herbs usually planted in gardens."

John Josselyn used the seed for "Pleurisie, adding coriander seed and pulverised harts horn."

Carduus benedictus is one of the ingredients in George Baker's "pretious oil" — *Oleum magistrale.*

CARROT *Daucous carota sativa*   Everyone
                    " 1 li carrett seed 12d per li" – J.W.Jr.
CARROT, WILD *Daucus carota*
                    (Naturalized)

In Gerard's time the garden carrot and the wild were more nearly alike in leaf and bloom than they seem today, with our garden carrot perfected to a mere preliminary tuft of leaves, and the wild, long known as "Bird'snest" in England and here now called "Queen Anne's Lace," beautifying our meadows and distressing farmers.

Gerard has little to say of garden carrots, however. The root is "commonly boiled with fat flesh and eaten . . . the nourishment . . . is not much, and not verie good . . . something windie, but no so much as be the Turneps, and doth not so soon as they pass through the bodie." The seed "breaketh and consumeth windinesse and provoketh urine, as doth that of the wild Carrot."

Parkinson entertains no carrots, but Culpeper says Galen highly commended garden carrots to break wind, though experience teaches they breed it first "and we may thank nature for expelling it, not they. The seeds of them expel wind, indeed, and so mend what the root marreth."

Both Gerard and Culpeper have more respect for, and devote more space to the wild carrot, as did Galen who said it was more "effectual" and contained in it "a certain force to procure lust." Pliny quoted Orpheus as saying it "winneth love." Gerard thought it "helpeth conception" and Dioscorides felt that those who had first eaten of it would not be hurt by "all manner of venomous beasts."

Culpeper says it belongs to Mercury and "therefor will expel wind . . . remove stitches in the sides, provoke urine and women's courses and remove the stone. . . . Boiled in wine it helpeth conception. The leaves applied with honey cleanse running sores."

CATMINT *Nepeta catarai* J.J.
(Naturalized)

While Josselyn lists "Cat-mint" among "such Plants as are common with us in England," it is generally considered to have been brought here from England and by the time Josselyn saw it to have made a handsome escape into "this wildernesse."

It would seem to have been a common household remedy such as any prudent housewife might have brought with her to be planted upon settling.

Gerard calls it "Nep or Cat Mint," says it has the faculties of the Calamints or "Mountaine Mints," hot and dry. It is "commended against cold paines of the head, stomacke and matrix, and those diseases that grow of flegme and raw humours, and of winde." With wine or "mede" it is good for those fallen from high places and much bruised. It is used in baths for women to sit over, "to bring down their sicknesse and to make them fruitfull."

Parkinson calls it *Nepeta* or Nep, and says it is "much used" as above, in baths, and also it is fried into Tansies, as is "Clarie," to strengthen women's backs. He also mentions its use, in wine, for those who have had a fall, and the adds that a decoction of Nep cures scab on the head or the body.

Culpeper says that Nep or Catmint is an herb of Venus, which we might have suspected by now. He reiterates all the above and adds only that the bruised herb applied to piles will ease the pain arising therefrom.

No one tells us what cats see in it.

CELANDINE   *Chelidonium majus*   J.J.
SWALLOW WORT   Hammond
                                                        (Naturalized)

Josselyn lists this among the "Garden Herbs amongst us" as Celandine, "by the West Country men called Kenning Wort." He says it grows "but slowly." We now know how surely as one hundred years later Cutler called it "common by fences and amongst rubbish."

For once, the Latin name Gerard applies to "Great Celandine," *Chelidonium majus*, is the same as our own. His familiar names are Swallow-woort and Tetterwoort. The "Kenning" in Josselyn's name applies to seeing and the reputed restorative power of this herb to injured eyes. The "swallow" name also refers to the same powers, the "ancients" having declared that the mother swallow restores the sight of her fledglings, if their eyes have been poked out, with this herb." Gerard and Culpeper, who seem to have poked out some baby swallows' eyes in their time, are sure that the sight was restored, but if by this herb they could not say. Gerard says the juice is "good to sharpen the sight." The root cures jaundice and is reported to be good against toothache. The root is also an excellent remedy for whatever ails hawks, says Gerard, which is an interesting association of eyesight lore.

Culpeper says this is an herb of the Sun, and as eyes are subject to the luminaries this is the best cure for the eyes there is. He agrees with all Gerard has said, and adds that it is a cure for tetters and ringworm and will remove warts. Outwardly applied with oil of camomile, the bruised herb relieves all inward pains. The powder of the root placed in an aching hollow tooth will cause it to fall out.

So much for the Great Celandine. The Lesser Celandine, naturalized in Virginia, is called Pilewort, having the "signature" for that disease which it "readily cures" upon its roots. But this is an aside.

CHERRIES   *Prunus virginiana* (Choke-cherry)

(Native)

Prunus pensylvanica (Pin cherry)

(Native)

Prunus avium (Bird- or Sweet-cherry)

(Naturalized)

Prunus cerasus (Sour or Pie Cherry)

(Naturalized)

When William Wood, in his *New Englands Prospect*, wrote a poem to "recite the most usefull" trees found here, he gave two lines to the only two fruit trees truly native:

> Within this Indian Orchard fruites be some,
> The ruddie Cherrie and the jettie Plumbe.

Which makes it sound better than his later description of eating cherries "which so furre the mouth that the tongue will cleave to the roof." "English ordering," he says, "may bring them to be an English Cherrie, but yet they are as wild as the Indians." This must have been our choke cherry. "English ordering" appears to have brought in scions of the "sweet" and "sour" cherries of the old country, of which Gerard said good tarts are made and from the "gum" of which a good medicine for old coughs could be had. Today they are all growing together in our thickets and woods and roadsides.

CHERVIL   *Anthriscus cerefolium*
CHARVIL   J.J.
"1/2 oz Charnill seed 3d per oz" –J.W.Jr.
CHERVIL, SWEET   *Myrrhus odorata*
SWEET CICELY   *Osmorhiza longistylis*

Having hunted Winthrop's "Charnill" to no conclusion, I heard an Englishwoman call Sweet Cicely "Charvil" and realized mistaking a *v* for an *n* would be easy on such a list as Robert Hill's. However, this decision did not clear away all confusion. Considering that both chervils are mentioned in the same breath by both Gerard and Parkinson, although they are now given separate genuses, it is hard to know which one Winthrop intended. And as we have a native wild Sweet Cicely, *Osmorhiza longistylis*, called also "Sweet Jarvil" and "Anise-root," we do not know which of the three Josselyn listed as thriving in New England gardens.

As Gerard describes, "the leaves of Chervill are slender, and diversly cut, something Hairy, of a whitish green; the stalks be short, slender, round, and hollow within . . . floures be white and grow upon scattered tufts . . ." This is "Common Chervill," "one of the pot herbes . . . pleasant to the stomacke and taste." Inwardly taken or outwardly applied, it affects the bladder. It has "a certain windiness, by meanes whereof it provoketh lust." The Dutch make a "hotchpot" of it, "a kinde of Loblolly."

Gerard's "Great Chervill or Myrrh," called "Sweet Cicely," has a stalk that can rise to two cubits and leaves like "Hemlocks," [by which he means *Conium maculatum*, the undoing of Socrates]. He says it grows in his garden and in the "gardens of other men who have been dilgent in these matters." Its leaves are "exceeding good, wholesome, and pleasant among other sallad herbs, giving the taste of Annise-seed unto the rest." The seeds and roots are also excellent in salad "dressed as the cunning Cooke knoweth how. . . ." But Gerard himself eats them with oil and vinegar after they have been boiled, "which is very good for old people that are dull and without courage: it rejoyceth and comforteth the heart, and increaseth their lust and strength."

Parkinson entertains both chervils, the sweet and the ordinary, in his garden and seems to depend here entirely upon Gerard, salad experiences and all. He does add the French to the Dutch as liking to make a loblolly of ordinary chervil, "used as a pot herbe with us." To the uses of the "great or sweet Chervill which of some is called Sweet Cicely," he adds candying the roots as of "singular good use to warme and comfort a cold flegmaticke stomack, and . . . thought to be a good preservative in the time of the plague."

Culpeper places all chervils under Jupiter and attributes to them expelling powers and the ability to "procure an appetite to meat." The juice will heal ulcers. In his *London Dispensatory* he says, "Chervil water distilled about the end of May, helps Ruptures, breaks the Stone, dissolves congealed blood, strengthens the Heart and Stomach."

Evelyn says chervil must never be wanted in salads, being exceedingly wholesome and "chearing the spirits." He recommends "the Roots . . .

boiled and eaten cold . . . for Aged Persons." He says that chervil, like spinach, is used in tarts and can serve alone "for divers causes."

C H I C O R Y *Cichorium intybus*
                              "1/2 oz Cicory seeds at 3d per oz" – J.W.Jr.
C I C O R Y   (Naturalized)
S U C C O R Y
E N D I V E *Cichorium endiva*
                              "1/2 oz endiue seed 3d per oz" – J.W.Jr.

Parkinson solves our problem of how to treat chicory and/or endive, as Gerard has included them in one chapter, "Of Garden Succorie," by stating quite directly in *his* chapter, "Succorie and Endive," that he "put both Succorie and Endive into one chapter and description, because they are both of one kindred; and although they differ a little the one from the other, yet they agree both in this, that they are eaten eyther greene or whited, of many."

Thanking Parkinson, we can go back now to begin with Gerard, who lists first "Garden Succorie" and then "Garden Endive," describing the first as having leaves, narrow, deeply cut and gashed, and "little blew floures consisting of many small leaves." The second is like it but has larger and broader leaves and white flowers "tending to blewnesse." Curiously enough, for us today, considering chicory, he firmly says they are grown only in gardens. Both are cold and dry, somewhat binding, and, being bitter, "they doe also clense and open," excellent medicines for a hot liver, as Galen said. They help in jaundice and prevent the stone. Succory leaves are good for inflamed eyes, bruised and laid on. (Where eyes are predominantly blue, all blue flowers seem to bear the same signature.) In salads they "comfort the weake and feeble stomacke."

Parkinson recommends burying them both in sand, to lose bitterness, and he quotes Horace to this effect.

Culpeper follows Gerard almost exactly in his *Physicians Library* adding that the root boiled in wine is "very harmless."

Evelyn finds chicory "more grateful to the stomach than the Palate" but says of Endive that while the stalk of Endive blanched "eats firm," some prefer the larger leaves to those of lettuce.

In Queen Elizabeth's last illness she took only "succory broth," as noted by Geoffrey Grigson in *An Englishman's Flora*.

1 *Cichorium sativum.*
Garden Succorie.

3 *Intybus sativa.*
Garden Endiue.

CHIVES *Allium schoenoprasum*		Everyone
CIVES		(Introduced)
CIBOLLS		
CHEBOLS		
GARLIC *Allium sativum*		C.M.
		(Introduced)
GARLIC, WILD *Allium canadense*		Everyone

I have put these members of the *Allium* family together here, because they were so often joined in the early settlers' and sailors' naming of familiar herbs they saw growing wild, or thought they did, and because they seem to have arrived early with everyone. Gerard says chives are hot and dry, open and make thin, "ingender hot and grosse vapours," are harmfull to the eyes and brain, and cause troublesome dreams. Garlic, Gerard says, kills worms, helps an old cough and hoarseness, is a preservative against pestilential air, aids women who sit over it in a bath, cures ringworms and scabbed heads, and is good to be laid on the bitings of the shrew mouse. In fact, it is good for so many things that Galen named it the rustic's treacle.

Parkinson repeats this idea by saying that garlic in many countries is known as the poor man's treacle, or a remedy against all diseases. Boiled in salt broth, it is "often eaten of them that have strong stomackes." It is never eaten raw by anyone.

Culpeper puts both chives and garlic under Mars, warns of the bad dreams, attributes many virtues to garlic and warns of many vices, and adds that a green bean or two chewed after eating garlic will remove "the disagreeable smell of the breath proceeding therefrom."

Evelyn refers them all to Onions.

CINQUEFOIL *Potentilla reptans*   Hammond
                                                    (Adventive)

When Captain Hammond concocted his "Excellent Water for ye Sight" and included "Cinquefoyle" with sage, fennel, vervain, betony, eyebright, celandine, "Herb of Grass" or rue, and pimpernel, he was taking no chances of missing any of the herbs which, singly, had been considered to have special properties toward this end in Old England. When he said cinquefoil he may have meant only *Potentilla reptans*, the little plant which Gray calls an "adventive from Europe." On the other hand, with the wealth of native cinquefoils, this may have been one of those plants which the early settlers rejoiced to find in the wilds and to which they attributed the same properties as they felt were in the Old World variety. In any case it was *Potentilla reptans* which was supposed to have supernatural powers in pre-Gerard times.

Gerard recognizes that "most vain and frivolous" things have been said of "Cinquefoile or Five Finger Grasse" of which he presents several sorts, all apparently with the same properties, but he "willingly withstands" all these, even those handed on by Dioscorides. Cinquefoil, Gerard allows, is good for the bloody flux, against all poisons, for toothache, falling sickness, agues, ruptures, ulcers, diseases of the liver and the lungs. No wonder it was brought over.

Parkinson does not give it garden room, but Culpeper in the *Physicians Library* outdoes Gerard. The root, he says, will stop blood flowing in any part of the body. It helps infirmities of the liver and lungs and "appeaseth the rage" of fretting sores. Boiled in vinegar, the root is good against the shingles. In fact, "You may safely take half a dram at a time in any convenient liquor." It is "an herb of Jupiter" and strengthens the parts of the body ruled by him.

Avens and cinquefoil were said by Culpeper to have the same properties in medicine.

CLARY   *Salvia sclarea*

"1/2 oz Clary seeds at 3d per oz" – J.W.Jr.

Gerard says the first kind of "Clarie, which is the right," has two-foot, square stalks with large leaves growing "by two and two," whitish and hairy, like the stalks. The flowers are like those of sage or dead nettle, "of white, out of a light blew." Apothecaries call it *Gallitricum*, some *Tota bona*, others *Sclarea*, and the English call it "Clarie or Cleere eie." Hot and dry in the third degree, its seed, powdered and mixed with honey, clears the sight. The seed stamped, infused or steeped, and the slimy substance applied "plaisterwise" will draw out splinters and thorns and dissolve swellings in the joints. The seed powdered and drunk in wine "stirreth up bodily lust. The leaves taken in any way help weak backs, especially when fried with eggs in manner of a Tansie."

Parkinson has Garden Clary in his kitchen garden and says the most common use is in Tansies to help weak backs. The leaves "taken dry, and dipped into a batter made of the yolkes of egges, flower and a little milke, and then fryed with butter until they be crispe, serve for a dish of meate accepted with manie, unpleasant to none." The seed, Parkinson says, is used by some to put into a corner of the eye, if a mote or other thing happens to be in it. "But assuredly although this may peradventure doe some good," he says — and we expect a patronizing conclusion about ignorant people — but he surprises us by concluding firmly, "yet the seeds of the wilde will doe much more."

Pursuing the wild clary back to Gerard, we find he calls is *Oculus Christi* and attributes all the properties of the garden variety to it. Culpeper treats of them both as under the dominion of the moon, agrees with all above, adds that the powder provokes sneezing and the seed or leaves, taken in wine, provoke venery.

Evelyn likes clary in "Omlets" made with cream, fried in butter, and eaten with sugar and lemon juice.

CLOTBUR   *Arctium lappa*
LESSER CLOTBUR   *Arctium minus*   J.J.
BURDOCK   (Naturalized)

Gerard treats of "the great Burre Docke" and "the lesse Burre Docke" as having a long list of "vertues" in common. The ancients declared the roots beaten with salt would cure the bites of mad dogs and serpents, drawing out the poison of the viper. The juice of the leaves with honey takes away pains of the bladder. The stalks, peeled before the burrs come, are nourishing in broth and also increase seed and lust. The root, stamped and strained with a good draught of ale, is good for a cold or windy stomach. A decoction of the root and the seed are good for toothache and kibed heels and, taken in wine, is good for pain in the hip. And, finally, the root in Malmesey strengthens the backs of women and aids them.

Culpeper begins at this point and, ascribing burdocks to Venus, claims that by applying burdock to the head, feet or navel one may control the womb, drawing it upward or downward or staying it, with the child in it.

CLOVE GILLYFLOWERS   *Dianthus caryophyllus*
                                                        J.J.
                                                      Whipple

Josselyn says "Gilly Flowers will continue two Years" in his section on "such Garden Herbs (amongst us) as do thrive." Elsewhere he notes that "Gilliflowers thrive exceedingly there and are very large. The collibuy, or humming-bird, is much pleased with them. Our English dames make a

Syrup of them without fire, they steep them in Wine till it be of a deep colour, and then put to it spirit of Vitriol, it will keep as long as the other." And again, when Josselyn was treating "a neighbour of mine in Haytime, having overheat himself and melted his grease, with striving to outmowe another man," he gave him a decoction of avens roots and leaves in water and wine, sweetening it with "Syrup of Clove Gilliflowers."

The Whipple House 1683 inventory lists "five bottles of syrup of clove gilly flowers."

Obviously, clove gillyflowers or clove pinks were staple in seventeenth-century New England gardens, like roses and various members of the borage family. Called gillyflowers because they bloomed in July, the name was also applied to garden stocks and wallflowers, both of which appear on the Winthrop seed list. The absence there of the clove gillyflower can be easily explained by the fact that roots would be the obvious way of bringing these favorite household flowers across the Atlantic.

The whole family of pinks and carnations was cherished by English gardeners on the eve of understanding hybridization. One Mr. Fairchild was to achieve a cross between a pink and a Sweet William by mixing the "farina" of one with the other. But, on the whole, the varieties seem to have arrived by happy overcrowding of all sorts and colors, as would seem to have been the case with "Mistress Tuggy" in Westminster, whose garden had a collection which "in the excellencie and varietie of these delights" exceeded all that Thomas Johnson had ever seen.

Gerard says the "conserve made of the floures of the Clove Gillofloure and sugar is exceedingly cordial and wonderfully above measure doth comfort the heart." He also ventures to say it will expel the "poyson and furie" of pestilential fevers and that other and even wild varieties, while most esteemed for their use in garlands and nosegays, may be commended "against infection of the plague."

Parkinson speaks of the "red or Clove Gilloflower," separating carnations and "gilloflowers" with some difficulty, as having the most clovelike scent of his nearly fifty sorts. "Pinkes," he calls "wild, small Gilloflowers" and gives us a good number of them.

The "red or Clove Gilloflower" is, however, the one "most used in Physicke in our Apothecarie shops" — although he believes all of them might serve, though not to give so "gallant a tincture to a Syrupe as the ordinary red will do." It is "accounted to be very Cordiall."

1 *Cariophyllus maximus multiplex.*
The great double Carnation.

2 *Cariophyllus multiplex.*
The double Cloue Gilloфloure.

CLOVER GRASS *Trifolium pratense*   J.J.
(Naturalized)

When Josselyn comments upon how well the English "clover grass"
does, he is referring to the red clover which was so highly thought of
among farmers in England at this time as a ground-conditioner, sowed
upon worn-out fields and plowed in later. It was a substitute for animal
manure when that was hard to come by, as it certainly would be in a new
country with barely enough domestic animals to feed the settlers. We see
this old idea suitably revived in the middle of the next century by Jared
Eliot, first truly American author of a book upon American husbandry.

COLEWORT *Brassica oleracea*
See CABBAGE

COLTSFOOT   *Tussilago farfara*   C.M.
                                                    (Introduced)

The name of this early spring plant honors its ancient reputation as a remedy for a cough, or *tussis*. Gerard says it may be Englished as "Coughwort," and explains that the *Farfara* name comes from the resemblance of the leaves to those of the white poplar tree. The yellow flowers of coltsfoot come before the leaves and fade quickly, but the leaves stay green all summer and are to be seen beside ditches and upon waste places almost everywhere.

Gerard says the fresh leaves have a drying quality and are good for ulcers. The dried leaves are hot and biting. Burned upon coals and the "fume" taken through a funnel, the dried leaves help those short of breath and with "impostumes of the brest." They are also helpful when "taken in manner as they take Tobacco." The green leaves and roots, in a decoction or a syrup, are good for a cough. The green leaves pounded with honey are good for hot inflammations, like St. Anthony's fire.

Parkinson does not give it garden room.

Culpeper puts it under Venus and agrees with what Gerard has said.

Cotton Mather prescribes, in his "Angel of Bethesda": "If a cough be very troublesome, lett the drink be a pectoral Decoction of Hysop, Coltsfoot, Liquorice, and the like, sweetened with Syrup of Poppy-Heads." And again, in the same vein, discussing Consumptions, he tentatively suggests a change of air and "sweetening the blood" with "Theas or Decoctions or Infusions of Sassafras, of Sarsaparilla . . . of Colts-foot and the like."

Coltsfoot also figures in the Abram How cure for measles.

So we know why it is so commonly seen along the roadside.

COLUMBINE   *Aquilegia vulgaris*
            Aquilegia vulgaris flore pleno (Double)
                      "1 oz Cullumbine seeds 3d" – J.W.Jr.
COLUMBINE, WILD   *Aquilegia canadensis*   J.J.

Gerard says the "blew Columbine hath leaves like the great Celandine . . . blewish greene. . . ." and every slender "sprig" brings forth "one floure with five little hollow hornes . . . with small leaves standing upright, of the shape of little birds . . ." There is also a red columbine, like it except for the color, and some are purple or of a horseflesh color. The "double Columbine" has many of these little flowers (having the form of birds) "one thrust into the belly of another sometimes blew, often white, and other whiles of mixt colours, as nature list to play with her little ones. . . ."

Gerard says the virtues are not yet sufficiently known, columbines being "used especially to decke the gardens of the curious, garlands, and houses." However, a few have recommended columbines for the stopping of the liver, jaundice, and soreness of the throat, although the ancient writers said nothing about it.

Parkinson lists the same sorts of columbine as Gerard: Single, Double, Double Inverted Rose, and Degenerate, and gives the colors as blue, violet, flesh, white, purplish, reddish and some "party coloured," as blue and white. He says both single and double columbines are "nursed up in our gardens for the delight of both their forms and colours." He lists the same uses mentioned by Gerard but with less diffidence, namely: for stone in the kidneys, women in travail, swoonings and for the jaundice.

Culpeper puts columbines under Venus, says they are grown in almost every garden, and lists the above uses also.

C O M F R E Y *Symphytum officinale* J.J.
How C.M.
(Introduced)

Gerard treats "Of Comfrey, or great Consound" after the Borage, Bugloss, Alkanet, Hounds-tongue sequence and just before Cowslips of Jerusalem, so he is treating very soundly of the large borage family, of which Comfrey must loom as the largest for any garden. Comfrey, he says, grows a yard high with a hollow stalk like that of a sow thistle and long, broad, hairy, pricking leaves like borage, but pointed like those of elecampane. So we have all the old favorites compared. He says the flowers are light red; Josselyn saw white. Some are blue. The roots of comfrey, however, are the sole source of its virtues, having a "clammy juice." John Josselyn announces with an air of producing good news that "Compherie" grows well. Gerard gives several concoctions of the root boiled with sugar and licorice, combined with "Folefoot" (coltsfoot), mallow-seeds and poppy-seeds to be used in over-straining of the back for various reasons,

1 *Confolida maior flore purpureo.*
Comfrey with purple floures.

‡ 3 *Symphytum tuberofum.*
Comfrey with the knobbed root.

and for gonorrhea. Likewise a medicine made of comfrey roots, knotgrass and clary leaves boiled in wine for the same end. The juice alone drunk with wine will stop excessive flowings of practically anything, being power-fully "glutenative." This to such an extent that if it is added to meat cut in pieces and seething in a pot, they will stick together and become one "lumpe." Which, of course, makes it good for green wounds.

Parkinson refers to comfrey only for comparison of other borages, as when he says that lungwort is like a spotted comfrey or the ever-living borage has leaves like comfrey, which places comfrey quite possibly in the ditches outside his gardens.

Culpeper echoes Gerard in all the troubles that comfrey is good for, spit-ting blood and all inward hurts and bruises and so on. He, too, believes the pieces of meat in a pot story.

CORIANDER *Coriandrum sativum* J.J.
(Introduced)

Coriander, says Gerard, is a "very stinking herbe" with faint green leaves like Chervill or Parsley at first and then more like those of Fumitorie, though smaller. The flowers are white and grow in "round tassels like unto Dill." The leaves are unwholesome, but the seed, dried, is warm and "convenient to sundry purposes." Such as: the seed covered with sugar, as comfits, and taken after meat, helps digestion; drunk with wine, it kills worms and stops all extraordinary issues of blood. Since it closes the stomach against sending up fumes, it aids digestion and prevents gout. It also takes away sounds in the ears, dries up rheum and eases "squinancy."

Josselyn lists "Coriander" with dill among "such Garden Herbs (amongst us) as do thrive there."

CORN *Zea mays* Higginson
J.J. R.W.

When the early settlers speak of "corn," generally they mean various grains brought with them from the old country, as when we refer today to "John Barleycorn." What we call corn today was "Indian corn," or the Indians' staple crop, just as oats, wheat, barley, rye and so on made up the basic food supplies of the newcomers. Gerard begins his section on "Corne" with wheat, which he says grows in almost all the countries of the world "which are inhabited and manured." Johnson adds the story about transmutation started by Theophrastus and Vergil and only barely disproved in modern Russia, where one sort changes to another— wheat to darnel and oats and so on. John Josselyn, listing the "Garden Herbs" which do well, names wheat, rye, barley and oats and notes that barley "commonly degenerates into oats." "Rie" follows wheat in Gerard and then "Spelt Corne," or what the lazy English women liked, then "Starch Corne," Barley of various sorts, "Otes," wild and otherwise, "Darnell," Rice, Millet, and, finally, "Turkie Corne." This he promptly deals with as a Turkey wheat, although his first entry is of "Corne of Asia." There is a handsome illustration of "Turkey Wheate" with a picture of two ears, and then pictures of the Yellow, Red and "Blew" Turkey wheat, in ear also. While this may seem confusing, there is no confusion in Gerard's mind. "The graine is of sundry colours, sometimes red, and sometimes white, and yellow, as myselfe have seene in myne owne garden where it hath come to ripeness." So who should know better than he that it is not really from Turkey, but was brought into Spain first "out of America and the Islands adjoining, as out of Florida and Virginia, or Norumbega, where they used to sow or set it, and to make bread of it. . . ." Probably unknown to Greek and Roman authors, he says, it is known to the inhabitants of America and the Islands, as "Mais." The Virginians call it "Pagatowr." The bread made from it is a "More convenient food for swine than for men."

CORN SALLET *Valerianella olitoria*
                                      "1/2 oz Corn sallet at 2d" –J.W.Jr.
LAMB'S LETTUCE   (Introduced)

Gerard says that this plant, called the "white pot herbe," has been con-sidered a kind of valerian by some, but he places it as a lettuce, because it resembles it, especially in being edible, which valerian is not. For once, the temperature and virtues are all in one sentence. "This herbe is cold and something moist, and not unlike in facultie and temperature to the garden Lettuce; in stead whereof, in Winter and in the first moneths of the Springe it serves for a sallad herbe, and is with pleasure eaten with vinegar, salt and oile, as other sallads be, amongst which it is none of the worst."

Parkinson finds "Lambes Lettice . . . wholly spent for sallets, in the begin-ning of the yeare . . . before any almost of the other sorts of Lettice are to be had." It has "small bleake blew flowers" and is "of a waterish taste, almost insipide."

Evelyn says it is "loosening and refreshing: The Tops and Leaves are a Sallet of themselves, seasonably eaten with other Salleting the whole Winter long and early Spring." He adds that "the French call them Salad de Preter, for their generally being eaten in Lent."

COSTMARY *Chrysanthemum balsamita*
  See MAUDLIN

COWSLIP   *Primula veris* (English)   C.M.
          Mertensia virginica (Virginia)
          Dodecatheon meadia (American)
          Caltha palustris (Early American)
          Pulmonaria officinalis (Jerusalem)

When Cotton Mather, in one of his cures for smallpox, recommends giving *Diacodium* (made from poppy heads) in "cowslip Water," he is probably referring to the pretty yellow marsh marigold, *Caltha palustris*, which many of us were taught to call cowslip, thereby confusing future associations and English friends. These plants have been eaten as salads in rural New England. We think this may be one of the plants Cotton Mather sent to the Royal Society as being not in any herbal and unique in New England.

Gerard, Parkinson and Culpeper all treat of the common English primrose, *Primula veris*, as cowslip, oxlip and peagle or paigle. Famed and favored as a complexion aid and for curing palsy, it is not hardy as far north as the early settlers may have wanted to take it.

On the other hand, the little plant which Gerard inserts between comfrey and burdock, two of the great plants used for home medicine, is *Pulmonaria officinalis*, which he calls "Cowslips of Jerusalem" or spotted comfrey, an ancient "lungwort" or remedy for lung troubles, bearing its signature of spots for spotted lungs. And here he appears to include also what we now call Virginia cowslips, or *Mertensia virginica*.

The "American cowslip" of the handsome picture in Thornton's dramatization of special flowers is *Dodecatheon meadia*, also known as shooting star, which is not common north of Pennsylvania.

CRESS, GARDEN   *Lepidium sativum*
                                "1 oz cresses seed 3d" –J.W.Jr.
   PEPPERGRASS   (Introduced)
CRESS, WATER   *Nasturtium Officinale*
                                   (Naturalized)
CRESS, WINTER   *Barbarea vulgaris*
ST. BARBARA'S WORT   *Barbarea*
                                   (Naturalized)

Gerard begins to list cresses in Chapter 8, *Of Winter Cresses*, where he places Herbe Saint Barbara, that beautiful yellow infester of our New England fields. He says the seed "causeth one to make water, and driveth forth gravell, and helpeth the strangurie." The juice in an unguent, with wax, oil and turpentine, "mundifieth" corrupt ulcers. "In winter when salad herbes be scarce, this herb is thought to be equall with Cresses of the Garden, or Rocket." And it helps the scurvey when boiled with scurvy grass, called in Latin, *Cochlearia*.

Gerard lists as Garden Cresses *Nasturtium hortense*, *N. Hispanicum* and *N. Petreum*, that grows on rocks. He says they all have small, narrow, jagged leaves, round stalks a cubit high, and many small white flowers with seeds like those of shepherd's purse. One kind from Paris, sent by "Robinus," has leaves like a curled fan of feathers. Spanish Cress has leaves like basil. The seeds of Stone Cress are contained in pouches like those of treacle mustard, or *Thlaspi*. Gerard reports the seed is more biting than the herb, whose sharpness is allayed if it is eaten young. He quotes Galen in saying that cresses may be eaten with bread, as ancient Spartans used to do, and he says low country men feed on cresses with bread and butter. It is eaten with other salad herbs, such as tarragon and rocket, and is chiefly sown for this use. Its virtues lie in its use against scurvy. Dioscorides thought it harmful to the stomach. Since Gerard lists it as killing the child in the mother's womb, this may well be. However, the seed is a remedy against sciatica, stamped and applied with vinegar and malt. Inwardly taken, it is a remedy for those who have fallen from high places, since, apparently, it prevents blood from clotting.

Gerard then mentions Indian Cresses, *Nasturtium Indicum*, sent him by John Robin in Paris. Having no certain knowledge of "this beautiful plant," he is content to refer it to the kinds of cresses for further consideration. Gerard says *Nasturtium Indicum* has yellow flowers with a crossed star "overthwart the inside" of a deep orange color. Gerard lists also Sciatica Cress, called *Iberis* and *Cardamantica*.

Parkinson lists only Garden Cresses, *Nasturtium hortense*, as the two-foot-high, white-flowered, torn-leaved plant pictured in Gerard. He says of its uses that the Dutchmen eat cresses with their bread and butter, and also stewed in a "Hotch potch" with other herbs. "We," says Parkinson, "doe eate it mixed among Lettice or Purslane, and sometimes with Tarragon or Rocket, with oyl and vinegar and a little salt, and in that manner it is very savoury to some mens stomackes." "Physically," he says, the uses are against spots, worms, phlegm, and pains in the breast.

Culpeper has three cresses: First, Black-cress "not much unlike wild mustard is under the dominion of Mars." Its seed strengthens the brain almost as well as mustard seed. Beaten up with honey, it is good for the cough, jaundice and sciatica. Boiled, it is a good poultice for inflammations in women's breasts and men's testicles. Second, Sciatica Cresses, are a "Saturnine plant" and are used in a poultice of the leaves or roots in a salve of old hog's grease, to be applied to the places pained, four hours for a man, two for a woman, and the places then to be rubbed with wine and oil and wrapped with wool or skins to sweat a little. This will also remove blemishes of the skin, and Culpeper adds, "Esteem this a valuable secret." If the skin is ulcerated by this treatment, it can be treated with a salve made of oil and wax. Culpeper adds in a footnote that cresses either boiled or eaten in salads are very wholesome. And then he cannot forbear to add immediately that garden-cresses beaten up with lard applied to children's scales or scalded heads will make the scales fall in a day. Water cresses, says Culpeper, are under dominion of the Moon.

We do not know what Winthrop brought. We suspect it was Herb St. Barbara.

On the other hand, Evelyn makes much of "Cresses, *Nasturtium*, Garden Cresses" sown monthly; the *Indian* sort are especially aromatic and quickening to the brain and spirits. The leaves and flowers can be mixed with colder plants and the buds can be candied. The "vulgar watercress, proper in the Spring" is good for cold stomachs but has little nourishment.

John Josselyn found "watercresses," probably the *Nasturtium officinale*, already naturalized.

CUCUMBER *Cucumis Sativus*   W.W.
                                                Higginson
                                                J.J.

Higginson appears to have thought "cowcumbers" one of the indigenous blessings of the country, with "pompions." Josselyn lists cucumbers after musk melons and before pompions, among which he seems to include squashes.

Gerard hedges a bit on Cucumbers: "There be divers sorts of Cucumbers; some greater, others lesser: some of the garden, some wilde; some of one fashion, and some of another. . . ." All, however, he says, are cold and moist in the second degree. They "putrifie soon in the stomacke and yeeld . . . a cold and moist nourishment . . . very little, and the same not good." All must be eaten green. The seed is cold, the fruit colder. It unstops the liver, helps inflamed chests, and, outwardly applied, it makes the skin smooth and fair. It is good for "other parts troubled with heat," such as, for instance, "red and shining fierie noses (as red as red Roses) with pimples, pumples, rubies and such like . . ." provided the face be bathed at the same time with a liquor made up of white vine vinegar, iris roots, brimstone, camphor, almonds, oak apples and lemons set in the sun for ten days. ("Oke apples are much of the nature of Galls, yet are they far inferiour to them, and of lesser force.")

Parkinson speaks of the "cowcumber," whose use is mainly as a salad, or pickled for a sauce, or preserved in brine. They are fittest for the hottest season.

Culpeper says "Cowcumbers" is the pronunciation of the vulgar. Cucumbers are under the Moon, and good for the complexion and ulcers of the bladder.

Evelyn finds cucumbers the most approved salad, alone or "in composition" to sharpen the appetite and cool the liver. He gives various ways of treating it, not too much oiled, "macerated" in its own juice, cloves and onion added, sometimes dressed with the juice of lemon, orange, or vinegar, and salt and pepper. Pickled, they are for winter.

CUDWEED   *Gnaphalium obtusifolium*   J.J.
EVERLASTING   *Gnaphalium uliginosum*

Culpeper's picture is of the English kind with little flowers all the way up the stalks, "with every leaf standeth a small flower."

Gerard says of the Vertues "Gnaphalium boiled . . . cleanseth the haire from nits and lice; also the herbe being laid in wardrobes and presses keepth apparel from moths." It is good against worms when it is boiled in wine "and drunken" and it also "prevaileth against bites and stings." "The fume or smoke of the herbe dried and taken with a funnel, being burned therein, and received in such manner as we do use to take the fume of Tabaco, that is, with a crooked pipe made for the same purpose by the Potter, prevaileth against the cough of the lungs, the great ache or pain of the head, and cleanseth the brest and inward parts."

Culpeper says "Venus is lady of it," and the plants are all "astringent or binding or drying, and therefore profitable for defluxions of rheum from the head, and to stay fluxes of blood wheresoever. The green leaves laid to a green wound stay the bleeding and make it heal quickly. The decoction in red wine and drunk is good for "inward and outward wounds." And Pliny said that the juice of the herb in wine and milk is a "sovereign remedy against the mumps and quinsey" and, so taken, prevents a recurrence of either disease. Culpeper has seen the tops beaten up with sugar and taken, a pill the size of a pea at a time, with success against the "chin-cough" that "almost incurable disease."

Parkinson knows *"Gnaphalium Americanum,"* too, as Live Long or Life everlasting and calls it "This silver tuft or Indian Cotton weede." On the whole, "all of them may to some good purpose be applied to rheumaticke heads. . . . They are also layd in chests and wardrobes, to keep garments from moths; and are worn in the heads and armes of Gentiles and others, for their beautiful aspect."

CULIFLOWER *Brassica oleracea*
"2 oz Culiflower seed 2s 6d per oz" –J.W.Jr.
See **CABBAGE**

CURRANTS *Ribes rubrum* (Red)   Everyone
Ribes nigrum (Black)   J.J.

Shall we say, like all our authors, that currants are so well known? Some were brought over. Some were found here. They are, like many fruits, so commonly accepted and cultivated as to need no comment from us now.

DAFFODILL, BASTARD *Erythronium americanum*   J.J.

Unfortunately, there seem to be no references to purely ornamental, bulbous plants in early New England gardens. The squills and stars of Bethlehem and saffron crocuses had their uses and excuses for being here. While Gerard's and Parkinson's daffodils fill pages, as do their tulips, the settlers appear to have disregarded them as garden necessities. All we have is Josselyn's mention of our trout-lily or dogtooth violet, as a "bastard daffodill."

DAISIE, NEW ENGLAND *Saxifraga virginiana*   J.J.

DAISY (OX-EYE) *Chrysanthemum leucanthemum*
(Naturalized)

Of the cure worked with the "New England Daisie or Primrose," which Johnson-upon-Gerard places among the navelworts, we have seen in Josselyn's *Rarities*. In Gerard it is called "Rose Penniwort" and is recommended for inflammations, breaking the stone, and for hot tumors.

Our field daisy, which seems so like a native wildflower today, appears in Gerard as "The great Daisie, or Maudelenwoort," excellent in "Vulnerarie drinks," and as a poultice for the "cruell torments of the gout" when boiled in butter with mallows and applied to the joint. Called Ox-eye, according to Geoffrey Grigson, because it was once hoped to be the plant Dioscorides called *bouthalmon*, it is interesting to see Gerard recommending it for inflammation of the eyes. See MAUDLIN.

DANDELION *Taraxacum officinale*   J.J.
(Naturalized)

Gerard says of the dandelion, or, according to him, *Dens Leonis*, that it is rather like "Succorie, that is to say wilde Endive," cold but "drieth more." It cleanses and opens "by reason of the bitternesse." Boiled, it strengthens a weak stomach. Raw, it "stops the bellie and helps the Dysentery," especially when boiled with lentils. The juice helps incontinence. The juice drunk is good against "the involuntary effusion of seed"; boiled in vinegar, it is good for troubles of the bladder. A decoction of the whole plant helps the yellow jaundice.

1 *Bellis maior.*
The great Daifie.

Culpeper says dandelion is "vulgarly called piss-a-beds" and is under the dominion of Venus. Of an opening and cleansing quality it opens obstructions of the liver, gall, spleen, and wonderfully "openeth the passage of urine." It also helps to bring rest and sleep to those afflicted with ague fits. The French and Dutch often eat dandelions in the spring.

Evelyn recognizes it in *Acetaria*. Macerated in several waters to extract the bitterness, he says that, "tho somewhat opening, it is very wholesome and little inferior to Succory, Endive, etc. The French Country-People eat the Roots and 'twas with this homely Sallet the good-wife Hecate entertained Theseus."

DILL   *Anethum gravelolens*

"1/2 oz dillseed 3d per oz" – J.W.Jr.

Gerard describes dill as having a stalk, high, round and jointed with little leaves, cut like those of fennel, and little yellow flowers "standing in a spokie tuft or rundle," a perfect picture of the dill flower. Gerard gives the Latin as "*Anethum.*" Being hot and dry, according to Galen, a decoction of the dried tops and also the seed increases milk in nurses, provokes urine, stays gripings, increases seed, cures ulcers and "stayeth the yeox, hicket, or hisquet," according to Dioscorides. This last "vertue" can be realized by the smelling of the seed boiled in wine, more especially in wormwood wine or wine with branches of wormwood and roses in it, and the bathing of the stomach with this. Common oil in which dill is boiled or sunned, as "we do oyle of roses," helps digestion, mitigates pain, procures sleep and provokes lust.

Parkinson is briefer and brisker. He agrees with Gerard that it is usually found in gardens. The leaves, he says, are used "as they do Fennel" with fish, but are so strong some people refuse it. With pickled cucumbers it agrees very well by giving the cold fruit a spicy taste. "Some," he says, "use to eat the seed to stay the Hickocke."

Culpeper says dill is under Mercury, so it strengthens the brain. He repeats Gerard's points, but ties it in a cloth after being boiled in wine to be smelled to stop "the hiccough." He recommends women troubled with pains of the mother to make a decoction to sit in. In a plaster it is excellent for old ulcers or sores.

DITTANDER   *Lepidium latifolium*
PEPPERWORT   J.J.
                                                (Naturalized)

Josselyn found "Dittander or Pepper Wort flourisheth notably," in New England gardens.

The senior Winthrop promises his son some roots. "To my loving sonne Mr. John Winthrop of Ipswich, Sonne, I received your letter and doe blesse the Lorde for your recovery and the wellfare of your family, you must be very careful of taking colde about your loynes; and when the ground is open I will send you some pepper-worte roots."

Gerard's "Dittander and Pepperwort" come second in his chapter on Horse Radish." Gerard's third "wild radish" is "Annuall Dittander," *Lepidium Annum*, by his definition, although the others have been *Raphanus rusticanus* and *R. sylvestris*. I have used Geoffrey Grigson's definition, although we have also *Lepidium campestre*, or "poor man's mustard" and "Mithridate mustard" from the *Book of Weeds*, or perhaps *Lepidium virginicum* "poor man's pepper," in Gray. In any case these all have "extreme hot" roots and leaves. Gerard's third "wild radish" has seeds which taste like "Thlaspi, or Treacle mustard" which Pliny said was one of the scorching and blistering simples and would mend the skin of the face after scars and scabs. Gerard says all these "wild radishes" are hot and dry and have a cleansing and opening quality.

Parkinson says Dittander or Pepperwort is used as a sauce or salad for "some cold churlish stomackes" but it is too hot, bitter and strong for weak and tender stomachs.

Culpeper says of "Pepper-wort, or Dittander" that it is "under the direction of Mars." He quotes Pliny and Paulus Aeginetus as saying it is effectual against pains in the joints, gout, sciatica "or any other inveterate grief," the leaves being bruised and mixed with "old hogs-lard." This applied to the place "four hours in men, and two hours in women, the place being afterwards bathed with wine and oil mixed together, and then wrapped with wool or skins after they have sweat a little." He says it also "amendeth the deformities or discolourings of the skin," and that the juice used to be given in ale to women with child to procure a speedy delivery.

Gerard has included Dittany as an English name for the above, but as he also has a chapter on Dittany concerning two quite other plants, to avoid confusion we make a note here of:

DITTANY   *Origanum dictamnus*

Called also Dittany of Candie, candy mustard and dittany of Crete, which is famous because deer and goats wounded by arrows heal themselves by eating this herb which has strong healing qualities. And

DITTANY, FALSE *Dictamnus albus*
FRAXINELLA

According to Parkinson, who calls this the Bastard Dittanie, and at-tributes to it also many strong qualities as an anti-poison and anti-pestilence.

DOCK *Rumex sanguineus* (Bloodwort)
       R. patientia (Monks rhubarb)
       R. acetosella (Sorrel)   How
       R. crispus (Yellow dock)   C.M.
                                 J.J.

Under Bloodwort we included Gerard's quote from Dioscorides about "Dockes: that there are four kinds, wild or sharp-pointed, garden dock, round-leafed dock, and the sour docks called sorrel." Culpeper says that all docks are under Jupiter, and have a cooling drying quality. The seeds of most kinds stay fluxes of all sorts. The roots boiled in vinegar help the itch and cure freckles. All docks boiled with meat make it boil the sooner. And he praises bloodwort above all others. We shall meet docks again under SORREL.

(Burdock is *Arctium lappa*. See also CLOTBUR.)

DOGSTONES *Orchis*   J.J.

Which of the native orchids Josselyn saw being made into an "amorous cup" which wrought the desired effect, we do not know. This is obviously a sort of doctrine of signatures affair, and indicates a plant with what Gray calls "roundish tuberoids."

DOVESFOOT *Geranium columbinum* J.J.
(Naturalized)

Here Gerard becomes very personal. Dovesfoot or cranesbill, he says, is good for green and bleeding wounds and lessens inflammations. The herb and root dried and powdered, taken a half-teaspoonful at a time in red wine for twenty-one days, cures "miraculously ruptures or burstings, as my selfe have often proved whereby I have gotten crownes and credit." If the patients are elderly, he adds a concoction of dried red snails (those without shells). The decoction of the herb made in wine is excellent for inward wounds "as myselfe have likewise proved."

Culpeper says it is "gentle though martial" and agrees with all the uses above, adding also that it is good for stone in the kidneys.

DOVESDUNG *Ornithogalum umbellatum*
STAR OF BETHLEHEM Boorde
(Naturalized)

Gerard says "there be sundry sorts of wilde field Onions called 'Starres of Bethlehem.'" He quotes the ancients as saying the roots roasted in hot embers and applied will heal old ulcers. Dioscorides says the roots are eaten both raw and boiled. Parkinson says they are so common in all countries as to need no description and are eaten in Italy "as well for necessary food as for delight."

This is one of the plants referred to by Andrew Boorde in his *Breviary of Health* when, after a carbuncle has been "fired," a poultice of stamped Doves Dung be applied. I have retained his capitals, as he capitalizes all flowers and not the sort of thing we might assume this to be, were it not capitalized. In the Bible, II Kings, 6:25, in the great famine in Samaria, "doves dung" brought a great price. And no less a person than Linnaeus, among others, believed this to be the small bulbous wild flower which spatters the rocky hills of that part of the world with splashes of white and was also called "bird-milk." It is edible when roasted.

Its familiar names, in both England and New England, have something to do with the clock, because it shuts up at night. Four o'clocks, for example, in New England.

DRAGONS　*Arisaema draconitum*
DRAGON-ARUM
INDIAN TURNIP　*Arisaema triphyllum*
JACK-IN-THE-PULPIT
WATER ARUM　*Calla palustris*　　　　　　　　　　　J.J.
　　　　　　　　　　　　　　　　　　　　　　　　　　　C.M.

Gerard gives us a collection of somewhat similar plants called Great Dragons, Small Dragons, Cuckow Pint and Wake Robins, hooded Cuckow Pint or Friars Cowle, and a Water Arum which seems to be *Calla palustris*. In any case, they all bear a heavy head of red berried fruit in the autumn and have a bloom which suggests different things to different people, as evidenced in the almost countless names. At any rate, they all belong to the *Arum* family. Their "vertues" are similar and all powerful, scouring and cleansing the entrails and the lungs, curing old ulcers, good against serpent bites and web in the eyes. The scent of the flowers is harmful to pregnant women. Starch made from the roots blisters the hands of laundresses.

Parkinson treats of Dragons as in Gerard's first picture, with "leaves very much divided on all sides," certainly not in the least like our Jack-in-the-pulpit which seems to represent the dragon family here. And Parkinson claims the roots, like those of Arum or Wakerobin, are given with treacle "to expell noisome and pestilentiall vapours from the heart."

Culpeper says the root of Dragons resembles a snake, the plant is under the dominion of Mars and "is not without its obnoxious qualities." He agrees with all the uses according to Gerard and adds that Pliny and Dioscorides affirm that no serpent will approach anyone who carries this herb. "Indian Turnip" explains itself.

EARTHNUT　*Aralia*　　　　　　　　　　　　　　　　　　J.J.
　　　　　　　Apios tuberosa

Josselyn describes the earthnut in New England under plants "common also in England" and says there are "divers kinds, one bearing very beautiful flowers." Tuckerman thinks he means *Apios tuberosa* and that this is also

what William Morrell meant in his Latin poem about New England, quoted
in translation in Chapter II. From the picture of "Earth-nut, Earth Chestnut, or Kipper-nut" in Gerard, however, the plant Josselyn is referring to
would appear to be not a vine or wild bean, but a plant similar to the
"Hemlocks, or herbe Bennet" which it follows in order. Gerard describes it
as having "small even crested stalkes a foot or somewhat more high:
whereon do grow next the ground leaves like those of Parsley and those that
doe grow higher like to those of Dill; the white flowers doe stand on the
top of the stalkes in spokie rundles, like the tops of Dill. . . ." Gerard says
of the root, "The Dutch people doe use to eate them boyled and buttered,
as we doe Parsneps and Carrots, which so eaten comfort the stomacke, and
yeeld nourishment that is good for the bladder and kidneyes."

The *Aralia* include several familiar species such as the "five-leaved
ginseng," the "dwarf ginseng" and "wild sarsaparilla," all with edible
roots.

See GROUNDNUT.

ELDER   *Sambucus canadensis*   J.J.
        Sambucus ebulus (Dwarf elder Danewort)   C.M.

The elder, both the shrub and the related dwarf elder, are apparently
among the really ancient remedies of mankind. Gerard ascribes much the
same "vertues" to both: powerful purges, and good for dropsy, mad dog
bites, infirmities of the matrix, gout, and inflammations of the mouth. The
juice of the dwarf elder will dye hair black.

Culpeper devotes all his attention to the dwarf elder, although he says
both sorts are under the dominion of Venus. He uses all parts of the common elder; the shoots or stalks or leaves boiled in broth to "expel phlegm
and choler"; the middle bark for a purge; the berries for the dropsy; the
juice of the root to provoke vomiting. The root will cure bites of adders
and mad dogs. He knows about the hair dye and all Gerard's uses, and
adds that the distilled water of the flowers is good for cleansing the skin.
The water is also good for baths, redness of the eyes and the palsy.

Helenium.
Elecampane.

ELECAMPANE *Inula helenium*   Boorde
ENULA CAMPANA   Hammond   How
                                              "Cases" – J.W.Jr.

Gerard describes "Elecampane" as "bringing forth from the root great
white leaves . . . almost like those of great Comfrey, but soft, and covered
with a hairie downe, of a whitish greene colour . . . the stalke a yard and a
half long . . . great floures broad and round . . . are yellow . . . also the middle ball or circle. . . ." In shops he says it is called *Enula campana.* In English it is called "Elecampane, and Scab-woort, and Horse-heale; some report that this plant took the name *Helenium* of Helen wife to Menelaus,
who had her hands full of it when Paris stole her away into Phrygia."

Gerard says its "vertues" are against shortness of breath, old coughs,
"and for such as cannot breathe unless they hold their necks upright."
Given in a "looch, which is a medicine to be licked on," it will help void
"thicke and clammie humours which stick in the chest and lungs." It is
good to mix with counterpoisons. The juice of the root will drive worms

from the belly. The root chewed will "fasten the teeth." The root preserved aids digestion. Galen recommended it for "divers passions of the huckle-bones, called the Sciatica." The root boiled soft and mixed with butter and powdered ginger makes an excellent ointment against the itch, scabs and manginess.

Parkinson does not grow it in his gardens.

Culpeper says it is under Mercury and agrees with all the above, adding only that the roots and herbage put into new ale or old beer and drunk daily will strengthen and quicken the eyesight.

ENDIVE *Cichorium endiva*

"1/2 oz endive seed 3d per oz" –J.W.Jr.

Gerard's chapter on "Garden Succorie" begins by listing "sundry sorts of plants comprehended under the title of *Cichoracea*, that is to say Cichorie, Endive, Dandelion, etc." As we have seen under CHICORY, both "Succorie and Endive" are put in one chapter by Parkinson and considered mainly as salad material, although they have "vertue to cool the hot burning of the liver, to helpe the stopping of the gall, yellow jaundice, lacke of sleepe, stopping of urine, and hot and burning feavers." The white endive espe-cially "doth comfort the weake and feeble stomacke, and cooleth and re-fresheth the stomacke overmuch heated."

Evelyn says of Endive, "the largest, whitest and tenderest Leaves best boil'd, and less crude. It is naturally Cold, profitable for hot Stomachs; Incisive and opening Obstructions of the Liver. . . . the middle part of the Blanch'd Stalk separated, eats firm; and the ampler Leaves by many pre-ferred before Lettuce."

EYEBRIGHT *Euphrasia officinalis* (Naturalized)

Gerard says, "Euphrasia or Eye-bright is a small low herbe not above two handfuls high, full of branches covered with little blackish leaves dented or snipt about the edges like a saw, the flowers are small and white, sprinkled and powdered on the inner side, with yellow and purple specks mixed therewith."

Josselyn said the "Stitchwort" growing in New England, defined as *Stellaria graminea* by Tuckerman, was taken by ignorant people for eyebright. But apparently not for long as the *Euphrasia officinalis* now grows along the New England coast, and those seeking the true eyebright can find it.

Gerard says, "It is very much commended for the eyes. Being taken itselfe alone, or any way else, it preserves the sight . . . stamped and laid upon the eyes, or the juice thereof mixed with white Wine and dropped into the eyes, or the distilled water, taketh away the darknesse and dimnesse of the eyes, and cleareth the sight." It also "comforteth the memorie."

Parkinson does not consider it a garden plant.

Culpeper agrees in all matters with Gerard, adding only that if it is "tunned up with strong beer that it may work together and drunk . . ." it will "restore the loss of sight through age."

FEATHERFEW   *Chrysanthemum parthenium*
FETHERFEW   J.J.
FEVERFEW   C.M.
(Introduced)

Gerard describes "Fetherfew" as having "many little round stalkes divided into certaine branches. The leaves are tender, diversly torne and jagged, and nickt on the edges. . . . The floures stand on the tops of the branches with a small pale of white leaves, set round about a yellow ball or button, like the wilde field Daisie. . . . The whole plant is of a light whitish greene colour, of a strong smell and bitter taste."

Gerard says it is hot in the third degree, dry in the second and "fully performeth all that bitter things can do." In diseases of the matrix it can be drunk, sat over in a hot bath or applied in a poultice. Both leaves and flowers can be applied to hot inflammations, according to Dioscorides, who also says the powder drunk with "Oxymell, or syrup of Vineger, or wine for want of the others, draweth away flegme and melancholy." The powder taken in wine is also good for those with "Vertigo, that is a swimming and turning in the head. Also it is good for such as be melancholike, sad, pensive and without speech."

Parkinson entertains only the double-flowered kind of "Featherfew" and says it is used for women's diseases and is reputed to help those who have taken too much opium. In Italy it is fried with eggs.

Culpeper says, "Featherfew" is governed by Venus and is of "general utility to the fair sex."

Josselyn remarks that "Fetherfew prospereth exceedingly" when he lists it among the English garden herbs which "thrive there." Cotton Mather advises it for toothache, in a bag "bedewed with Rum," applied hot.

Geoffrey Grigson has called it "the housewife's aspirin."

FENNEL   *Foenticulum vulgare*   J.J.   Dr. Oliver
                                    (Introduced)   (Naturalized)
                                "1 oz fennell seed 1d" – J.W.Jr.

FENNEL, SWEET   *Foeniculum dulce*
                                "1 oz sweet fennell seed 1d" – J.W.Jr.

Gerard says, "Of Fennell" that the first kind is so well known as to need no description. The second kind, he says, is "Sweet Fennel, so called because the seeds thereof are in taste sweet like unto Annise seeds." It resembles the common fennel "saving that the leaves are larger and fatter, or more oleus; the seed greater and whiter, and the whole plant in each respect greater." The seed is hot and dry in the third degree. The powder of the seed preserves the eyesight, as witness this verse, given in Latin and translated by Gerard:

> Of Fennell, Roses, Vervain, Rue and Celandine,
> Is made a water good to cleere the sight of eine.

The leaves eaten or the seeds drunken in a "Ptisan" fill women's breasts with milk. The decoction of the same is good for the kidneys and against the "wambling" of the stomach. The whole plant is good for the kidneys, lungs and liver.

Parkinson knows three sorts of fennel, of which two are sweet. It is "of great use to trimme up and strowe upon fish" and to flavor cucumbers pickled "and other fruits." The roots boiled with parsley will open obstructions. The seed is put into "Pippin pies and divers other such baked fruits, as also into bread to give it the better relish." When Sir Henry Wotton sent the sweet fennel to John Tradescant, he directed him to "white it," which makes it the more delightful, says Parkinson, to those "accustomed to feed on greene herbes."

Culpeper takes the charm from Parkinson's fish dish by saying fennel is good to be boiled with fish as it "consumeth the phlegmatic humour arising therefrom." Covered by Mercury under Virgo, it "beareth antipathy to Pisces." He finds it good for increasing milk, cleansing the blood and for gout, cramp, stone and jaundice. He thinks it is good for those bitten by a serpent or who have eaten poisonous herbs. (What a pity he does not name these.)

Evelyn says the sweetest fennel comes from "Bolognia." Aromatic, hot and dry, it expels wind, sharpens the sight and "recreates the Brain." The stalks peeled when young are to be dressed like celery. The "tender tufts" are eaten alone with vinegar, oil and pepper. He warns of a small green worm "which sometimes lodges in the Stemm of this Plant, which is to be taken out. . . ."

FERN   J.J.
   B R A K E S   *Pteridium aquilinum*
   M A I D E N H A I R   *Adiantum pedatum*
   M A L E   *Dryopteris filix-mas*
   S W E E T   *Comptonia*

Unlike as these members of the fern family may be, we list them here as a whole. John Josselyn mentioned "Fearn" followed by "Brakes" in his list of plants common to both Old and New England. In Gerard's chapter on ferns, the male fern, "*Filix mas*," comes first and then the "Female Ferne, or Brakes." Both are hot, bitter, dry and somewhat binding. The male fern roots taken in honey drive out worms. Taken with scammony or black hellebore, after eating garlic, it will drive out worms and kill the unborn child. The female fern operates in the same way, bringing "barrennesse, especially to women . . . and causeth women to be delivered before their time."

Of maidenhair fern, Josselyn comments on the quantities found in New

England and says scathingly that there apothecaries need no substitute for the genuine article. This is an interesting aside, as the true maidenhair is *Adiantum capillus-veneris*, not very like our native New England variety, which resembles Gerard's "Wall rue, Or Rue Maidenhair." This Gerard considers superior medicinally to the first true maidenhair. He says it is used instead of it in "Flanders and Germanie." In any case, maidenhair fern generally is good for the stone, phlegm in the lungs, coughs, stitches and pains in the side. It reduces swellings, makes hair grow again, and is good against ruptures in young children.

The sweet fern, *Comptonia*, which is not really a fern, is named for the horticulturally inclined Henry Compton, Bishop of London, who stimulated plant exploration in Virginia especially. This was one of the plants which surprised the early settlers. Soldiers marching through it on hot days felt it made them faint by its pervading sweetness. Ink was made from it. Josselyn reports, "Sweet fern [*New-England's Rarities*] the tops and nudicaments of sweet fern boiled in water or milk and drunk helpeth all manner of fluxes, being boiled in water it makes an excellent liquor for Inck." He also reports pigs liking the roots.

FLOWER DE LUCE   J.J.
  FLAG, BLUE   *Iris prismatica*
  FLAG, YELLOW   *Irish pseudacorus* (See under ACORUS)
  FLAG, SWEET   *Acorus calamus* (See under ACORUS)

Josselyn observed a "Blew Flower de Luce; the roots are knobby but long and straight and very white, with a multitude of strings. He found it "excellent for to provoke vomiting and for Bruises on the Feet or Face. They flower in June and grow upon sandy hills as well as wet grounds." (So he says.)

Gerard treats, under "Floure-de-luce," a charming quantity of irises, the first being the common variety to which he ascribes a great many "vertues" including the ones Josselyn mentioned, for bruises on the face. Johnson, editing, summed it all up in one sentence. "It bindes, strengthens, and condenses; it is good in bloddy flixies and stays the Courses."

Considering that Gerard has a splendid assortment of irises recorded and Parkinson an even greater one, recommended as definitive by no less than

Thomas Johnson in one of his notes in Gerard's *Herball*, it is strange that only one would seem to have been brought in by the early settlers. Or perhaps not strange, since Culpeper waxes so enthusiastic about "Flower-de-Luce," which he says also "beareth the name of yellow water-flag." Flag is, he says, under the dominion of the Moon. The root is astringent, cooling and drying, so it helps all kinds of fluxes. The distilled water of the whole plant is a "sovereign remedy for weak eyes," "dropped therein" or merely applied to the forehead with a wet sponge. It is also good for ulcers and swellings. An ointment made of the flowers is best for external applications.

The legend that it is the origin of the French Fleur-de-Lis, having showed by its growth the existence of a ford where French soldiers under Clovis were able to escape the Goths, so far antedates its becoming an English wildflower that its "vertues" are all English ones.

There is a suggestion in Culpeper's text of its use for syphilitic sores. The yellow flag also dyes yellow and, with the sulphate of iron, black.

Parkinson considers "Flowerdeluce" the familiar name for all irises and endows them all with the "vertues" of purging both "inward, as well as outward." Some use the green roots to cleanse the skin, but carefully, and the dried roots powdered to make "Orris" for sweet powders to perfume linen. The juice in the nostrils induces sneezing and, taken inwardly it causes "vomiting very strongly."

F L A X   *Linum usitatissimum*   J.J.
                                                                  (Introduced)

Gerard's "Garden flaxe" he calls *Linum sativum* which means only that it is the cultivated sort. It "risith up with slender and round stalks. The leaves thereof bee long, narrow and sharpe pointed; on the tops of the sprigs are faire blew floures . . ." Virgil "testifieth in his Georgickes," as translated for us by Gerard:

> Flaxe and Otes sowne consume
> The moisture of a fertile field;
> The same worketh Poppie, whose
> Juice a deadly sleepe doth yeeld.

Which is given us to show the ancients were right in advising that it be sown in "fat and fruitfull soile, in moist and not drie places." Sown in the spring, it flowers in June, and after it is cut down the stalks are put out into water and, with some weight laid upon them, exposed to the heat of the sun. The looseness of the rind indicates when it is ready to be taken out and dried in the sun, and after "used as most huswives can tell better than any selfe."

Linseed oil is infinitely useful and has the same properties as fenugreek (*Trigonella foenum-graecum*) or "Greek hay." The seeds of either or both made into a powder, boiled with mallows, violet leaves, smallage, and chickweed, then stamped with hog grease, will "appease" all manner of paine, reduce and soften swellings and bring apostumes to suppuration, especially when applied warm.

Parkinson apparently considers flax a field crop. In his garden he entertains only toad flax. Culpeper likewise confines himself to this false flax and calls it "Flaxweed, called likewise toad-flax," which see under T.

FUMITORY *Fumaria officinalis*   (Adventive)

Gerard says fumitory is a "very tender little herbe." It is good for "all them that have either scabs or any other filth growing on the skinne, and for them also that have the French disease." It removes stoppings of the liver and the spleen, purifies the blood, and is "a most singular digester of salt and pituitous humors."

It is beneath Parkinson's garden notice.

Culpeper says it is governed by Saturn and is effectual for the liver and the spleen, clarifying the blood from "saltish, choleric and malignant humours" which cause scabs, tetters, itch and such like. It is good against melancholy and prevents infection by the plague when taken with a good treacle. The juice alone is good for the eyes and, when combined with vinegar, makes a good wash for the skin.

GALE *Myrica gale*   J.J.
GAUL

Josselyn observed "Gaul or noble Mirtle" growing in New England as

well as in Old. Gerard calls it "Gaule, sweet willow or Dutch Myrtle tree,"
and says the "fruit is troublesome to the brain; being put into beere or aile
whist it is in boiling (which many use to do) it maketh the same heady, fit
to make a man quickly drunke. The whole shrub, fruit and all, being laied
among clothes, keepeth them from moths and worms."

The "bayberry" or "candleberry" from which candles were made belongs
to the wax-myrtles and is *Myrica pensylvanica*, of which, curiously enough,
there seems to be no mention, unless it was confused with the "Gaul" observed by Josselyn.

GALINGAL

Of which Endecott sent a root to be used for the relief of Mistress Winthrop, grows, says Gerard, in the East Indies and China and is "good
against all cold diseases." Obviously it was one of the apothecaries' remedies brought over with the settlers for which they tried to find native sub-
stitutes. Culpeper says it is seldom seen in England, belongs to Mars and
was approved by Pliny.

GARGLIA

Was mentioned in Sewall's diary as having been found by him, as well as
yarrow. Perhaps his writing was misread and this might just possibly be
"garget" or pokeweed, *Phytolacca decandra* or *Americana*. The young
sprouts were used as greens; the roots are poisonous and may have been
used as a narcotic: the berries have medicinal properties. One wonders.

GARLIC   *Allium sativum*   C.M.   J.J.
                                                            Hammond
                                                            (Introduced)

Gerard says Galen found the wild garlic stronger than the garden variety,
that the leaves may be eaten in the spring with butter "by such as are of a

strong constitution and labouring men," and that the distilled water is good against the stone and to provoke urine.

Parkinson also mentions it as eaten by those with strong stomachs, though always "sodden," never raw. And he says it is considered in many countries "the poor mans Treacle, that is, a remedy for all diseases."

Culpeper says "Mars owns this herb." It provokes urine and women's courses, helps mad dog bites, kills worms in children, cleanses the skin of blemishes, eases ear pains, removes "all inconveniences" from drinking stagnant water or eating such poisonous plants as henbane, wolfbane or hemlock, is good against the falling sickness, but bad for those "oppressed with melancholy." Cummin seeds or green beans chewed after eating garlic remove the disagreeable smell from the breath.

Evelyn has quite a bit to say upon the subject.

"Garlick, Allium; dry towards Excess; and tho by both Spaniards and Italians and the more Southern People familiarly eaten with almost everything, and esteemed of such singular Vertue to help Concoction, and thought a Charm against all Infection and Poyson (by which it has obtain'd the Name of the Country-man's Theriacle) we yet think it more proper for our Northern Rustics, especially living in Uliginous and moist places, or such as use the Sea; Whilst we absolutely forbid it entrance into our Salleting, by reason of its intolerable Rankness, and which made it so detested of old that the eating of it was (as we read) part of the Punishment for such as had committed the horrid'st Crimes."

While the garlic referred to in our references above was probably the garlic they knew in England, or *Allium sativum*, the cultivated variety, we remember the early voyagers and settlers finding garlic growing wild which would be *A. canadense*. *A. oleraceum* (meaning for the pot) and *A. vineale* (meaning from vineyards) or a field garlic, are also both introduced and naturalized.

Hammond put a clove of garlic dipped in honey in the ear before going to bed "to comfort the head and brain." Cotton Mather quotes Sydenham as binding it to the feet of smallpox patients. Josselyn gave it combined with hyssop to croupy hens. As Josselyn himself would say, *Probatum est*.

GENTIAN   *Gentiana crinata*   J.J.
FELWORT
BALDMONEY

Gerard says that Gentius, King of Illyria, was the "first finder of this herbe" and apothecaries call it after him "unto this day." In England it is called Felwort or Baldmoney, and there are several sorts. One called the "Autumne Bel-floure" looks enough like our fringed gentian to pass for it, in a woodcut at least.

Gentians appear to be one of those ancient remedies whose long-standing is evidenced by the emphasis put upon their values as counterpoisons and their uses against mad dog and snake bites. Gentians are also good to rub on the udders of cows so bitten and for cattle who are broken-winded or have a cough. It helps those who are fallen from high places, or have "evil" livers or stomachs. The decoction is good against all cold diseases of the inward parts.

Parkinson says that undoubtedly if the bitterness was not so great, the "wonderfull wholesomeness of Gentian" would be known and would work admirable cures as a counter-poison and for the stomach, liver and lungs.

Culpeper does not notice the bitterness as a deterrent and recommends gentian, which is under the dominion of Mercury, for almost every conceivable use. It strengthens the brain and apprehension, is good for coughs, the spleen and dropsy. It is useful for all women's troubles, so much so that pregnant women must beware of it. It expels worms, cleanses old sores, and is good against all diseases of the brain, such as headache and melancholy.

Cotton Mather advised it among "easy Purges" taken with rhubarb and anise.

Gentian was one of the ingredients of Hammond's prescription for "the Megrum," mixed with mugwort, sage and chamomile in honey and applied to the head, warm.

GILLYFLOWERS
   See CLOVE GILLYFLOWERS   *Dianthus caryophyllus*
                                                               J.J.
   See STOCK GILLYFLOWERS   *Matthiola incana*
                                                             J.W.Jr.

GLASSWORT *Salicornia virginica* J.J.
SAMPHIRE
CORAIL
LEADGRASS

Gerard deals in his chapter on "Glasse Saltwoort" with "Glassewoort" which resembles a branch of coral, and is called *Kali* and *Alkali* by the Arabians. "The salt that is made from the ashes is *Sal Alkali*. Stones are beaten to powder and mixed with ashes, which beeing melted together become the matter whereof glasses are made. Which while it is made red hot in the furnace, and is melted, becomming fit to work upon doth yeald as it were a fat floting aloft which when it is cold, waxeth as hard as a stone, yet it is brittle and quickly broken."

Josselyn speaks of "Soda bariglia, or massacote, the Ashes of Soda, of which they make Glasses," and follows it with listing "Glass-wort, here called Berrelia, it grows abundantly in Salt Marshes."

Gerard says that "a little quantitie of the herbe taken inwardly . . . doth provoke urine . . . calleth forth the dead childe. It draweth forth by siege waterish humours, and purgeth away the dropsie." "A great quantitie taken is mischievous and deadly."

That the plant was used medicinally in spite of these rather dire warnings in Gerard, we know from Josselyn who says, "Glasswort, a little quantity of this plant you may take for the Dropsie but be very careful that you take not too much, for it worketh impetuously. . . ."

GOATS-BEARD See **VIPERS GRASS**

GROUND IVY See **ALE HOOF**

GROUNDNUT *Apios tuberosa* Bradford
Morrell

While the dwarf ginseng, *Panax trifolium*, is a tempting guess at what the early settlers meant by groundnut, it is probably safer to believe Professor

Tuckerman and picture them subsisting in the New World on the climbing and twining "wild bean" or "potato bean," *Apios tuberosa* or *Apios americana*. Both native American plants are still called "groundnut." See also EARTHNUT.

HELLEBORE   *Veratrum viride*   J.J.
ELLEBORE   C.M.

The true hellebore, *Helleborus niger*, or Christmas rose, has no relationship with the False Hellebore, White Hellebore, Indian Poke, Tabac du Diable or American Hellebore which confronted the early settlers. John Josselyn's report is graphic. He reports that hellebore grows in such quantities in New England that a man may gather a cartload of it without any trouble. He does not explain what the man would do with a whole cartload. Its uses seem to be somewhat diverse.

Josselyn says, "White Hellebore is used for the Scurvie by the English. A friend of mine gave them first a purge, then conserve of Bear-berries, then fumed their leggs with vinegar, sprinkled upon a piece of mill-stone made hot and applied to the sores white Hellebore leaves; drink made of Orpine and Sorrel were given likewise with it, and Sea-scurvi-gras. To kill lice, boil the roots of Hellebore in milk, and anoint the hair of the head therewith or other places." In another place, with reference to "the Sassafras tree," he says the leaves are good to lay upon the legs of those with scurvy with "the leaves of white Hellebore." In still another passage, he describes a ritual among young Indian males who sit about in a circle with birch bowls and, having eaten hellebore and vomited, strive with each other to eat again and retain it. The one who lasts the longest is head of the others for one year.

Cotton Mather announces in his chapter on the founding of Harvard College that anyone who does not believe it to be great needs hellebore. Taken how, he does not say, perhaps only as snuff to awaken him from the near-dead and restore his wits.

Governor Endecott, we know, recommended tobacco up the nose for Mistress Winthrop as better than hellebore.

Samuel Sewall writes in his Diary that "Mr. Cook scrapes white Hellebore which he snuffs up and sneezes 30 times and yet wakes not nor opens his eyes."

HEMP *Cannabia sativa* (Cultivated hemp)   J.J.
      Acnida cannabina (Water hemp)   W.W.
      Apocynum cannabinum (Indian hemp)   J.W.Jr.

Gerard deals with several hemps, the male and female of the first variety, and then with the wild, bastard and water hemp, "Called Water Agrimony." The temperature and "vertues" of the "Manured hempe," the first variety, are fairly rugged. Galen says the seed is "hard of digestion, hurtful to the stomacke and head, and containeth an ill juice. . . . it consumeth winde . . . and is so great a drier as that it drieth up the seed if too much be eaten of it." The pulp of the hemp seed is given to those with yellow jaundice, often with success. (Which explains the use of the name agrimony.) Mathiolus is quoted as saying the "seed given to hens causeth them to lay eggs more plentifully."

Higginson lists among the herbs found growing in the country, "two kinds of Herbes that beare two kinds of Flowers very sweet, which they say, are as good to make Cordage or Cloath as any Hempe or Flaxe we have."

The Massachusetts General Court in 1641 tried to encourage harvesting wild hemp, with what success we do not know.

HERB BENNET See AVENS

HERB OF GRACE See RUE

HERB PARIS *Cornus canadensis*   J.J.

Our native "bunchberry," which attracted the attention of John Josselyn toward two entries and a sketch, is very like its English counterpart, called "One-berry or Herbe True-Love" and *Herba Paris* by Gerard; *Paris quadri-folia* by Linnaeus.

Gerard says, "Herbe Paris is exceedingly cold whereby it represses the rage and force of poison." The berries have been given with success to those who become "peevish or without understanding." From experiments made upon dogs in Paris it was proved an antidote to poison, since of two dogs down whose throats poison had been rammed, the one given herb Paris immediately afterward recovered. The other "died incontinently." As antidote to poison would seem to be its sole use, but for the reports of people in Germany using its leaves for green wounds.

Curiously enough, Gerard in one chapter couples Herbe Paris with "Moone-wort" *Botrychium lunaris*, which seems to give it an occult association. He avers of moonwort that "Alchymists and witches doe wonders withall, who say that it will loose lockes and make them to fall from the feet of horses that grase where it doth grow . . . ," which Gerard says are "all but drowsie dreames and illusions." Actually moonwort was considered good for wounds — probably his only reason for including this fern with Herbe Paris.

HERB ROBERT *Gerannium robertianum*   J.J.

Gerard says "Herbe Robert" is somewhat cold and yet both scouring and somewhat binding, in fact, "participating of mixt faculties." The total "vertues" are contained in one sentence: "It is good for wounds and ulcers of the dugs and secret parts, it is thought to staunch bloud . . . to heale up bloudy wounds."

HERB TWOPENCE *Lysimachia nummulari*   (Introduced)
(Naturalized)

Perhaps there is no better proof of the early popularity of this little plant which now carpets large areas of streamsides and meadow edges near old settlements, than to quote Gerard: "Boiled with wine and honie it cureth the wounds of the inward parts and ulcers of the lungs and in a word there is not a better wound herbe, no not Tabaco it selfe, nor any other whatsoever." And again, "The herbe boiled in wine with a little honie or meade prevaileth much against the cough in children called the Chinne cough."

HOLLYHOCK *Althaea rosea*
"1/2 oz hollyhock seeds at 2d" –J.W.Jr.

Gerard's chapter is titled "Of the Garden Mallow called Hollihocke" (which he gives in Latin as *Malva hortensis*). He says there are several mal-

3 *Malua purpurea multiplex.*
Double purple Hollihocke.

lows, some of the garden, some of the marsh or seashore, "others of the field, and both wilde."

"And first of the Garden Mallow or Hollihocke."

"Of a height of foure or six cubits . . . slender foot-stalks . . . single floures not much unlike to the wilde Mallow, but greater . . . sometimes white or red, now and then of a deep purple colour, varying diversly, as Nature list to play with it." Double hollyhocks he believes to be all red or purple. The "vertues" are those of the wild mallows "which are more commonly used." Gerard notes that one name for the hollyhock was "*Rosa ultramarina,* or outlandish Rose."

Parkinson lists eight mallows, of which the eighth is *Malva hortensis rosea simplex et multiplex diversorum colorum*, which, he agrees with Gerard, is found only in gardens. Parkinson says all mallows have a viscous or slimy quality which, used inwardly, helps make the body "soluble" and, used outwardly, will "mollifie hard tumours."

Culpeper treats of "Mallows and Marsh Mallows" and says Venus owns them both. So see under Mallow, as Winthrop ordered Mallow seed "1/2 oz mallow seed at 1 d." One wonders if perhaps his "hollihoks" were double. After seeing Parkinson's botanical name for the garden hollihock, one can be grateful for the new name for it, *Althaea rosea*, from the Greek word meaning to cure.

HONESTY   *Lunaria annua*
MOONWORT   *Lunaria pediviva*   J.J.
WHITE SATTIN FLOURE

Gerard describes "Sattin floure" as having "hard and round stalkes, dividing themselves into many other small branches, beset with leaves like Dames Violets, or Queenes Gillofloures . . . in fashion almost like Sauce alone, or Jacke by the hedge . . . floures like the common stocke Gillofloure, of a purple colour . . . the seed cometh forth contained in a flat thin cod . . . in fashion of the Moone . . . composed of three filmes . . . wherein the innermost is cleere shining, like a shred of white Sattin newly cut from the peece." By a "barbarous name" they are called, "Bolbonac . . . or as it pleaseth most Herbarists, *Viola peregrina* . . ." though "later Herbarists doe call it Lunaria." In English Gerard says it is "Penny-floure or Money floure, Silver Plate, Pricke-songwoort; in Norfolke, Sattin, and White Sattin, and among our women it is called Honestie." It seems to be the "second Treacle mustard" of the "old Herbarists," its seed being hot and dry "like in taste and force to the seed of Treacle Mustard." The roots are eaten "with sallads as certaine other roots are." And "a certaine Chirurgian of the Helvetians composed a most singular unguent for wounds of the leaves of Bolbonac and Sanicle stamped together, adding thereto oile and waxe. The seed is greatly commended against the falling sicknesse." Parkinson lists "*Viola Lunaris sive Bolbanach*, The Sattin flower," and quotes most of Gerard's comments, except that he has never seen it growing wild himself, as Gerard reported he had. Parkinson says "some do eate the young rootes hereof . . . as Rampions are eaten with Vinegar and Oyle; but we know no Physicall use."

Culpeper seems to know of no physical use either. Here may be the exact place for one of those little word-of-mouth clues which are often most revealing. Miss Emma Safford of Ipswich, Massachusetts, a direct descendant of Massasoit, said that she had been told as a child on Cape Cod that the only seeds the Pilgrims brought with them were the seeds of this plant. Can this not well be a very old joke about the almost total improvidence of the arriving Pilgrims?

HOPS *Humulus lupulus* Gookin
"Escaped"

Gerard says there are two kinds of hops, one the manured or garden kind, the other "of the hedge." They have the same properties. Similarly, Asa Gray recognizes our native hop vine, and that which the early settlers brought with them which has escaped and which he calls "introduced from Europe, long cultivated and established in waste places, fence rows, old house-sites, etc." Hops would seem to rank somewhere with the domestic cat as an indispensable adjunct to any household. Used both in raising bread and preserving and flavoring beer, its functions are as humble and cozy as those of the cat, and as ancient. The early voyagers were pleased to see hop vines on the banks of Maine rivers. The early settlers brought their own hop seeds with them. One of the earliest (and almost the only) account of organized Indian industry is Gookin's report of the Indians' help in raising hops and building stone walls.

Gerard says the flowers are gathered in late summer and "reserved to be used in beere." The buds, he says, are used for salads in the spring. The flowers, he says, "make bread light, and the lumpe to be sooner and easier leavened, if the meale be tempered with liquor wherein they have been boyled." A decoction of the juice cleanses the blood, the liver and the kidneys and "driveth forth yellow and choleric humours." In fact, "the manifold vertues of Hops do manifestly argue the wholesomeness of beere above ale; for the hops rather make it a physicall drinke to keep the bodie in health, than an ordinary drinke for the quenching of our thirst."

This last would seem to be a blow to be struck in a long argument which had started when Henry VI doubted the wholesome value of hops as a crop to be used in ale. Evelyn, even after Gerard, had his doubts about "this one ingredient, by some suspected not unworthily," which "transmuted our wholesome ale into beer, which doubtless much alters its constitution . . . preserves the drink, indeed, but repays the pleasure in tormenting diseases and a shorter life."

Hops boiled in the water in which potatoes have been boiled and set aside to ferment, with a little live yeast added, is still used to raise large batches of bread in rural areas today.

Parkinson does not include them in his garden.

Culpeper puts hops under the dominion of Mars and "consequently its operations are obvious" . . . "opening obstructions of the liver and spleen, cleansing the blood, loosening the belly, expelling the gravel . . ." and so on. Obviously a vine one should not be without. Incidentally it is one of those plants of which one must have both sexes.

Evelyn allows hops an entry in his book on salads, as "hot and moist, rather Medicinal than fit for a Sallet, the Buds and young Tendrels excepted." These he says may be eaten raw "but more conveniently being boiled and cold like Asparagus." He adds, however, "They are Diuretic, depurate the Blood, and open Obstructions."

HOREHOUND   *Marrubium vulgare*   Boorde
                                                       (Naturalized)

Pliny named horehound from the Hebrew *marroh*, meaning a bitter juice, which places the herb in time. Gerard's list of its "vertues" is comprehensive and confirms its ancient values to mankind. Curiously, I have found no early references to its cultivation in New England and the term "naturalized" will have to stand for its usefulness to the early settlers.

Gerard says that, boiled in water, common horehound, "openeth the liver and spleen, cleanseth the brest and lungs, and prevailes greatly against an old cough, the paine of the side, spitting of bloud, the ptysicke, and ulceration of the lungs." Boiled in wine, it eases "sore and hard labour in childe-bearing." In syrup it eases those with consumption of the lungs, "as hath beene often proved by the learned Physitions of our London Colledge." With honey the leaves cleanse ulcers. The juice clears eyesight, dropped into the eyes or drawn up through the nostrils. The plant is good against poisons and the bites of serpents.

Culpeper repeats all this, announcing it a "herb of Mercury."

HORSETAIL   *Equisetum hyemale*   J.J.
SCOURING RUSH

Horsetail's botanical name means horse bristle. As an ancient household aid to pot-cleaning it is called, says Gerard, "Pewterwoort" in England since it is "not unknowne to women who scoure their pewter and wooden things of the kitchin therewith."

It appears in Josselyn's drawing of the skunk cabbage, to show the size of the heretofore unknown plant against a very well-known one.

Its "vertues," according to Gerard, are to cure wounds, inward and outward, nosebleed and "fluxes of the belly." Boiled in wine, it is very profitable for ulcers of the kidneys and bladder, cough and "difficulties of breathing."

HOUSELEEK   *Sempervivum tectorum*   Boorde
                                                                  J.J.

Gerard speaks of "Houseleeke or Sengreene" as classified by Dioscorides as three sorts into: great, small, and biting Stonecrop or Wallpepper. The first is called Jupiter's beard, he says, and is known in many countries, growing in walls and upon roofs especially, "continually greene, and alwaies flourisheth." Houseleeks are good against "St. Anthonies fire, the shingles, and other creeping ulcers and inflammations. . . . They take away fire in burnings and scaldings . . . the pains of the gout." The juice in a pessary or held in the mouth quenches burning fevers. The juice takes away corns. The decoction is good against inflammation of the eyes.

Parkinson does not entertain this obviously invaluable countryman's herb in his garden.

Culpeper, however, is very loyal to this ancient herb and is not above giving us one of the best-known superstitions about it — that it keeps off lightning. Jupiter, says Culpeper, claims dominion over it "from which it is fabulously reported that it preserves whatever it grows upon from fire and lightning." It is good, he says, "for all inward and outward heats." Besides all its remedies noted by Gerard, Culpeper adds that the leaves applied to the temples ease headaches. And the leaves being rubbed on the stings of nettles or bees take away the pain and prevent blisters.

Geoffrey Grigson has an interesting note in his *Englishman's Flora*. He says that houseleek's name in New England, hen-and-chickens, is a West of England name that came to the New World with it, and that the plant itself has so long been around dwellings as to have no known wild existence anywhere.

Josselyn reported, "Houseleek prospereth notably."

HYSSOP *Hyssopus officinalis*   J.J.
"1 oz hyssop seed 2d" –J.W.Jr.

Gerard says he will follow the example of Dioscorides, who "gave so many rules for the knowledge of simples" and yet left hyssop undescribed as being too well known to need identification. Gerard thoroughly approves of this and announces he will follow this example "not onely in this plant, but in many others which bee common, to avoid tediousnesse to the Reader." There then results a difference in hyssops between Gerard and Johnson, who interrupts Gerard's listing of "Hyssope with blew floures; Hyssope with reddish floures; white floured Hyssope; Thinne leafed Hyssope" to say that Gerard's next hyssop, the "Dwarfe narrow leaved Hyssope," is really the hyssop judged by most writers to have been the one used by the Arabian physicians. Johnson feels the hyssop used by the Greeks was nearer to "Origanum and Marjerome." The list of "vertues" treats of hyssop mainly compounded with figs, gargled in the throat, drunk with honey and rue added to cure trouble in the lungs, and taken in a juice to expel worms. In fact, Gerard has left hyssop still in its Asia Minor husk.

Parkinson is more definite and decided. He treats only of the "Garden Hyssope" which has "blewish purple gaping flowers one above another in a long spike or eare." And its uses are many. In "Ptisans and other drinkes to help to expectorate phlegme." It is "many Countrey peoples medicine for a cut or greene wound, being bruised with sugar and applyed." It is taken in pills against the falling sickness, in a decoction with oil to kill lice. An oil of the leaves and flowers is good for anointing benumbed joints. And it is accounted a special remedy against an adder's sting if the place be rubbed with hyssop bruised and mixed with honey, salt and cummin seed. In his *Theatrum*, Parkinson lists it with his "Sweet smelling Plants" and quotes Dioscorides as recommending it with rue and honey for coughs.

Culpeper warns that all hedge hyssops are under the dominion of Mars and are unsafe to take inwardly except when "rectified" by an expert, as they are violent purgers, especially of choler and phlegm. Well prepared, they are good for gout, sciatica and dropsy. Externally applied in ointments, they will — merely rubbed upon the belly — destroy worms, and are excellent for old sores.

Culpeper places hyssop pure and simple under Jupiter and the sign Cancer. He quotes Dioscorides as giving it with rue and honey for coughs, and includes a few fig recipes, while he also gives all those noted by Parkinson. To all these he adds one new use, letting the steam from the decoction funnel into the ears to ease inflammations in them and stop singing noises. And he also uses hyssop with sugar for cuts.

John Josselyn has his own special use for hyssop — in treating poultry in the New World. "Pipe or Roupe . . . a common disease amongst their poultry, infecting one another," he feels is due to "a cold moisture of the brain," and he says the best cure for this trouble, with which "they will be very sleepie," is garlic "and smoaking of them with dryed Hysope."

Evelyn has a bit to say about hyssop in his *Acetaria* which raises the tone a bit. Evelyn groups it with thyme and marjoram and "Mary-gold" (by which he means *Calendula* and not the poisonous African marigolds) as being "of Faculty to Comfort, and strengthen, prevalent against Melancoly and Phlegm." He declares that "Plants like these, going under the Names of Pot Herbs, are much more proper for Broths and Decoctions than the tender Sallet; Yet the Tops and Flowers reduc'd to Powder, are by some reserved for Strewings, upon the colder Ingredients; communicating no ungrateful Fragrancy."

IRIS See **FLAG**

I V Y   See   **A L E H O O F,   G R O U N D   I V Y**
Josselyn reported "No Ivy" in New England.

J U N I P E R   *Juniperis communis*   J.J.
Samuel Sewall

Josselyn reports, "Juniper growes for the most part by the seaside, it bears abundance of skie coloured berries fed upon by partridges and hath a woodie root . . . ," which induced him to believe it was the plant mentioned in Job 30.4. Juniper coals, he says, last longer than any others, yet the Indians never burn it. The Englishwomen, he reports, dry their sheets upon the bushes. Samuel Sewall planted a hedge, as recommended by the best authorities, of one eglantine to two of juniper. And we remember the young man searching through his friend's herbal to see if juniper berries were an abortifacient.

K N O T   B E R R Y   O R   C L O U D   B E R R Y   See   **B E R R I E S**

K N O T G R A S S   *Illecebrum verticillatum*   J.J.

Although Gerard puts his "sundry sorts of Knot-Grasses" all under *Polygonum* and says a good deal about their resembling *Herniaria*, it seems best for us to accept the Reverend W. Keble Martin's botanical name for the knot-grass, which does, indeed, seem to have been one of the rupture worts. Culpeper says its uses are legion. Under Saturn, though some say under the Sun, it is good for all kinds of bleedings, bites, fluxes, broken joints and ailments of the bladder, stomach and ears.

L A N G   D E   B E E F E   *Picris echioides*
"1/2 oz Lang de beefe at 1d" – J.W.Jr.

Under "Buglosse," Gerard includes as his second sort, "*Buglossum luteum,*

Lang de beefe." He says that it is a kind of bugloss, "altogether lesser, but the leaves thereof are rougher, like the rough tongue of an oxe or cow, whereof it took its name." He says the stalk is commonly red, and the tops of the branches carry "floures in scaly rough heads . . . composed of many small yellow leaves in manner of those of Dandelion, and flie away in down like as they do." As for its "vertues": "as well Buglosse as Lang-de-Beefe . . . to keep the belly soluble."

Parkinson gives "*Buglossum luteum sive Lingua Bovis.* Langdebeefe" a whole chapter in his Kitchen Garden and begins by saying this is really where he should have considered borage and bugloss also, but he is firm about langdebeefe not belonging to them. They, he feels, should have the ox-tongue name and langdebeefe should be considered as an "*Hieratium* Hawkeweed." (In all justice to Gerard he listed one as name *Buglossum Luteum Hieracio cognatum.*) Anyway, after all these fumblings, everyone agrees that Langdebeefe has but one use — the leaves only rank as a potherb "among others, and it is thought to bee good to soften the belly."

Culpeper would seem to ignore it altogether, even combined with borage and bugloss.

LAVENDER *Lavandula vera*
Lavandula spica J.J.

Gerard's chapter is "Of Lavender Spike." The distilled water of lavender is first considered as among the "vertues," whether it be "smelt unto, or the temples and forehead bathed therewith." In cases of a "light Migram" or for those that "use to swoune much," it is "refreshing." But combined with other "herbes, floures or seeds and certaine spices" and given in cases where there is "abundance of humours especially mixed with bloud, it is not then to be used safely," especially when purging or bloodletting has not preceded the treatment. "Some unlearned physitions . . . overbold Apothecaries, and other foolish women" err in this respect, giving such concoctions to those with Apoplexy. The flowers, "I meane the blew part," made into a powder and drunk in the distilled water help "the panting and passion of the heart," as does the conserve of the flowers made with sugar. The oil of lavender is good for those with the palsy.

Parkinson treats of "Lavender Spike" as "Little used in inward physicke, but outwardly; the oil for cold and benumbed parts . . . to perfume linen, apparrell, gloves, leather, & the dryed flowers to comfort and dry up the moisture of a cold braine."

Culpeper says Mercury owns this herb "and it carries its effects very patiently." It is of especial use in "paine of the head and brain that proceed of a cold cause," for falling sickness or giddiness of the brain, as a gargle for loss of voice, for faintings and swoonings. The oil, called "oil of spike," must be cautiously used.

Josselyn lamented that "lavender is not for the climate."

LAVENDER COTTON *Santolina chamaecyparissus* J.J.

Gerard says, "Lavender Cotton groweth in gardens almost everywhere" in England, and Josselyn reported it as flourishing in gardens in New England. Its "vertues" according to Gerard are so short and definite that we may quote them in full: "Pliny saith that the herb *Chamaecyoarissus* being drunke in wine is a good medicine against the poysons of all serpents and venomous beasts. It killeth wormes either given greene or dry, and the seed hath the same vertue against wormes, but avoideth them with greater force. It is thought to be equall with the usuall worme-seed."

Parkinson approves of lavender cotton especially "to border knots with" as it does well being cut. He places it "among other hot herbes, eyther into baths, ointments, or other things that are used for cold causes. The seede is also much used for the wormes."

Culpeper says it is under Mercury, resists poison and putrefaction and helps cure the bitings of venomous beasts. It is good against worms and, in a decoction, cures scabs and itch.

LEEKS *Allium porrum*

"1 oz leekes seeds at 3d" –J.W.Jr.

Gerard calls them *Porrum Capitatum*. He says that Nero was fond of "this root," and therefore he was called in scorn *Porrophagus*. He says, "The leeke is hot and dry, and doth attenuate and make thin as doth the onion." "Being boyled" he says, "it is less hurtfull." (Gerard does not think much of leeks. It is the one of the very few cases where he appends "The Hurts" to "The vertues.") The vertues are not many: "it concocteth and bringeth up raw humours that lie in the chest. Some affirm it to be good in a loch or licking medicine, to clense the pipes of the lungs." The juice, taken with honey, is "profitable" against the bite of venomous beasts, and, with oil of roses, dropped into the ears "mitigateth their pain." The seed with an equal weight of myrtle berries will keep wine from souring, or, if it is already sour "amend the same, as divers write." Gerard now gives "The hurts" in full: "It heateth the body, ingendreth naughty blood, causeth troublesome and terrible dreames, offendeth the eyes, dulleth the sight, hurteth those that are by nature hot and cholericke, and is noysome to the stomacke, and breedeth windinesse."

Incidentally, Gerard speaks of "Cives or Chives, and wilde Leekes" and says chives "are like in Facultie unto the Leeke, hot and dry" and "worke all the effects that a leek doth."

Parkinson is kinder to leeks. Under "The Use of Leekes" which he calls *Porrum*, he begins: "The old World, as wee finde in Scripture, in the time of the children of Israels being in Egypt, and no doubt long before, fed much upon Leekes, Onions and Garlicke boyled with flesh; and the antiquity of the Gentiles relate the same manner of feeding on them, to be in all Countries the like, which howsoever our dainty age now refuseth wholly, in all sorts except the poorest. . . . They are used with us also sometimes in Lent to make pottage, and is a great and generall feeding in Wales with the vulgar Gentlemen." Leeks, Parkinson says, "are held to feed the chest and lungs from much corruption . . . that sticketh fast therein . . . as also for them that through hoarseness have lost their voice, if they be eyther taken rawe or boyled with broth of barley. . . . And baked under hot embers is a remedy against a surfeit of Mushromes. . . . The greene blades of Leekes being boyled and applied warme to the *Hemorrhoides* give a great deal of ease."

Culpeper dismisses leeks fairly briefly, after onions, saying that they participate in the same quality as onions, though not in so great a degree. He mentions the antidote to a surfeit of mushrooms — "baked under the embers and then taken when sufficiently cool to be eaten," and again, "boyled, and applied warm they help the piles."

LEEKS, WILD   *Allium canadense*   J.J.

Josselyn mentioned "Wild-Leekes, which the Indians eat much with their fish."

"Homer's Molley" which Josselyn also lists is probably *Allium tricoccum*, according to Tuckerman.

LETTUCE   *Lactuca sativa*

"3 oz lettice seeds at 2d per oz" – J.W.Jr.

Gerard is very firm about lettuce, saying that the ancients considered there were but two sorts, the wild and the tame; but "time, with the industrie of later writers have found others both wilde and tame, as also artificiall." These he proposes to "lay down." His first observation after describing the "naturall" or Garden Lettuce, is to say that "by manuring, transplanting, and having a regard to the Moone and other circumstances, the leaves of the artificiall Lettuce are oftentimes transformed into another shape; for either they are curled, or else so drawne together, as they seeme to be like a Cabbage or headed Colewort, and the leaves which be within and in the middest are something white, tending to a very light yellow." Gerard shows four illustrations: Garden Lettuce, Curled Lettuce, Cabbage Lettuce, and Lumbard Lettuce.

Lettuce, says Gerard, is a "cold and moist pot-herbe." It cools the heat of the stomach, called "heart-burning," quenches thirst, causes sleep, makes plenty of milk in nurses who "through heate and drinesses grow barren and drie of milke," and makes a pleasant salad, served raw with vinegar, oil and a little salt. Boiled, it is sooner digested and is more nourishing. Gerard says that in his day and country it is served at the "beginning of supper, and eaten first before any other meate," as Martial says was done in his time, as translated for us:

> Tell me why Lettuce, which our Grandsires last did eate,
> Is now of late become, to be the first of meate?

However, says Gerard, lettuce may now be eaten at both times "to the health of the body; for being taken before meat it doth many times stir up appetite; and eaten after supper it keepeth away drunkennesse which commeth by the wine; and that is by reason that it stayeth the vapours from rising up into the head." And finally, for Gerard, "The juice of Lettuce cooleth and quencheth the naturall seed if it be too much used, but procureth sleepe."

Parkinson says there are so many sorts of lettuce that he doubts he will "scarce be beleeved of a great many." For he intends to reckon up eleven or twelve sorts: some of little use, some of more; some excellent, some rare, some for each season. The "Cabbage kindes" are well known and the "Romane or red Lettice is the best and greatest of all the rest." It was brought into England by John Tradescant. There is a white Romane which must be "whited" and a reddish-leaved Virginia lettuce which is "not of any regard," and loose Lumbard lettuce, and several curled kinds, and, finally, "our winter Lettice." Parkinson says that all sorts are "spent in sallets, with oyle and vinegar, or as every one please . . . while they are fresh and greene, or whited. . . . They are also boyled. . . . They all cool a hot and fainting stomacke." The juice applied with oil of roses to the foreheads of those sick and weak and sleepless, "procureth rest, and taketh away the paines of the head: bound likewise to the cods, it helpeth. . . . If a little camphire be added, it restraineth immoderate lust; but it is hurtfull to such as are troubled with the shortnesse of breath."

Culpeper says the Moon owns lettuce, and lists all the above uses, adding to the remedy for lust that, outwardly applied over the heart or other organs of the body, it will comfort and strengthen them. He says Galen advised old men to use lettuce with spices or with "mint, rocket and such-like hot herbs . . . to abate the cold of one and the heat of the other."

Evelyn could be the sole advocate of lettuce and it would never need another. He says that lettuce, in spite of its "*Soperiferous* quality, ever was and still continues the principal Foundation of the universal Tribe of Sallets, which is to Cool and Refresh, besides its other Properties." He lists all the cures and uses stated by Gerard and Parkinson and Culpeper, as above, and puts one use more prettily by referring to "the effect it has upon the Morals, Temperance and Chastity." Lettuce was, says Evelyn, Galen's "beloved Sallet" and was believed by him to breed "the most laudable Blood." Augustus attributed his recovery from a dangerous sickness to lettuces, and he is reputed to have erected a "Statue and built an Altar to this noble Plant."

Tacitus was extremely fond of lettuce, although he spent "almost nothing at his frugal Table in other Dainties." And Galen, in celebrating lettuce, said he began a meal with lettuce in his younger days and then, as he grew old, concluded the meal with it.

Evelyn says lettuce exceeds all other materials for salads in that it is wholesome both alone and in combination, with vinegar, oil, pepper, salt and sometimes with a little orange or lemon juice and sugar. With "unblanched Endive, Succory, Purselan," it was served up at the "best Tables," with "indeed little other variety," "Sugar and Vinegar being the constant Vehicles." Now, however, says Evelyn, sugar is almost wholly banished from salads except for the "more effeminate Palates, as too much palling, and taking from the grateful Acid now in use."

LICORICE   *Glycyrrhiza glabra*   C.M.
WILD LICORICE   *Galium circaezans*   J.J.

Cotton Mather's "Liquorice" may have been the imported root, or a plant grown here. When Josselyn brewed his splendid beer for the ailing Indians and included elecampane, aniseed, sassafras, fennel, catmint, sowthistle and licorice, it is likely that he had licorice growing to hand since he scraped the root, bruised it and cut it into pieces, and that it was *Glycyrrhiza glabra*, which Gerard had in his garden and Parkinson in his.

Gerard gives licorice its ancient due as a remedy for afflictions of the throat and lungs, and mentions the tradition that a bit of licorice held in the mouth enables a man to endure thirst for many days, which is an indication of its desert uses long before it was brought to England through Europe from its Scythian origins. Culpeper notices it as a commercially profitable crop in the England of his time. "Pontefract root" is a name for it still heard in England today. Parkinson grows it and recognizes its value in asthmatic afflictions, the juice made up with hyssop water, and into a "Lohoc or licking medicine" with "Gum Tragacanth in Rose Water." All three authorities agree that it is good for the kidneys and bladder. Culpeper says it is under the dominion of Mercury.

LILIES   *Convallaria majalis*
LILLY CONVALLIE
LILY OF THE VALLEY   J.J.
(Introduced)   (Naturalized)

As Gerard seems to be the entire source for both Parkinson's and Culpeper's reports upon the "May lily" or "Convall Lily or Lilly of the Valley," we give Gerard's account of its "vertues" in full.

"The floures of the Valley Lillie distilled with wine and drunke the quantitie of a spoonful restoreth speech unto those that have the dum palsie and that are falne into the Apoplexie, and is good against the gout, and comforteth the heart. The water aforesad doth strengthen the memorie that is weakened and diminished, it helpeth also the inflammation of the eies, being dropped thereinto. The floures of May Lillies put into a glasse and set in a hill of antes close stopped for the space of a moneth and then taken

1 *Lilium convallium.*
Conuall Lillies.

2 *Lilium convallium floribus suaue-rubentibus.*
Red Conuall Lillies.

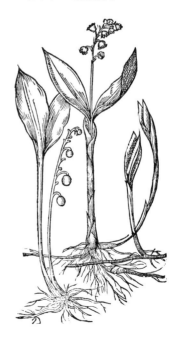

out, therein you shall find a liquor that appeaseth the paine and griefe of the gout, being outwardly applied, which is commended to be most excellent."

Josselyn saw these growing and considered them as among the plants common to both countries.

FAIR WHITE LILY *Lilium candidum*   Bradford

Bradford's "fair white lily" in the Plymouth gardens is the ancient "white lily" of Gerard which is full of "vertues." The root stamped with honey glues together cut sinews, and takes away all scurviness of the beard and face. Stamped with vinegar, barley meal and the leaves of henbane, it cures tumors. With oil it restores hair to burned places. Altogether the lily, flower, leaves and root, combined with wine, is good against old wounds, ulcers, and sores "coming of venery and such like." The distilled water brings easy delivery in childbirth.

Parkinson is brief but positive in his listing of the uses of the white lily. Above all other sorts, he says, it has "a mollifying, digesting and cleansing quality helping to suppurate tumours. . . . The water of the flowers distilled is of excellent vertue for women in travail . . . also . . . to cleanse the skin . . ."

Culpeper adds to a review of the above only by saying that white lilies are under the dominion of the Moon and by antipathy to Mars expel poisons and are useful in pestilential fevers.

Bradford welcomed the "fair white lily" and the "fragrant rose" in Plymouth gardens in the early thirties of the seventeenth century.

MOUNTAIN LILIES *Lilium canadense*   J.J.
                   Lilium superbum

Josselyn noted, "Mountain-lillies, bearing many yellow Flowers, turning up their leaves like the Martagon, or Turks Cap, spotted with small spots as deep as Saffron, they Flower in July." This sounds like *canadense* although Tuckerman settled for *superbum*, and neither grows in "Mountains." However, this may be responsible for the great mountain gorges behind Dr. Thornton's *Lilium superbum* portrait nearly two hundred years later.

RED LILIES *Lilium philadelphicum*   J.J.
                 Lilium canadense

John Josselyn noticed red lilies growing in New England, one very common and growing everywhere in the bushes, which would seem to have been our native *Lilium philadelphicum.*

WATER LILIES *Nuphar microphylium*   J.J.
                   Nymphaea odorata
                   Nuphar variegatum

John Josselyn remarked upon water lilies as Indian food, and says that "the black roots dryed and pulverized are wondrous effectual in the stopping of all manner of fluxes of the belly drunk with wine or water." But he does not say if he means the yellow pond-lily or the fragrant water-lily, both native plants.

Elsewhere he specifies "Water Lilly, with yellow Flowers," and says, "the Indians eat the Roots, which are long a boiling, they tast like the Liver of a Sheep. . . ." "The moose feed upon them and are killed by the Indians when their heads are under water."

LIVE FOR EVER *Antennaria canadensis*
LADIES-TOBACCO
PUSSY'S TOES
IMMORTELLE   J.J.

Josselyn speaks of "Live for ever . . . a kind of Cud-weed, flourisheth all summer long till cold weather comes in, it growes now plentifully in our English Gardens. It is good for cough of the lungs, and to cleanse the breast taken as you do Tobacco and for pain in the head. The decoction or the juice strained and drunk in Beer, Wine or Aqua vitae killeth worms. The Fishermen when they want Tobacco take this herb being cut and dryed."

Tuckerman's Linnaean name *Antennaria margaritaceae* meaning "pearl-like" is now called "Pearly Everlasting," *Anaphalis margaritacea*, an introduced plant. It seems more likely that the fishermen's substitute for tobacco is the plant with a familiar name still retaining an association with smoking.

Here may be the place to remind ourselves that Gerard has several cud-weeds under the name *Gnaphalium*, though his *Gnaphalium americanum*, he says, came from the Mediterranean regions. The name puzzled Johnson until he surmised it might be from Brazil. Parkinson knew it was from the West Indies. In any case, these cudweeds kept away moths, cleansed hair of nits and lice, were good against worms and venomous bites, and when smoked, cured coughs and headaches. As seen under CUDWEED.

L I V E R W O R T   *Hepatica americana*
N O B L E   L I V E R W O R T   J.J.

Gerard speaks of Noble Liver-wort or *Hepatica trifolia, Herba trinitatis*, and "in English, Golden Trefoile," and says it is reported to be "good against the weaknesse of the liver which proceedeth of an hot cause for it cooleth and strengtheneth not a little."

Parkinson offers a variety of "Hepatica or Noble Liverwort" but says, "These are thought to cool and strengthen the liver, the name importing as such; but I never saw any great use of them by any of the Physitians of our London Colledge, or effect by them that have used them in Physicke in our Country."

Culpeper, however, appears to find it very useful indeed. Under Jupiter and Cancer, it is good for all diseases of the liver both to cool and cleanse it. Bruised and drunk in small beer, it cools the liver and kidneys, stays the spreading of sores and scabs and is a remedy for those whose livers are cor-rupted by surfeits. In short, it "fortifies the liver exceedingly and makes it impregnable."

Josselyn noted, "Noble Liver-wort, one sort with white flowers the other with blew."

But it is to Edward Taylor, the clergyman-physician-poet, we owe the mention of *Herba trinitatis* in his poem, part of which is quoted at the end of this book, where he specifies a whole list of garden flowers including "Herba Trinitatis in my breast."

1 *Hepaticum trifolium*.
Noble Liuerwort.

LOOSE-STRIFE   *Epilobium angustifolium* (Rose bay Willow-herb)
Lysimachia nummularia (Moneywort)
Lysimachia punctata (Introduced)
Lysimachia terrestris (Swamp Candles)
Lysimachia vulgaris (Introduced)
Lythrum salicaria (Spiked Loosestrife)
Oenothera biennia (Tree Primrose)   J.J.

Josselyn noted "several kinds of Lysimachus or loose-strife" and the above assortment would appear sufficient to cover what he saw, since upon opening his edition of Johnson-upon-Gerard we find the chapter, "Of Willow-herbe, or Loose-strife," setting forth thirteen assorted Lysimachias and including the "Rose bay Willow herbe" and even a splendid "Tree Primrose" inserted by Johnson as *Lysimachia lutea virginiana* which we know was noted by Parkinson. The latter confounds confusion by saying that he "stood long in suspense" as to the "tribe" to which he should refer this plant, but as he made no mention of "any other Lysimachia in this work" he placed it with *Hesperis* or Dames Violets.

However, Josselyn is specific about one or two. Of "Lysimachus or Loose-strife" he says in *Two Voyages*, "there are several kinds, but the most noted is the yellow Lysimachus of Virginia." He then quotes a description straight from Johnson adding to Gerard, mentioning on his own only "this . . . is taken by the English for Scabious." In *New-Englands Rarities* he notes "Tree Primrose, taken by the Ignorant for Scabious" and then: "Lysimachus or Loose Strife, it grows in dry grounds in the open Sun four foot high, Flowers from the middle of the Plant to the top, the Flowers purple, standing upon a small sheath or cod, which when it is ripe breaks and puts forth a white silken doun, the stalk is red and as big as ones Finger."

Gerard explains that the name Lysimachia came from its special virtue in appeasing the strife and unruliness of oxen plowing if it is put about their yokes, a discovery made by King Lysimachus. In English it is called Willow-herb and Loose-strife. The yellow Lysimachia is the best for physic, being cold, dry and very astringent. The juice is good against the bloody flix, for green wounds and nose-bleed. The smoke of the burned herb drives away serpents and venomous beasts, and kills gnats and flies in a house.

We have seen Parkinson offering the Tree Primrose as his Lysimachia entry.

Culpeper deals with two kinds of "loose-strife or Willow Herb." The first is the "common yellow loose strife, good for all manner of bleeding," whose smoke "driveth away flies and gnats which are used in the night-time to infest the habitations of people dwelling near marshes and in the fenny countries." The second is the "Loose strife with spiked heads of flowers branching out like the spikes of lavender, of a purple violet color," and growing in wet grounds. This is "an herb of the moon," better than eye-bright for sore eyes, an aid to the blind, excellent for wounds and ulcers, as a gargle and to cleanse the skin. This is Culpeper on his own, as Gerard refers only to the yellow loosestrife medicinally.

LOVAGE *Levisticum officinale*

"1 oz lovadj seeds at 2d" – J.W.Jr.

Common lovage, says Gerard, is *Levisticum vulgare* and is sometimes called *Ligusticum*. It is hot and dry in the third degree. It has large and broad leaves, "like to smallage" and hollow, knotty stalks, three cubits high, having "spoky tufts, or bushing rundles."

We can quote almost all: "The roots . . . are good for all inward diseases. The seed warmeth the stomachs, helpeth digestion, wherefore the people of Gennes in times past did use it in their meates, as we doe pepper. . . ." "The distilled water . . . cleareth the sight, and putteth away all spots, . . . freckles . . . and redness of the face, if they be often washed therewith."

Parkinson seems to omit it from his garden.

Culpeper is enthusiastic and loyal. He says it has "umbels of yellow flowers," is a herb of the sun under the sign Taurus, and if Saturn offend the throat, "this is your cure." Culpeper also recommends it . . . "it openeth, cutteth, and digesteth humours." It is good in gargles and will help in fevers as well as anything "occasioned by cold." He agrees that it helps blotches and boils and spots or freckles.

LUNGWORT *Pulmonaria officinalis: Mertensia virginica*

"Spotted Cowslips of Jerusalem" is Gerard's name for the lungwort, which does indeed resemble our native *Mertensia virginica* also called a cow-slip. Gerard explains that the flowers look like cowslip flowers.

Parkinson has rather diffidently placed "Lungwort, or Cowslips of Jeru-salem" after the true cowslips, although these are "generally more used as Pot-herbs for the Kitchin, then as flowers for delight." Still, they are "of like form." Sometimes called "Spotted Comfrey" and "as fitly called spotted Buglosse," Parkinson says it is much commended as good for ulcered lungs and for those who spit blood. "It is of great use in the pot, being generally held to be good both for the lungs and the heart." Culpeper speaks of "Lung-flower," "generally called autumn gentians." As see J.J. and his "Autumn bellflower."

MADDER *Rubia tinctorum*   "roots sent"
MADDER, FIELD *Sherardia arvensis*   (Adventive)

Among the first plants "to be sent to New England" was madder. Gerard says, "There is but one kinde of Madder onely which is manured or set for use, but if all those that are like unto it in leaves and manner of growing were referred thereto, there should be many sorts as Goose-grasse, soft Cliver, our Ladies Bedstraw, Woodroofe, and Crosse-woort, all which are like to Madder in leaves, and therefore they be thought to be wilde kinds thereof." The garden or manured madder, he says, has trailing branches full of joints and at every joint there is a star of green rough leaves, set round about. The flowers grow at the top of the branches and are of a faint yellow color.

Gerard had a quarrel with a "great Physition (I do not say the great learned)" over the temperature of madder, whether it binds or opens or does both. Upon the "vertues" there is equal dispute, even with Galen and Dioscorides, but Gerard thinks madder is astringent and binding, so it is good against inward and outward bleeding. On the other hand, the root boiled opens the stopping of the liver and kidneys and is good for jaundice. It colors urine red, which deceived Dioscorides into thinking it was blood.

According to Pliny, the stalk and leaves can be used against serpents. Parkinson does not grow it in his garden.

Culpeper resolves all differences by saying that, as a herb of Mars it has an opening quality but afterward binds and strengthens. He adds to Gerard's "vertues" that it colors urine instantly when merely held in the hand.

They do not mention it as a domestic dye, which is very likely the chief reason for its "sending over." Governor Winthrop's suggestion, in 1631, that his son import a quantity of sheepskins (two or three hundred) "dyed redd" as a saleable commodity, together with the "coursest woollen clothe" "some redd" would point up the need for madder as a dye in New England. The so-called "field madder" was also used as a dye, though inferior to the "garden or manured madder."

MAIDENHAIR See FERN

MALE FUELLIN See VERONICA

M A L L O W S   *Althaea officinalis* (Marshmallow)   C.M.
              Malva crispa (French mallows)   J.J.
              Malva moschata (Musk mallow)

Considering that John Winthrop, Junior, ordered both hollyhocks and mallow seeds, it is not likely that his "mallow" was marshmallow, or *Althaea officinalis*, since hollyhocks, *Althaea rose*, were considered to have the same properties. Nor would the musk mallow, *Malva moschata*, be likely as his choice, although Josselyn saw it growing in gardens in New England. It seems more probable that Winthrop ordered *Malva crispa*, or French mallows, which Josselyn also saw growing in gardens.

However, let us begin with Gerard, who puts "The Garden Mallow called Hollihock" first, as being not wild, though he says it "serves for all those things, for which the wild Mallowes do, which are more commonly and familiarly used." He then takes up an assortment of different mallows, including Marshmallow and a French curled mallow. Of them all, the virtues of the marshmallow are foremost and prodigious. Parkinson said, "All sorts of Mallowes by reason of their viscous or slimie quality . . . make the body soluble . . . outwardly applied they mollifie hard tumours . . . ease paines . . . those that are the most use are the most common." But Culpeper, in his chapter, "Mallows and Marshmallows," gives a very long list of virtues and says the marshmallows are the most effective against all the diseases mentioned: hot agues, pleurisy, excoriations of the guts, falling-sickness, constipation, bee-stings, hardness of the liver and spleen, swelling of the testicles, roughness of the skin, "dandriff," loss of hair, head-colds and splinters. The boiled leaves produce milk in nurses and a speedy delivery. In fact, says Culpeper, "A sovereign remedy."

Gerard mentioned the "crispe or curled Mallow, called of the vulgar sort French Mallowes" as "an excellent pot-herbe, for the which cause it is sowne in gardens, and is not to be found wilde that I know of."

Parkinson has "*Malva crispa* or French mallow" in his Kitchen Garden, "much used as a pot-herbe, especially when there is cause to move the belly downward . . . boyled or stewed, eyther by it selfe with butter, or with other herbes . . ."

Evelyn considers "Mallow, *Malva*, the curled, emollient, and friendly to the Ventricle, and so rather Medicinal; yet may the Tops well boiled, be ad-

I *Althæa Ibiscus.*
Marſh Mallow.

mitted." Evelyn thinks both the garden and the wild mallow "proper rather for the Pot than Sallet. . . ."

With all the "introduced" and "naturalized" mallows flourishing about our countryside today, to know what Winthrop intended and brought is difficult. I leave you the choice between the marshmallow and the French mallow.

We have no less an authority than Cotton Mather for growing a good plot of marshmallows, Culpeper's "sovereign remedy."

MARIGOLDS   *Calendula officinalis*   How
                                                                                                J.J.
                                                      "1/2 oz marigold at 2d" –J.W.Jr.

Gerard says it is *Calendula* because, "It is to be seene in floure in the Calends almost of everie moneth." He describes "the greatest double

Marigold" as having "broad leaves springing immediately . . . the upper
sides . . . of a deep greene, and the lower side of a more light and shining
greene . . . stalkes somewhat hairie . . . floures . . . beautiful, round, very
large and double, something sweet, with a certain strong smell, of a light
saffron colour, or like pure gold. . . ." He says the flower is hot almost in
the second degree, especially when it is dry. In this regard, he says, "The
yellow leaves of the floures are dried and kept throughout Dutchland
against Winter, to put into broths, in physicall potions, and for divers other
purposes, in such quantity that in some Grocers or Spice-sellers houses are
to be found barrels filled with them . . . insomuch that no broths are well
made without dried Marigolds." Gerard says that conserve made of the
flowers and sugar and taken "in the morning fasting cureth the trembling
of the heart, and is also given in time of plague or pestilence, or corruption
of the aire." How calls them "merrigoulds." Josselyn saw them thriving
in gardens.

1. 2. *Calendula major polyanthos.*   4 *Calendula multiflora orbiculata.*
    The great double Marigold.   Double globe Marigold.

Parkinson says, in toto, "The herbe and flowers are of great use with us among other pot-herbes. The flowers either green or dried are often used in possets, broths, and drinkes, as a comforter of the heart and spirits, and to expel any malignant or pestilential quality, gathered neere thereunto. The Syrup and Conserve made of the fresh flowers are used for the same purposes to good effect." He agrees that they seem to bloom most freely at the beginnings of the months.

Culpeper does not go into this, but says it is "an herb of the Sun, and under Leo." They "strengthen the heart" and are "little less effectual in the small-pox and measles, than saffron." He says the juice with vinegar "instantly giveth ease" to a hot swelling. And, besides the uses mentioned above, he mentions that a "plaister" of the dry powder, hogs-grease, turpentine and resin outwardly applied strengthens the heart in fevers.

MARJORAM   *Origanum onites*
                                    " 1 oz pott majoran at 4d" – J.W.Jr.
MARJORAM, SWEET   *Origanum majorana*
                                    " 1 oz sweet majoran at 8d" – J.W.Jr.
MARJORAM, WILD   *Origanum vulgare*   How
                                                              J.J.
                                                         (Naturalized)

Gerard describes "Pot Marierome or Winter Marierome" as having ". . . divers small branches whereon are placed such leaves as . . . 'sweet marjoram' . . .but not so hoarie, nor yet so sweet of smell, bearing at the top of the branches tufts of white floures tending to purple. The whole plant is of long continuance and keepeth greene all the Winter whereupon our English women have called it, and that very properly, Winter Marierome."

Sweet marjoram, Gerard says, is "a low and shrubby plant, of a whitish colour and marvelous sweet smell, a foot or somewhat more high. The stalkes are slender . . . parted into divers branches, about which grow forth little leaves soft and hoarie; the floures grow at the top in scaly or chaffie spiked eares, of a white colour. . . . The whole plant . . . is of a most pleasant taste and aromaticall smell, and it perisheth at the first approach of Winter."

Wild marjoram, says Gerard, or "Bastard Marjerome groweth straight up with little round stalkes of a reddish colour full of branches . . . leaves be broad . . . whitish greene . . . on the top of the branches stand long spikie sealed eares, out of which shoot forth little white floures like the flouring of wheate. The whole plant is of a sweet smell and sharp biting taste."

In temperature Gerard says all the "Organies . . . do cut, attenuate or make thin, and heate . . . in the third degree." Galen said wild marjoram is the more "forceable." As for "vertues" they are many: "Organy given in wine is a remedy against the bitings and stingings of venomous beasts . . . cureth them that have drunk Opium, or the juice of the blacke poppy, or hemlockes. . . . The decoction of Organy provoketh urine, bringeth downe the monthly course, and is given with good success to those that have the dropsie. . . . In a looch, or medicine to be licked, against an old cough and stuffing of the lungs . . . it healeth scabs . . . the juice mixed with a little milke, being poured into the eares, mitigateth the paines thereof . . . the same mixed with . . . the rootes of the white Florentine Floure de Luce, drawne up into the nostrils . . . draweth down flegme. . . . The herbe strowed upon the ground driveth away serpents. . . . The decoction looseth the belly . . . drunke with vineger helpeth . . . the spleen, drunke in wine helpeth against all mortall poisons. . . . These plants are easie to be taken in potions, therefore . . . they may be used . . . unto such as cannot brooke their meate . . . and have a sowre and sqamish and watery stomack . . . as also against the swouning of the heart."

Gerard adds another marjoram chapter: "Goats Maierome" or "Tragoriganum" which is especially good against the "wamblings of the stomacke . . . the soure belchings of the same, and stayeth the desire to vomit, especially at sea."

Sweet marjoram, Parkinson says, is sometimes called "knotted Marierome" because the flower heads are "like unto knots." The sweet marjorams are "not onely much used to please the outward senses in nosegays, and in the windows of houses, as also in sweete powders, sweete bags, and sweet washing waters, but are also of much use in Physicke, both to comfort outward members or parts of the body and the inward also. . . ."

Culpeper announces Wild Marjoram, called also Origanum or "bastard marjoram," is under the dominion of Mercury and that "there is a deadly antipathy between this herb and the adder." He says, "there is scarce a better remedy growing for such as are troubled with a sour humour in their stomach." It restores appetite, helps the cough, remedies infirmities of the spleen, and — in fact — does all that Gerard has mentioned. Sweet marjoram, Culpeper says, is an herb of Mercury under Ariel, and is therefore an excellent remedy for the brain . . . and other parts of the body and mind under the dominion of the same planet. It is "comfortable in cold diseases" and "helpeth the loss of speech by resolution of the tongue." Its oil makes joints supple. And it is used in odoriferous waters.

John Josselyn used "a little sweet marjoram" in his eel recipe.

MASTERWORT *Astrantia major*   J.J.
COW PARSNIP *Heracleum maximum*

The Masterwort of Gerard he calls *Imperatoria* or *Astrantia* and its "vertues" are long and many : the roots and leaves stamped will cure carbuncles : the root drunk in wine cures the cold fit of agues and provokes sweat; strengthens the stomache, cures mad dog bites, restores the appetite, helps those fallen from high places, cures hemorrhoids. It is hot and dry in the third degree.

In the New World the cow parsnip or *Heracleum maximum*, also an impressive plant in its own right, inherited the name of Masterwort, how deservedly we do not know.

MAUDELEIN *Herba Giulia* (of the Italians)
Balsamita foemina (Gerard)
Ageratum (Johnson)
"1 oz maudlin seed at 2d" – J.W.Jr.
COSTMARY ⎫
BIBLE-LEAF ⎬ *Chrysanthemum balsamita*
ALE-COST ⎭
MAUDLIN *Chrysanthemum leucanthemum*
(in Wiltshire according to Grigson)

Gerard and Parkinson treat of "Costmary and Maudelein" together — or, as Parkinson spells it "Maudeline." Gerard says, spelling it still another way, "Maudleine is somewhat like to Costmary, whereof it is a kinde, in colour, smell, taste, and in the golden floures." Smaller than Costmary . . . "but of a better smell," it looks like Tansie, according to Parkinson. Costmary grows to three feet, Maudlin to two. According to Gerard, "they grow everywhere in gardens and are cherished for their sweet floures and leaves." Costmary is called Alecoast, Parkinson thinks, because it is of special use in the Spring to make Sage ale, used with other "such like" herbs. Maudlin especially, according to Gerard, is effectual in troubles with the bladder. The conserve made with Costmary leaves and sugar "doth warme and drie the braine" and stops catarrhs. The seeds expel worms. And "the leaves of Maudleine and Adders tongues stamped and boiled in Oile Olive, adding thereto a little wax, rosin, and a little turpentine, maketh an excellent healing unguent or incarnative salve to raise or bring up flesh from a deepe and hollow wound or ulcer, whereof I have had long experience," says Gerard.

Parkinson mentions most of this, but has something especially to say of "Maudeline." It is "much used with Costmary and other sweet herbes to make sweet washing water, whereof there is great store spent." "The flowers also are tyed up with small bundles of lavender tops, these being put in the middle of them, to lye upon the toppes of beds, presses etc. for the sweet sent and savour it casteth."

Culpeper says Maudlin is so like Costmary it cannot be treated except as Costmary and then says Costmary is called also alecost or balsam herb and is in "almost every garden." It is under the dominion of Jupiter and "provoketh urine abundantly . . . gently purgeth choler and phlegm." In fact, it does good to the liver, stomach, head, and expels worms. He also mentions the salve.

And finally, maudlin is the common name for our field daisy in Wiltshire. Who is to say what the seeds were that John Winthrop, Jr., ordered as "maudlin"? See DAISY.

MAYWEED   *Anthemis cotula*   J.J.

John Josselyn lists this under "such plants as have sprung up since the English planted and kept cattle in New England." He noted: "Mayweed, excellent for the Mother; some of our English Housewives call it Iron Wort and make a good unguent for old sores."

Gerard calls it "Mayweed, or Wilde Cammomill" and says it is not used for "meate nor medicine" as it is "not at all agreeing with mans nature notwithstanding it is commended against the infirmities of the mother, seeing all stinking things are good against those diseases." It grows as a weed among corn and blisters the hands of weeders.

Parkinson does not admit Mayweed to his Garden of Pleasant flowers.

Culpeper says of Mayweed that there are three sorts, one with flowers nearly as large as those of the double camomile. Culpeper says it is used for the same purposes as camomile and then adds, "it is also good for women whose matrix is fallen, by washing their feet with a decoction thereof. It is likewise good to be given to smell to by such as are troubled with the rising or suffocation of the matrix."

MELON, MUSK   *Cucumis melo*
CUCUMBERS   *Cucumis sativus*   J.J.

Josselyn claimed that "Musk Melons are better than our English and Cucumbers," all in the same breath, which was obviously the way cucumbers and melons were considered by many; so we treat of them together here.

Gerard says the meat of the "Muske Melon" is "harder of digestion than is any of the rest of Cucumbers." Parkinson says that melons, though formerly eaten only by great personages, since the fruit was "not only delicate but rare," are now made more common by "divers others that have skill and convenience of ground for them." Both Gerard and Parkinson say the seed of melons and cucumbers operate the same physically, which is to say, cooling for "parts troubled with heat."

Evelyn says that "Melon, *Melo*," has been reckoned among fruits rather than salads, but he likes the pulp mingled with the salad. Or it can be eaten separately with salt and pepper. A melon which requires sugar "wants of Perfection." In hot climates they drink water after eating melons, but in colder climates wine is better. He does not hold with the idea, put forward by Gerard as well known, that the pulp is apt to "corrupt in the stomach." Parkinson insists that a melon must be eaten with a "good store of wine, or else it will hardly digest."

MINT, SPEAR- *Mentha spicata* J.J.

While Gerard has a large assortment of mints, from "Red Garden Mints" through "Crosse or curled," "Speare," "Heart," "Balsam," "Cat-," "Horse" to "Mountain Mint or Calamint," we are confining ourselves here to Spearmint as that is, according to Culpeper, the most useful of all the mints, and what is said about it will suffice for the rest. (See CATMINT and CALAMINT also.)

Gerard deals broadly with Garden Mint, a "marvelous wholesome for the stomacke . . ." "It staieth the Hicket . . . applied to the forehead . . . doth take away headache . . ." In a broth it stops fluxes. It is good against watering eyes and is a sure remedy for children's sore heads. It is applied with salt to the bitings of mad dogs. It will not allow milk to curdle in the stomach, which "vertue" leads some to wonder if it is not "an enemy to generation by overthickening of the seed." Dioscorides taught that applied inwardly by women "it hindreth conception." It is very good laid upon the stings of wasps and bees. It is poured into the ears with honey water. And: "It is taken inwardly against Scolopenders, Beare-wormes, Sea-scorpions, and serpents."

Parkinson grows the Red, Speare and "Party coloured" mints. Succinctly he says mints are used in baths with "Baulme" and other herbs to comfort and strengthen the nerves and sinews. He agrees with Gerard that it is good for weak stomachs "much given to casting" and for feminine fluxes. It is boiled in milk for those whose stomachs are apt to cause milk to curdle. Parkinson also applies it with salt to mad dog bites. It is good to be boiled with "Mackerell and other fish." When dried it is used with "Penniroyal to bee put into puddings . . . also among pease that are boyled for pottage." Where no docks are ready at hand they bruise mint leaves and lay them upon stings "to good purpose."

Culpeper says mint is "an herb of Venus." He quotes Dioscorides, saying it has a heating, binding, drying quality, so it stays bleeding. It is an incentive to venery. Culpeper quotes all the virtues noted by Gerard and Parkinson which make mint "a very powerful stomachic." He also notes the cures for mad dog bites. He quotes Pliny as saying mint exhilarates the mind and is "therefore proper to the studious." And he says it is very useful against poisonous creatures, will dry up excess of milk in nurses and, taken in wine, helps women in child-bearing.

Evelyn also concentrated upon spearmint: "dry and warm, very fragrant a little pressed, friendly to a weak stomach and powerful against all Nervous Crudities."

Josselyn reported spearmint thriving in New England gardens.

MONEY-WOORT *Lysimachia nummularia*
HERBE TWO PENCE   J.J.

Gerard describes it as "having stalkes trailing upon the ground set by couples at certaine spaces with smooth greene leaves somewhat round, whereof it took his name, from the bottom of which leaves shoote forth small tender foot-stalkes whereon do grow little yellow floures, like those of Cinquefoil." He says that some people call it *Serpentaria* because wounded serpents are said to heal themselves with it, and some call it *Centummorbia* because of its wonderful effect in curing diseases. One can tell it is dry by its binding taste and it is also moderately cold. The flowers and leaves stamped and laid upon wounds and ulcers will cure them, most effectually when boiled in olive oil and combined with rosin, wax and turpentine. "Boiled with wine and honie it cureth the wounds of the inward parts and ulcers of the lungs. In a word, there is not a better wound herbe, no not Tobaco it selfe, nor any other whatsoever." And then he adds what may have been the reason it carpets places near old settlements in Vermont. "The herbe boiled in wine with a little honie, or meade, prevaileth much against the cough in children, called the Chinne cough."

Parkinson ignores it but Culpeper says that Venus owns it and so it is good for all manner of fluxes. He agrees that it is good for all sorts of ulcers and wounds.

MONKSHOOD *Aconitum napellus*
"1/2 oz munkhoods seeds at 3d" – J.W.Jr.

Aconitum, says Gerard, is a "kinde of Wolfes-bane" and is called in English Helmet Floure — "so beautiful that a man would think they were of some excellent vertue," but not so. "Universally known in our London gardens," no cattle will touch it in the wild, but flies swarm upon it and create thereby the best possible antidote to be devoured if one has swallowed a bit of the plant, which is extremely poisonous. Others have recommended a mouse which has just eaten the plant, but Gerard says no one has ever really seen such a mouse, whereas the flies are readily available. Take twenty and a dram each of *Aristoclochia torunda* and *bole Armoniack.* It is deadly to man and beast. "Of late," in Antwerp the leaves were served up by ignorant persons in salads and all who ate thereof were "presently taken with the most cruel symptoms, and so died."

Parkinson includes all the "wolfesbanes" as *Aconitum.* "Blew Helmet flower or Monkes hood," he says, shows by its root why it was called Napellus by the ancients, as it resembles a small turnip. Parkinson offers a "counter poison Monkes hood" against the poison of the poisonful "Helmet Flower" but it is hard to define from his description and one had best remember about the flies. I do not find it in Culpeper.

Note. What was the poison the seventeenth-century lady of the manor used in England in the Civil War to kill off forty enemy soldiers quartered with her? Was it this?

MOTHERWORT *Leonurus cardiaca: Leonurus marrubiastrum*
(Naturalized)   (Adventive)

Gerard has two listings in his index for Motherwort. One refers us to Feverfew, which he calls *Matricaria.* The other is his chapter on Motherwort which he calls *Cardiaca.* Gerard says that some have called this Hercules Ironwort. We have seen the housewives of New England hopefully calling Mayweed Ironwort, so the name of Ironwort would seem to be purely complimentary and to indicate a really "sovereign remedy." In any case, Gerard's Motherwort and Culpeper's are the same plant: four-square stalks two cubits high of "an obscure or everworn red colour, the leaves somewhat black like those of nettles but greater and broader than the leaves of Horehound . . . purple flowers not unlike to those of dead Nettle . . . rank smell and bitter taste." Hot and dry in the second degree it is commended against infirmities of the heart. It is reported also to cure convulsions, cramps, and the palsy, to kill worms and open obstructions. It is also useful in women's ills and childbirth and for green wounds.

And then we have possibly the reason for its appearing along our country roadsides: "it is also a remedy for certain diseases in cattell, as the cough and murreine, and for that cause divers husbandmen oftentimes much desire it."

Culpeper says it is owned by Venus and under Leo and there is no herb better to drive melancholy vapours from the heart and to make the mind cheerful, blithe and merry. And he lists all the former virtues, missing only the cattle.

Parkinson overlooks it entirely, possibly having seen it where Gerard said it flourished besides in gardens — on rubbish heaps.

MOUSE-EAR   *Hieracium pilosella*   J.J.

Josselyn listed one of our hawkweeds as being common to both countries. It is to be hoped the Englishwomen welcomed it as one of the herbs they recognized upon landing, as its "vertues" were all useful ones. The decoction of what Gerard calls *Pylosella* when drunk would cure both inward and outward wounds, hernias, ruptures, or "burstings." The powdered leaves also cure wounds. A decoction of the juice was said to be so excellent that if red-hot, steel-edged tools were often drenched and cooled in it, they would be able to cut stone or iron without becoming dull.

Apothecaries in the Low Countries made a syrup of the juice which they used for coughs and consumption.

Culpeper agrees with all the above and adds that it cures jaundice. The Moon owns it.

MUGWORT   *Artemisia vulgaris*   Hammond
                                                          (Naturalized)

This was one of the ingredients of Captain Hammond's sovereign remedy for the headache where mugwort, sage, camomile and gentian were all boiled in honey and applied warm.

There was certainly something very special about this herb for a very long time as both Gerard and Culpeper are cautious about giving more than a few potent facts. Gerard quotes Pliny and Dioscorides as recommending it for all women's diseases. Gerard quotes Pliny as saying that the wayfaring man who has the herb tied about him is never weary or hurt by any wild beast. Gerard then interposes this rather sharp interruption: "Many other fantasticall devices invented by Poets are to be seene in the Works of the antient Writers tending to witchcraft and sorcerie, and the great dishonour of God; wherefore I do of purpose omit them as things unworthie of my recording, or your reviewing."

Culpeper says mugwort is under Venus and maintains the parts of the body she rules. He recounts several dramatic cures of female disorders. "This herb," he says rather sententiously, "has been famous for this from the earliest time and Providence has placed it every where about our doors, so that reason and authority, as well as the notice of our senses point it out for use."

Between them, Gerard and Culpeper contribute a very brief little interpretation of the emerging seventeenth-century's background philosophy for home medicine.

MULLEIN *Verbascum lychnitis*
Verbascum thapsus J.J.
(Adventive)
(Naturalized)

Josselyn saw mullein "with the white flowers" growing in New England, as brought in by the English and their cattle-keeping.

The yellow flowered *Verbascum thapsus*, called naturalized by Gray, is commoner, but apparently escaped his notice. Both would seem to have merited being transported, common roadside plants though they were in England. Mullein is good for coughs in cattle and in humans. It is an excellent remedy for inflamed swellings of the eyes, and all the ancient authorities recommend mullein especially as a cure for piles. Culpeper adds warts, gout and cramped sinews.

MUSTARD *Brassica alba* (Introduced)
Brassica nigra (Naturalized)

Whatever the mustard was which Culpeper and Parkinson and Evelyn grew, the four very dissimilar plants pictured in Gerard's mustard chapter do not give a clue. *Sinapi sativum* he calls his first one and that does, indeed, look very like the wild mustards in our fields today. In any case Parkinson says the best mustard in gardens was also found wild, "many rough divided long leaves, of an overworn green colour . . . the stalke is divided at the toppe into divers branches . . . divers pale yellow flowers. . . ." Parkinson grinds the seed with vinegar "to serve as sawce both for fish and flesh." The same is good for coughs, epilepsy and to quicken dull spirits.

Culpeper says it is an herb of Mars and lists almost all possible virtues, including taking out splinters, and clearing the throats of singers. He especially recommends it in a "blister" to draw the disease outward, like the splinter.

Evelyn describes it masterfully: "Mustard, *Sinapi*," he says, "exceedingly hot and mordicant, not only the seed but Leaf also . . . of incomparable effect to quicken and revive the Spirits, strengthening the Memory, expelling heaviness, preventing the Vertiginous Palsie, and is a laudable Cephalick. Besides it is an approv'd Antiscorbutick, aids Concoction cuts and dissapates Phlegmatick Humours. In short, 'tis the most noble Embamma, and so necessary an Ingredient to all cold and raw Salleting, that it is rarely if at all to be left out."

MUSTARD, KNAVES *Thlaspi arvense*   J.J.
   MITHRIDATE TREACLE
                                                (Naturalized)

When Josselyn saw a "plant like Knavers-Mustard called New England Mustard" Tuckerman thought he saw *Lepidium virginicum*, or Poor Man's Pepper. Gerard's assortment of pseudo-mustards is so widely ranged, however, as to allow of several interpretations. Called "Knaves mustard" because "it is too bad for honest men" and Treacle it would seem because of common use among the poor, Gerard's poor men's mustards are unflatteringly described "burning the tongue as doth mustard seed, leaving a taste or savour of Garlicke behinde for a farewell."

N E T T L E   *Urtica dioica*   J.J.   C.M.
                                                    Hammond

The nettle would seem to have been one of those ancient remedies which counteract poisons: hemlock, mushrooms, quicksilver, henbane, serpents and scorpions. It draws evil humors from the chest, stops nosebleed, stirs up lust, and helps those with pleurisy. ". . . grossely powned and drunke in white wine it is a most singular medicine against the stone in the bladder or in the reines as hath been often proved." All this from Gerard, who also recommends nettles for whooping-cough.

Parkinson likes only "the Hungarian dead nettle or the dragon flower," a stranger plant he includes with other strangers, all mulleins.

Culpeper is eloquent about nettles, which belong to Mars. The tops eaten in the spring "consume the phlegmatic superfluities in the body, which the coldness and moisture of winter have left behind." He follows Gerard on the anti-poison list and says also that nettles are especially good to expectorate tough phlegm, and to expel worms from children, and as a bath to strengthen weary limbs. The decoction or the seed has never been known to fail in expelling gravel and stone.

Josselyn says nettles were the first plant he noticed in New England, and he assumed it had come since the English planted and kept cattle there.

Captain Hammond said, "Nettle seeds bruised and drank in White Wine is Excellent for the Gravel." Which sounds as if he had got it straight from Gerard.

N I P P   *Nepeta eataria*

"1 oz nippseed at 2d" – J.W.Jr.

Gerard says in his Index, "Nep, see Cat mint," and then heads the chapter, "Of Nep, or Cat Mint." There seems to be a good deal of editorializing here between Johnson and Gerard . . . or between each and unidentified authorities. Anyway, it is established that, although "It is named of the Apothecaries, *Nepeta* . . . Nepeta is properly called, as we have said, wilde Pennyroyall." "Later herbarists do call it Herba Cataria, . . . because Cats are very much delighted. . . ." Nep is of a temperature hot and dry, and "hath the faculties of the Calamints." It is commended against cold pains of the head, stomacke, and matrix, and "those diseases that grow of flegme and raw humours and winde . . . for them . . . bursten inwardly . . . of some fall . . . from an high place." It is used in baths and decoctions for women to sit over . . . and to make them fruitful. In fact, it is good "against those diseases which the ordinarie mints do serve and are used."

Parkinson has "*Nepeta*, Nep" and says that while herbarists know three sorts of Nep, the usual one is that called Cat Mint, somewhat like Balme or Speare Mintes, with not so strong a scent as Clary. It is "used much by women, either in baths or drinkes, or, like Clarie, being fried in Tansies to strengthen their backs." It is good for those who fall from high places, and a decoction is available to cure scab.

Culpeper is very brief. He says it is an herb of Venus, generally used by women . . . in a decoction to bathe in or to sit over. Its frequent use takes away "barrenness, and wind, and pains of the mother."

OAK OF CAPPADOCIA *Ambrosia eliator*   J.J.
OAK OF JERUSALEM *Chenopodium botrys*   J.J.

As Gerard treats of these two plants together, so shall we. John Josselyn also seems to consider them much alike, saying only that they are "both much of a nature, but Oak of Jerusalem is stronger in operation; excellent for stuffing of the Lungs upon Colds, shortness of Wind and the Ptifick . . ."

Gerard says of Oak of Cappadocia, which he calls *Ambrosia* that "the fragrant smell that this kind of Ambrosia or Oke of Cappadocia yeeldeth hath moved the Poets to suppose that this herbe was meate and food for the gods. . . ." Of the other he gives us no description, saying only they were both brought from "beyond the seas" and that they are both: "Good to be boyled in wine, and ministered unto such as have their brests stopped, and are shortwinded, and cannot easily draw their breath, for they cut and waste grosse humours and tough flegme. . . ."

Then he says it "giveth a pleasant taste to flesh that is sodden with it, and eaten with the broth. It is dryed and layd among garments, not onely to make them smell sweet but also to preserve them from moths and other vermine, which thing it doth also perform." And that is all about these two herbs Josselyn saw growing and used for curing the Indians.

To clear up a point: Oak of Cappadocia is indeed a ragweed, but not *the* ragweed, which is *Ambrosia artemisifolia*. And the name of Jerusalem Oak lingers today only in the *Manual of Weeds*, by Ada Georgia, which describes it as "an unsightly object . . . strong-scented with an odor somewhat like turpentine . . . grazing animals, even sheep, usually leave it alone."

ONIONS *Allium*

"1 li new onion seed at 2s 3d" – J.W.Jr.

Gerard says the shops call them *Cepe*. They are hot and dry, "as Galen saith" and "move tears by the smell." They "do bite, attenuate, or make thinne, breake winde, provoke urine. . . . They be naught for those that are cholericke but good for such as are replete with raw and phlegmaticke humours . . . they open passages that are stopped . . . the juice sniffed up into the nose, purgeth the head . . . stamped with salt, rue and honey, and so applied, they are good against the biting of a mad dog . . . the juice . . . mixed with the decoction of penniriall and annointed upon the goutie member with a feather . . . easeth the same very much. The juice annointed upon a pild or bald head in the sunne, bringing the hair again very speedily. . . . The juice taketh away the heat of scalding with water or oyle, oil as also burning with fire and gun powder. . . . Onions sliced and dipped in the juice of Sorrell will take away a fit of the Ague. The hurts . . . causeth head-ache, hurteth the eyes, and maketh a man dimme sighted, dulleth the senses, engendereth windinesse and provoketh overmuch sleepe, especially being eaten raw."

Parkinson is more gracious. He calls them *Cepae* and says they are used many ways — "so many I cannot recount them, every one pleasing themselves, according to their order, manner, or delight." He says the juice is used "most familiar in the Countrey" where other remedies are not at hand, for burnings and scaldings. And he adds a useful hint. "The strong smell of onions and so also of Garlicke and Leekes, is quite taken away from offending the head or eyes, by the eating of Parsley leaves after them."

Culpeper says "Mars owns them" and proves his statement that "they possess the quality of drawing corruption to them" by saying that if you peel one and lay it upon a dung-hill it will rot in half a day by drawing putrefaction to it. So, he says, it is "natural to suppose they would have the same attractive power if applied to a plague-sore." He says they will cause warts to come out by the roots if applied with salt and honey. He lists all the aforementioned uses, including the burns, and adds his own recommendation that the juice, with vinegar, will remove blemishes from the skin. They also kill worms in children if the children drink only the water in which the onions were steeped all night. And they "do whet the appetite."

Evelyn's reference to onions is full of poetry and classical quotes. He says the best come from Spain and one weighing eight pounds has been recorded. "Choose therefore the large, round, white and thin skinned . . . eaten crude and alone with Oyl, Vinegar and Pepper, we own them in Sallet. . . . Boiled they give a kindly relish, raise appetites, corroborate the Stomach, cut Phlegm, and profit the Asthmatical. . . ." Then, he, too, lists the points against them, used to excess, and adds, "How this bulb was deified in Egypt we are told, and that whilst they were building the Pyramids there was spent in this Root" (according to Herodotus), "Ninety Tun of Gold among the Workmen. So lushious and tempting . . . the Israelites were ready to return to Slavery and Brick-making for the love of them."

So the comparative largeness of Winthrop's order need not surprise us.

ORACH See **ARACH**

ORGANIE *Mentha pulegium*   Endecott

While "Organie" and "Penniriall" are interchangeable names, the first we hear of this plant is in Endecott's letter to Governor Winthrop, and he speaks of it as "organie." However, as most people refer to Pennyroyal, please see it there.

ORPINE *Sedum purpureum*
STONE CROP
LIVE FOREVER   (Introduced)

Orpine seems to be one of those plants which follows and stays with man, like purslane, and plantain. Called Live For Ever, it seems to share with those plants called Everlasting a domestic decoration niche. Its place in medicine is not dealt with at any length or with much conviction by either Gerard or Culpeper. Still, it is here, and with it another camp follower, naturalized *Sedum acre*, which in Old England rejoiced in the name "Welcome home husband though never so drunk."

PARSLEY *Petroselinum hortense* J.J.
"4 oz parsley seed at 16d per li" – J.W.Jr.

Gerard calls it *Apium hortense* or "Garden Parsley" with leaves of a "beautiful greene." He lists also a kind with curled leaves and one he calls "Virginian" with "rounder . . . yellow-greene leaves." One of his unusual hints for good husbandry mentions that parsley "may oftentimes be cut." "The Apothecaries and common Herbarists name it *Petroselinum*," but Gerard feels it is not the true *Petroselinum* "which groweth among rocks and stones, whereupon it tooke his name." There are some who think "stone Parsley" does not differ much from garden parsley, although the latter is "of lesse force than the wilde; for wilde herbes are more strong in operation than those of the garden." (Which explains Josselyn's remarking upon New England herbs not proving too strong for English bodies.)

Gerard says parsley leaves "are pleasant in broth, in which, besides that they give a pleasant taste, they be also singular good to take away stoppings and to provoke urine. . . ." "The seeds are more profitable for medicine; they make thinne, open, provoke urine, dissolve the stone, breake and waste away winde, are good for suche as have the dropsie, draw down the menses, bring away the afterbirth; they be commended also against the cough, if they be mixed or boiled with medicines for that purpose. . . . They are also good to be put into clysters against the stone or torments of the guts."

The next chapter after Gerard's on parsley is called "Of Water Parsley or Smallage," and as Parkinson treats of them together, we let Gerard do so, too. "Smallage," Gerard says, does all that garden parsley will do but is not so safe and is known to harm those "with the falling sicknesse." It is good for ulcers, felons and old sores, used in a poultice, Pliny says, it "hath a peculiar vertue against the biting of venomous spiders."

Parkinson treats of "Parsley and Smalledge," and recognizes three kinds of parsley like Gerard adding that there is but one kind of "Smalledge." Of the latter, he says it resembles parsley and grows wild although "it is much planted in gardens." He mentions all Gerard's uses, except the Pliny quote.

Of parsley, the uses, according to Parkinson are: "in all sorts of meates, both boyled, roasted, fryed, stewed &c. and being green it serveth to lay upon sundry meates, as also to draw meat withall. It is also shred and stopped into poudered beefe as also into legges of mutton, with a little beefe suet among it &c." "The rootes . . . into broth to help open obstructions of the liver, reines and other parts. . . ." "The rootes likewise boiled and stewed with a legge of mutton stopped with Parsley as aforesaid, is very good meate and of very good relish, as I have proved by the taste, but the rootes must be young . . . and they will have their operation to cause urine." "The seed also is used for the same cause, when any are troubled by the stone."

Culpeper says parsley is "under the dominion of Mercury and is very comfortable to the stomach." He lists all the previously mentioned virtues and adds, "The distilled water of parsley is a familiar medicine which nurses give to children when they are troubled with wind in the stomach or belly, which they call the frets," and that it is "effectual against the venom of any poisonous creature."

Evelyn says, "Parsley, *Petroselinum* or *Apium hortense*, being hot and dry, opens Obstructions, is very Diuretic, yet nourishing, edulcorated in warm Water (the Roots especially) but of less Vertue than Alexanders; nor so convenient in our crude Sallet, as when decocted on a Medicinal Account. Some few tops of the tender leaves may yet be admitted. . . . In the meantime, there being nothing more proper for Stuffing. . . ."

PARSNIP *Pastinaca sativa*   J.J.   W.W.
"1 li new parsnipp seed at 20d" – J.W.Jr.

Gerard says "Of Parsneps or *Pastinaca latifolia sativa*" that they "nourish more than doe the Turneps or the Carrots, and the nourishment is somewhat thicker, but not faultie or bad, notwithstanding they be somewhat windy . . . they provoke urine and lust of the bodie: they be good for the stomacke, kidnies, bladder and lungs." "There is a good and pleasant food or bread made of the roots of Parsneps, as my friend Mr. Plat hath set forth in his booke of experimentes, which I have made no triall of, nor meane to do." Of Wilde Parsneps, or *Pastinaca latifolia sylvestris*, he says that by eating them, deer, according to Dioscorides, are preserved from the bitings of serpents. So the seed is given in wine "against the bitings and stingings of serpents."

Parkinson agrees with all this about the garden parsnip, says there is another called Pine Parsnep given him by John Tradescant, "to whom everyone is a debtor," with a root not quite so good, and adds that the wild parsnip can be made as good as the garden parsnip if it is brought in to be cultivated. The uses he gives are the same as above, adding that parsnips with butter are more used in Lent than at any other time although they are good all winter long.

Culpeper prefers the wild parsnip as "more medicinable" but agrees that the garden parsnip, under Venus, is "good and wholesome nourishment, though rather windy." He says, "it is said to provoke venery, notwithstanding which it fatteneth the body much if frequently used," and he agrees it is also "serviceable to the stomach and reins." But the wild parsnip, he says, has a "Cutting, attenuating, cleansing and opening quality" and "resisteth and helpeth the bitings of serpents." In fact, "The wild parsnip being preferable to that of the garden shews nature to be the best physician."

Evelyn says of the "Parsnep, Pastinaca, Carrot; first boil'd, being cold, is of itself a Winter-Sallet, eaten with Oyl, Vinegar etc. and having something of Spicy, is by some, thought more nourishing than the Turnep."

Josselyn noted "Parsnips of a prodigious size."

PATIENCE   *Rumex patientia*
MONKS RHUBARB   J.J.

Gerard (as see under "Bludwort") includes "Garden Patience" with a number of docks and devotes most of his space to this variety which "hath very strong stalks, furrowed or chamfered, of eight or nine foot high when it groweth in fertile ground, set about with great large leaves like to those of the water Docke, having alongst the stalkes toward the top floures of a light purple. . . . The root is verie great, browne without, and yellow within, in colour and taste like the true Rubarb." He says the leaves may be eaten

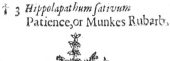

† 3 *Hippolapathum sativum*
Patience,or Munkes Rubarb,

and are "somewhat colde but more moist, and have withall a certaine clamminesse, by reason whereof they easily and quickly passe through the belly." Horace calls it a "short herb" because — to be fit to eat — it must be gathered before the stalk grows up. The decoction of the roots is drunk against "the bloudy flix, the laske, the wambling of the stomacke which cometh of choler, and also against the stinging of serpents. . . ." It is also "an excellent wholesome pot-herbe; for being put into the pottage in some reasonable quantitie, it doth loosen the belly, helpeth the jaunders, the timpany, and such like diseases proceeding of cold causes."

Parkinson says "Monkes Rubarbe or Patience" is a kind of dock in all its parts, but larger and taller than others, and, unlike them, it "beareth whitish flowers." Parkinson suspects this plant to be most like the "true Rubarbe of the Arabians" which "Merchants have brought." In any case, the leaves of Patience are "often and of many used for a pot-herbe, and seldom to other purpose; the roote is often used in Diet-beere, or ale, or in other drinkes made by decoction, to helpe to purge the liver and clense the blood."

Culpeper says it has some purging quality and leaves it there.

John Josselyn says it grows "very pleasantly."

PEACHES   *Prunus persica*   "stones sent"

We cannot know the kinds of peaches of which the stones were sent to the early settlers, but Johnson's notes in his edition of Gerard give "the names of the choice ones," as all good. They are: "two sorts of Nutmeg Peaches; The Queenes Peach; the Newington Peach; The Grand Carnation Peach; The Carnation Peach; The Blacke Peach: the Melocotone; The White; The Romane; The Alberza; The Island Peach; Peach du Troy." As the "Temperature and Vertues" as given by Gerard are all more briefly given by Parkinson, we will let him speake for them both.

Parkinson gives a long list of peaches including all the above with brief comments, such as that the Romane "is a very good Peach and well relished"; the Alberza is "late ripe" as is the "grand Carnation." Some of them sound rather hard and small, like the Nutmeg varieties. Parkinson places "the great white Peach" at the head of his list.

Of "The Use of Peaches" he says: "Those Peaches that are very moist and waterish (as many of them are) and not firme, doe soon putrefie in the stomacke causing surfeits often times . . . every one had neede be carefull . . . yet they are much and often well accepted with all the Gentry of the Kingdome. . . . The leaves, because of their bitternesse, serve well being boyled in Ale or Milke to be given unto children that have wormes. . . . The flowers have the like operation, that is, to purge the body somewhat more forceably than Damaske Roses; a Syrupe therefore made of the flowers is very good. The kernels of the Peach stones . . . openeth the stoppings of the uritory passages, whereby much ease ensueth."

Culpeper says, "Venus owns this tree. . . . Nothing is better to purge choler and the jaundice . . . than the leaves . . . made into a syrup or conserve." The powdered leaves stay the bleeding of open wounds. "The flowers steeped all night in a little warm wine . . . gently open the belly. . . .

The liquor which drops from the tree on its being wounded is given in the decoction of coltsfoot to those that are troubled with the cough." The bruised kernels boiled in vinegar are good for baldness. All the other virtues mentioned by Parkinson and Gerard are also mentioned by Culpeper.

Leonard Meager at the end of the century lists some forty peaches, including all those listed by Johnson-upon-Gerard and Parkinson.

PEARS *Pyrus communis*   J.W.
                                                        Endecott

Gerard, says Johnson, gave "eight figures with severall titles to them," to illustrate his chapter "Of the Peare Tree," but Johnson reduced them all to the "Katherine peare," the figure for which, he feels, "expresses the whole tree." Johnson then retires and leaves the Gerard text on the "Temperatures and Vertues" unedited by him. Gerard begins, "leaving the divers and sundry surnames of Peares . . ." (which, as we know, Johnson has deleted) "let us come to the faculties which the Physicians ought to know, which also varie according to the differences of their tastes; for some Peares are sweet, divers fat and unctuous, others soure, and most are harsh. . . . All Peares are cold, and all have a binding qualitie . . . stop the belly. . . . The harsh and austere Peares may with good successe be laied upon hot swellings . . . as may be the leaves of the tree. . . . Wine made of the juice . . . called in English, Perry, is soluble, purgeth . . . notwithstanding, it is as wholesome a drink being taken in small quantities as wine . . . comforteth and warmeth the stomacke and causeth good digestion."

Parkinson names for us more than sixty pears. He describes the "Catherine pear is known to all I thinke to be a yellow red sided pear, of a full waterish sweet taste, and ripe with the foremost." The best is "The ten pound peare, or the hundred pound peare, the truest and best, is the best Bon Chretien of Syon, so called, because the grafts cost the Master so much the fetching by the messengers expences when he brought nothing else."

(The Bon Chretien, an ancient French pear of enduring popularity is here today as our Bartlett pear.)

Parkinson's names include summer and winter varieties of such names as: "Bergomot . . . Warwick . . . Worster . . . Norwich . . . Popperin . . . Royall . . . Gilloflower . . . Morley . . . Gergonell . . . Warden. . . ." Of the uses he

3 *Pyrum Regale.*
The Peare Royall.

mentions first, as he did with apples, "to make an after-course for their masters table where the goodness of his Orchard is tried." The Warden is the best pear for baking or "to roast for the sicke or the sound." Perry is esteemed as a drink as good as cider, "of good use in long voyages." Perry made from wild pears, or "Choke Peares," in spite of the harshness and evil taste of the fruit, becomes in time as good as wine," and therefore we may admire the goodnesse of God that hath given such facility to so wilde fruits . . . to become useful . . . to the comfort of our soules and bodies." Briefly, as to the "Physicall properties,": "those that are harsh and sowre doe coole and binde, sweet doe nourish and warme. . . ."

Culpeper says pear trees, like apple trees, belong to Venus. He plagiarizes Parkinson on the "physical uses" and quotes Galen as saying they are good for green wounds. It has been said that after eating many pears a man should drink much wine, but "if a poor man find his stomach oppressed eating pears . . . working hard . . . wil have the same effect as drinking wine."

Meager lists over a hundred pears, many being different varities of Parkinson's list, but including several new ones, such as Warden named for

Parkinson himself — "Parkinson's-warden" — and many with purely descriptive names, such as "Pigs-tale" and "Ladies-buttock."

In 1634 Francis Kirby writes to John Winthrop, Junior, that Joseph Downing is sending one hundred apple and pear trees to the senior Winthrop. Joseph Downing writes himself to say he is sending "quodlin plants" packed in an oyster firkin with instructions for airing them on deck two or three days in the week, but he only wishes it was as easy to send pears from his "noucerie" as it had been to send them to Groton.

"Endecott's pear," however, stood as a living testimonial within the memory of many not long ago, and was even recorded by photography.

PEAS   *Pisum sativum*   Everyone

Gerard deals with "divers sorts of Peason," beginning with the "great Pease" or the "Rowncivall Pease," and running through various inferior sorts; field, tufted, skinless and wild. He reports Hippocrates as saying peas are less windy than beans and Galen as saying that peas are like beans and to be eaten in the same manner, although they are not so windy and lack "a clensing facultie." In fact, pease "have no effectuall qualitie manifest."

Parkinson names his pease: "The Rouncivall. The greene Hasting, The Sugar Pease, The Spotted Pease. The Gray Pease. The White Hasting. The Pease without skins. The Scottish or tufted . . . a good white Pease fit to be eaten. The early French Pease."

Parkinson describes the use of peas either as a "dish of meate for the table of the rich as well as the poore," when they are green. Dried they make a broth "wherein many doe put Tyme, Mints, Savory, or some other such hot herbes to give it the better relish." This, he says, is much used in Lent and "likewise at sea . . . a welcome diet."

Culpeper does not mention pease, obviously realizing their lack of an "effectuall qualitie."

Evelyn accepts pease in his salad in the shape of "the Pod of the Sugar-pease, when first beginning to appear, with the Rusk and Tendrels. . . ."

Leonard Meager lists: "Hot-spurs-pease, Redding pease, Sandwich pease, Sugar pease, Tufted or Rose pease, Gray Windsor pease, Great Maple pease, Great Bowling pease, Great Blew pease."

PELLITORY OF SPAIN *Pyrethrum* C.M.

Gerard calls it *Pyrethrum officinarum* and says it is not master-wort or *Imperatoria*, which some call by this name. Gerard describes it as having fennel-like leaves with flowers like the single white daisy. He says the French call it Pied d'Alexandre, or Alexander's foot. Its root, he says, is very hot and burning and so is good against the cold shivering of agues and "for all cold and continual infirmities of the head and sinews." Oil in which it has been boiled is good for anointing the body to procure sweating, and for parts that are blackened and bruised.

Parkinson lists only "the double wild pelletory," with flowers like the double featherfew but smaller. He says it provokes sneezing.

Culpeper says Pellitory of Spain is under the government of Mercury and is one of the "best purgers of the brain that grows." An ounce of the juice in wine will drive away the ague. Either the herb or root dried and chewed will purge the brain of "phlegmatic humours, thereby . . . easing paine in the head and teeth." The powder of the herb or root, snuffed up the nose "procureth sneezing." In an ointment it takes away black and blue spots and helps the gout and sciatica.

Cotton Mather quoted a remedy for toothache, "especially to chew Pellitory of Spain."

PENNY ROYAL *Mentha pulegium*
PENNIE ROYALL J.J. W.W.
PUDDING GRASSE Endecott
ORGANIE How

Gerard describes several kinds of "pennie royall," the second "this male kind" which "groweth upright of himself without creeping much like in shew unto wilde Marierome." This he has growing in his garden. The first is the common, the third a narrow-leaved one.

Gerard says Pennie Royal is "hot and drie in the third degree, and of subtill parts, as Galen saith." Boiled in wine and "drunken it is good for women's ills, including bringing forth the secondine, and the dead child and unnatural birth. It provoketh urine and breaketh the stone, especially of the kidneyes." Taken with honey, it "clenseth the lungs and cleareth the breast from all grosse and thicke humours." With honey and aloes, "it purgeth by stoole melancholic humours, helpeth the cramps and drawing together of sinewes." With water and vinegar it assuages the inordinate desire to vomit. Those at sea who have a great quantity of dried penny royal and cast it into "corrupt" water will find it helps. A garland of penny royal worn about the head is "of great force" against the swimming, pains and giddiness of the same. In a decoction it is good used in a bath or for women to sit over.

Parkinson does not give it garden room.

Culpeper says it is under Venus, and then quotes the ancient authorities. Dioscorides says that penny-royal makes tough phlegm thin, warms the coldness of any part to which it is applied, and digests raw or corrupt matter. Boiled and drunk, it "removeth the courses and expelleth the dead child and afterbirth." Mixed with salt and honey it voids phlegm from the lungs. Drunk with wine, it helps those stung or bitten by any venomous beast. Applied to the nostrils with vinegar, it revives persons fainting or swooning. Dried and burned, it strengthens the gums and helps those troubled by gout. As a plaster it takes carbuncles and splotches from the face. With fat it helps the splenetic or liverish. In a decoction it helps the itch and is good in baths. The green herb bruised and put into vinegar cleans ulcers and takes away bruises around the eyes and burns from the face. With honey and salt, it helps toothache. Bound to the afflicted spot, it takes away the pains of the joints.

Culpeper adds that Pliny said penny-royal mixed with mint in vinegar helped faintings and the falling-sickness and then gives us Dioscorides' description for shipboard use as: "put into unwholesome or stinking water that men must drink, as at sea, and where other can not be had, it maketh it less hurtful."

Endecott in his letter to Winthrop (quoted in full in the chapter on Medicine) calls it *organie* and lists it with mugwort as being "both good in this case of your wife. . . ."

Josselyn refers to "the upright pennyroyal."

P E O N Y *Paeonia officinalis*

The peony is one of the ancient plants handed on from garden to garden as a valuable aid in medicine, and yet so webbed about with superstition

Appendix

and misinformation that the early settlers may well have felt they did not need it as much as many other plants they brought to the New World. In any case, I do not dare to leave it out, considering its inevitable presence in old gardens in rural New England.

Gerard lists "Pyony and his kindes" in his index and treats "Of Peionie" in his Chapter 380. The Peony, he says, was named for a Greek physician, Paeon, who cured Pluto with this plant. As a medicine used on the gods, peonies became associated with all sorts of superstitions. Gerard recounts them all at length, but ends, ". . . it is no marvell that such kindes of trifles, and most superstitions and wicked ceremonies are found in the books of the most antient writers . . . cogged in for ostentation sake . . ."

Gerard, Parkinson and Culpeper, all three, treat of the peony as divided between the male and female variety, depending, they all agree, on the leaves being "divided and nicked" more or less, and in slight differences in the roots and blooms. They all allow single and double, red and white, and "blush." Parkinson says there is only one male sort but many female, and as his account of the "vertues" sums up the chief ones mentioned by them all, I quote him here: "The male Peony root is farre above all the rest a most singular approved remedy for all Epilepticall diseases . . . to be boyled and drunke, as also to hang about the neckes of the younger sort that are troubled herewith. . . . The seeds likewise is of especially use for women, for the rising of the mother. . . ." Parkinson disregards a use mentioned by Gerard and Culpeper — against "Ephilates or night Mare" and "melancholic dreams." Gerard says fifteen seeds taken in wine will help this. Culpeper recommends the same taken night and morning.

The peony chapter is the occasion for a rather nasty dig at "our Author," Gerard, by his editor, Johnson, who inserts after a statement by Gerard that a male peony is growing wild in Kent, "I have been told that our Author himselfe planted that Peionie there and afterwards seemed to find it there by accident. . . ."

PEPPERWORT *Lepidium sativum*   J.J.
See also **DITTANDER**   J.W.
(Introduced)

"Dittander or Pepper Wort, flouriseth exceedingly," says John Josselyn

in his list of herbs, "Of such garden herbs (amongst us) as do thrive there.
. . ." It is not on John Winthrop, Junior's seed list, probably because it is
easier to grow it from pieces of the roots. In any case, he could not have
been without it very long as his father writes to him in *c.* 1638, "To my loving sonne Mr. John Winthrop of Ipswich. Sonne, I received your letters
and doe blesse the Lorde for your recovery and the wellfare of your family,
you must be very careful of taking colds about your loynes; and when the
ground is open I will send you some pepper-worte roots, for the flux there
is no better medicine than the Cuppe used 2 or 3 times; and in case of sudden torments a Glyster of a quart of water boiled to a pint which with the
quantity of 2 or 3 nutmeggs of Saltpeeter boiled in it will give present ease.
for the pilles they are made of grated pepper made up with turpentine, very
stiff and some flouer withall and 4 or 5 taken fasting, and fast two hours
after but if there be any feaver with the flux, this must not be used until the
feaver be removed by the Cupp. this bearer is in great hast and so am I so
with our blessing to you and yours and salutations to all etc. I rest your
loving father   Jo. Winthrop."

Gerard says of Horse radish, Dittander, and Pepperwort, "These kindes of
wilde Radishes, are hot and drie in the third degree; they have a drying and
clensing quality, and somewhat digesting." He also notes that the roots of
Pepperwort are hotter than the leaves and that Pliny listed it among "the
number of scorching and blistering simples."

Parkinson rather hastily classes Dittander with the radishes and then says
that it might better have gone with the herbes, because the leaves and not
the roots are used. He says in full, "Dittander or Pepperworte is used for
some cold churlish stomackes as a sawce or sallet sometimes to their meate,
but it is too hot, bitter and strong for weak and tender stomackes."

Culpeper says of "Pepper-wort, or Dittander" that it is "under the direction of Mars." He quotes Pliny and Paulus Aeginetus as saying it is effectual against pains in the joints, gout, sciatica "or any other inveterate
grief," the leaves being bruised and mixed with "old hogs-lard." This applied to the place "four hours in men, and two hours in women, the place
being afterwards bathed with wine and oil mixed together, and then
wrapped with wool or skins after they have sweat a little." He says it also
"amendeth the deformities of discolourings of the skin," and that the
juice used to be given in ale to women with child to procure a speedy
delivery.

PERIWINKLE *Vinca minor* (Introduced)

Periwinkle would seem to be one of the familiar plants that followed everyone everywhere, like "hen and chickens" or house-leek. Gerard says the leaves boiled in wine and drunk will stop "the laske and bloudie flix." Parkinson recommends it for a ground cover to deck bowers in his chapter on clematis, advises that it be planted "where it may have roome to runne," and says it is good to stop bleeding. Culpeper adds to these uses that Venus owns it and that the leaves, eaten by man and wife "cause love between them."

PIMPERNEL *Anagallis arvensis* J.J.
Anagallis arvensis caerulea (Naturalized)

Gerard says "Of Pimpernell" that it is "like unto chickweed; the stalks are fouresquare, trailing here and there . . . broad leaves and sharp pointed . . . slender tendrils . . . small purple flowers tending to rednesse. . . . The female pimpernell differeth not from the male in any one point but in the colour of the floures, for as like as the former hath reddish floures, this plant bringeth forth floures of a most perfect blew. . . ." "They floure in Summer and especially in the moneth of August, what time the husbandmen having occasion to go unto their harvest worke, will first behold the floures of Pimpernell whereby they know the weather that shall follow the next day after; as for example, if the floures be shut close up, it betokeneth raine and foule weather; contrariwise, if they be spread abroad, fair weather." The "vertues" are varied. They "draw forth splinters and things fixed in the flesh" . . . "helpe the Kings Evill" and "are of power to mitigate paine." "The juice purgeth the head by gargarising . . . cures toothache being put up into the nosetrill, especially into the contrary nosetrill." The juice mixed with honey "cleanses the ulcers of the eye" . . . . "it is good against the stinging of vipers . . . prevaileth against infirmities of the liver and kidneyes, if the juice be drunk with wine." He adds that the pimpernell with the blue flower "helpeth up the fundament . . . and that red Pimpernell applied, contrariwise bringeth it downe."
Parkinson does not have it in his garden.

Culpeper describes it as Gerard did and says it is a solar herb. "This is of a cleansing and attractive quality . . ." drawing splinters and clearing the head. Applied both inwardly and outwardly it helps stingings and bitings of venomous beasts and dogs. The French use the distilled water to cleanse the skin. Culpeper says it helps toothache but advises it be dropped in the ear opposite to the pain. He recommends it against the stone and as "effectual to ease the pains of the hemorrhids." He uses it against wounds and ulcers and agrees that with honey and dropped into the eyes it "cleanseth them from cloudy mists."

Josselyn remarked upon seeing "The blew flowered pimpernell."

PLANTAIN *Plantago major*
WHITE MAN'S FOOT   J.J.
                                                C.M.   How
                                              (Naturalized)

Gerard says of all the plantains the "great" one is the best and gives a variety of uses . . . "good for ulcers, fluxes, rheumes and rottennesse." The root boiled in wine is good for the liver and kidneys and cures jaundice. The water makes a good wash and the juice is good for inflamed eyes.

Parkinson disregards it as a garden plant.

Culpeper says it is under the command of Venus and so also cures martial diseases by antipathy to Mars. He lists all the cures enumerated by Gerard and ends: "Briefly, the plantains are singular good wound-herbs to heal fresh or old wounds or sores, either inward or outward."

PLUMS *Prunus maritima* (Beach plum)   J.J.   W.W.
       Prunus nigra (Canada plum)   J.Lord
                                               "stones sent"

The "plumbs" which the early explorers and settlers found everywhere in quantity were soon joined by the varieties sent or brought from England. Of these Gerard says it would take a book to treat of all "particularly" and confines himself to "The Damson Tree" or what he calls *Prunus domestica* and to the "Mirobalane," the "Almond," and the "Sloe-tree." He mentions the "Bullesse and the Sloe" as wild plums. He says that dried plums, "commonly called Prunes" are more wholesome than the raw ones which are apt to "putrifie in the body." The leaves are good against swellings, the gum glues like the gums of the peach and the cherry, and the juice is good against "the lask and bloudy flix."

Parkinson lists over sixty sorts and recommends "My very good friend Master John Tradescante" as a source "for all the rerest fruits he can heare off in any place of Christendome."

Culpeper says all plums are under Venus; those that are sweet make the body soluble; those that are sour quench thirst and bind the belly; the firm are more nourishing than the moist and waterish; the juice is good as a gargle; the gum will break the stone; the oil from the plum's stone is good for piles; the dried fruit, stewed, is often used both in health and sickness to procure an appetite, open the belly and cool the stomach.

Leonard Meager's orchard advice lists about fifty varieties of plums, but, like Parkinson's names, they seem to contain a great many duplications due to the differences they note being in the colors of the same sorts. The chief sorts in both lists are: Mirabilons in Meager and Mirobabalane in Parkinson; Primoridans; Damascene or Damsons; Queen-Mother: Christian; and a variety of several of these in white or red or black or blue or green. Damsons and Bruneola plums are the best dried.

Parkinson says the great Damson is most used in medicine.

POMPION   *Cucurbita pepo*
PUMPKIN   J.J.
                                             "8 oz at 2s 8d per li" –J.W.Jr.
                                                                   Everyone

Gerard refers to "Melons or Pompions" and then deals only with pumpkins, which he calls in Latin *Pepo*, although he says there is a melon called "the Virginian water-melon" which he has not as yet seen. According to Galen, says Gerard, "Physitians" say this fruit should be called *Pepo Cucumeralis* or Cucumber Pompion. (And so it is today with our *Cucurbita pepo*.) "The fruit boiled in milk and buttered, is not onely a good wholesome meat for mans body but being so prepared is also a most physicall medicine for such as have an hot stomacke. ...." "The flesh or pulpe of the same sliced and fried in a pan with butter is also a good and wholesome meat, but baked with apples in an oven it doth fill the body with flatuous or windie belchings and is food utterly unwholesome for such as live idlely, but unto robustious and rustick people nothing hurteth that filleth the belly." The seed "clenseth more than the meat" and is good for those "troubled with the stone of the kidnies."

Parkinson agrees with the above recipes for boiling and buttering and adds, "They use likewise to take out the inner watery substance with the seedes and fill up the place with Pippins and having laid on the cover which they cut off from the toppe to take out the pulpe they bake them together and the poor of the citie as well as the country people doe eat thereof as of a dainty dish." He says the seeds, as well as that of "Cowcumbers and Melons, are cooling and serve for emulsions in the like manner for Almond milkes etc. for those are troubled with stone."

Culpeper says they are moist and under the dominion of the Moon, the seed is cooling and may serve well in emulsions, but the plant is rarely used in medicine.

We have seen Josselyn's recipe for pumpkin pie and heard his comments.

According to Higginson, pumpkins were the Indians' chief food after corn.

POPPY   *Papaver somniferum*   C.M.
                                          "1/2 oz popey seed at 2d" – J.W.Jr.

Gerard describes "Garden Poppies." The leaves are "long, broad, smooth, longer than the leaves of Lettuce, whiter and cut in the edges, the stem or stalke is straight and brittle, oftentimes a yard and a halfe high . . . on the top whereof grow . . . floures . . . white . . . purple . . . red . . . straked with some lines of purple . . . scarlet." Double and single. The shops keep the Latin name . . . *Papaver.* He quotes Galen saying "the seed, . . . is good to season bread . . . often used in comfits served at the table with other junketting dishes." Galen also added, "the same is cold and causeth sleepe." "*Opium* somewhat too plentifully taken doth also bring death, as Plinie truly writeth." "It mitigateth all kindes of paines, but it leaveth behinde it oftentimes a mischiefe worse than the disease . . ." "The leaves . . .

boiled in water with a little sugar . . . causeth sleep," as do the "heads boiled in water with sugar to a syrup." The leaves, "knops" and seeds stamped with vinegar, women's milk and saffron cure "Erysipelas, another kinde of St. Anthonies fire." The seed also "easeth the gout . . . sleepe . . . sleepe . . . sleepe." Gerard adds the "Wilde Poppy" is the one of which "the composition Diacodium is to be made."

Parkinson lists all the poppies, all double, "being sent from Italie and other places. The double wilde kindes came from Constantinople which whether it groweth neere unto it or further off, we cannot tell as yet." "Our English Gentlewomen . . . call it Ione silver pinne . . . Faire without and fowle within." Parkinson says, "It is not unknowne I suppose to any that Poppie procureth sleepe for which cause it is wholly and only used, as I think, but the water of the wilde poppyes . . . great use in Pleurisies . . . sovereign remedy against surfeits . . ."

Culpeper has three kinds of poppy: "white and black of the garden," and "the erratic wild poppy, or corn rose." Under "Government and virtues," he begins, "The herb is lunar, and the juice of it is made into opium." "Garden poppy heads, with the seed made into a syrup . . . procure rest and sleep to the sick and weak . . . stay catarrhs . . . hoarseness of the throat." "The black seed, boiled in wine and drunk, is also said to stay the flux of the belly and the senses." The empty shells of poppy heads and the leaves also "usually boiled in water" . . . "procure sleep." The oil . . . the green heads and leaves bruised and applied with a little vinegar . . . or with barley meal or hog's grease in a poultice "cooleth and tempereth all inflammations." "It is frequently put into a hollow tooth to ease the pain thereof."

He quotes Matthiolus that the wild poppy is good for those with the falling sickness. Distilled water of its flowers is more cooling than that of the garden poppies and is used against surfeits and against inflammations of all sorts. Galen says the seed "is dangerous to be used inwardly."

So we have Samuel Sewall in 1676 "watching" all night because his wife is ill and Dr. Brackenbury has advised "Diacodium to move rest."

P O T A T O   *Solanum tuberosum*   "sent"

While potatoes have been thought to have been fairly lately introduced as an article of North American diet, after a slow voyage through time cul-

minating in the Irish bringing the white potato to its present pitch of popularity, the early settlers of New England would appear to have been familiar with it as we know it today, called by both Gerard and Parkinson
"the Virginia potato."

Lion Gardiner, intrepid fort builder and holder of early Saybrook on the
Connecticut River, wrote in 1636 to John Winthrop, Junior, "I hear that
the bachelor" — a ship — "is to bring us provision. I pray you forget not
us when she comes from the bermudas with some potatoes for heare hath
beene some virginians that have taught us to plant them after another way
and I have put in practice and found it good."

Johnson-upon-Gerard in chapter 350 treats "Of Potatoes of Virginia,"
with a cut showing the potato we know. Gerard also knew another which
he refers to as the "common potato" which he says this much resembles.
He believes this Virginia potato to grow wild in that country as he has
received roots from there which now prosper in his garden in London. It
would seem to be one and the same plant, called Pappus by the Indians

Battata Virginiana ſiue Virginianorum, & Pappus.
Virginian Potatoes.

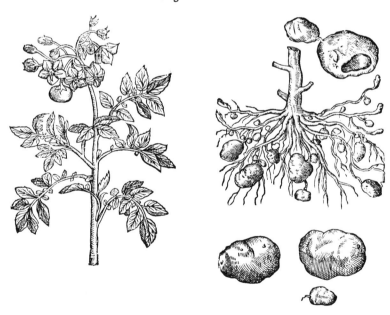

which is also their name for the common potato. Gerard says they are good either roasted in the embers or "boyled and eaten with oyle, vineger, and pepper, or dressed any other way by the hand of some cunning in cookerie."

Parkinson recognizes three potatoes: the Spanish (to which Gerard gave a separate chapter as it is not in the least like the other); the Virginia, "which some foolishly call the Apples of youth" and which he considers almost as delicate a dish as the Spanish, roasted or candied; and the Potato of Canada or Jerusalem artichoke (for which see Artichoke). In a letter from Edward Howes to John Winthrop, Junior, dated June, 1632, he sends a recipe which assumes they have plenty of potatoes.

"A receipt of a wholesome and savories drinke for such as are sick, weak or cannot drink water:

"5 or 6 gallons or quantum placet of water, put to every gallon a pinte of white wyne and a pretty quantitie of potatoe Rootes, which I suppose you have good store of, and after 2 or 3 days standing, drinke out halfe, and fill it up againe with fresh water, and the second drinke will be better than the first. Probat Mr. Thompson. This drinke Capt. Drake used very often to drinke of in his voyage about the world, and one of the voyage lately told it to me, with the manner as aforesaid."

PURSLANE *Portulaca oleracea*
J.J.
W.W.
" 1 oz pursland seed at 4d" – J.W.Jr.
(Naturalized)

Gerard says "Of Purslane" . . . "the stalkes of the great Purslane be round, thick, somewhat red, full of juice, smooth, glittering, and parted into certaine branches trailing upon the ground, the leaves be an inch long, something broad, fat, glib . . . the floures are little, of a faint yellow, and grow out of the bottoms of the leaves." The first he calls *Portulaca domestica*, Garden Purslane, and the second *Portulaca sylvestris*, Wilde Purslane. "The former is fitly sown in gardens, and in the waies and allies thereof being digged and dunged, it delighteth to grow in a fruitfull and fat soile not dry. The other cometh up of his own accord in allies of gardens and vineyards . . . let alone till the seed be ripe it doth easily spring up afresh for certain yeares after." "Purslane is cold . . . in the third degree, and moist in the second." "Rawe Purslane is much used in sallades, with oile, salt and vinegar: it cooleth an hot stomache and provoketh appetite . . . taketh away the paine of the teeth and fasteneth them . . . good for the bladder and kidnies, and allaieth the outrageous lust of the body . . . commended against the worms in young children, and is singularly good if they be feverish withall . . . stoppeth . . . all . . . fluxes whatsoever . . . thrown up with a mother syringe, cureth inflammations, frettings and ulcerations of the matrix . . . with a clyster pipe, helpeth the ulcerations and fluxe of the guts."

Parkinson is brief. He notes also that purslane delights to grow "in the alleyes of the Garden between the beds . . . or, as I have seene in some Gardens, upon those beds of dung that Gardiners have used to nourse up their Cowcumbers, Melons and Pompions, whereon after they have been taken away, they have sown Purslane. . . ." "It is used," says Parkinson, "as Lettice in sallets, to coole hot and faint stomackes in the hot time of the yeare, but afterwards if only for delight, it is not good to be too prodigal in the use thereof. The seed of Purslane doth coole much inflammation inward or outward, and doth a little binde withall."

"Purslain," Culpeper says, is "an herb of the moon." As such it "is good to cool any heat . . . stayeth hot and choleric fluxes . . . inflammations . . ." It is so useful in cooling all sorts of over-heatings, he says, that the "over frequent use of it extinguisheth the heat and virtue of natural procreation." Another somewhat unique use is that, when applied to the neck, it will take away "what is termed the crick in the neck." With oil of roses it can be so used and also against the burns of gunpowder and lightning blasts. "Applied also to the navels of children that are too prominent, it reduceth them." And "placed under the tongue, it assuageth thirst."

We have seen Wild Purcelane noted by Josselyn. In Champlain's time his men investigated Indian gardens and were astounded to see the Indians regarding the wild purslane, which covered the ground under their crops, as unusable, the French having been taught to regard it highly as a salad vegetable.

Evelyn says that "Purslain, *Portulaca* . . . quickens the Appetite, assuages Thirst, and is very profitable for hot and bilious tempers as well as Sanguine, and generally entertained in all our Sallets, mingled with the hotter Herbs; Tis likewise familiarly eaten alone with Oyl and Vinegar, but with moderation."

RADISH *Raphanus sativus*

J.J.
W.W.
"8 oz radish seed at 12d per li" – J.W.Jr.

Gerard describes many sorts of radishes: long and round, black and white, wild and tame, and some pear-shaped. Of the garden radish, or *Raphanus sativus*, that, while it may be sown ten months in the year, after the Summer Solstice is best for yielding good roots. Before then, they run to stalk and seed. They "manifestly heat and drie" and Galen "sheweth it is rather a sauce than a nourishment." The "rinde," given with "Oxymel, which is a syrup made with vineger and honey," . . . "cause vomitting." The seeds, likewise administered, "driveth forth worms." Raw radishes are eaten "with bread in stead of other food." The distillation is good against stones and gravel. "But for the most part they are used as a sauce with meates to procure appetite."

Horse radish, for Gerard, is *Raphanus rusticans*, closely related to his Dittander, and Pepperwort. He describes it as "having great leaves like those of the great garden Docke, called of some Monkes Rubarbe, of others Patience." He says the seed is so rarely seen, that it was reputed to have none.

Parkinson says of radishes that they "do serve usually as a *stimulum* before meat, and that the poor eat them alone with bread and salt." "The Horse Radish is used Physically, very much in Melacholicke, Speneticke and Scorbuticke diseases. And some use to make a kind of mustard with the rootes and eat it with fish."

Culpeper deals with "Raddish and Horse-Raddish" as both are under Mars. He lists all the uses mentioned above and ends with an observation, "Sleep not presently after the eating of raddish, for that will cause a stinking breath."

When Josselyn sees a radish "as big as a man's arm," does he mean Horse Radish?

Evelyn says that "Radish, *Raphanus*, albeit rather medicinal than so commendably accompanying our Sallets (wherein they often slice the larger Roots) are much inferior to the young Seedling Leaves and Roots, raised on the Monthly Hotbed almost the whole year round." He says that the bigger roots should be eaten when they are transparent and with salt only as they carry their pepper in them.

Evelyn recommends especially the "*Raphanus rusticanus*, the Spanish black Horse Radish, of a hotter quality and not so friendly to the Head, but a notable *Antiscorbutic*, which may be eaten all the Winter . . ." The thin shavings mingled with "our cold Herbs" are excellent ingredients. The fresh root grated into an earthen dish, tempered with vinegar and sugar becomes "a Sauce supplying Mustard to the Sallet and serving likewise for any Dish besides."

ROCKET   *Eruca sativa*

J.J.

"1/2 oz Rockett seed at 4d per oz" –J.W.Jr.

(Naturalized)

Gerard says "Of Rocket," "There be sundry kindes of Rocket, some tame, or of the garden, some wilde, or of the field, some of the water and of the sea." As Parkinson treats only of "Garden rocket, *Eruca sativa*," and Culpeper dismisses "the garden rocket" as "rather used as a sallad-herb than to any physical purposes" and proceeds to treat only of the "wild rocket," we may assume that Winthrop's seeds were of *Eruca sativa*, which is also Gerard's Latin for his first kind, or "Garden Rocket." "Garden rocket, or Rocket gentle," says Gerard, which he also calls "Romane Rocket," "hath leaves like those of Turneps, but not neere so great nor rough. The stalkes rise up of a cubit, and sometimes two cubits high, weak and brittle; at the top whereof grow the floures of a whitish colour, and sometimes yellowish, which being past, there do succeed long cods, which containe the seed . . ." Gerard reports Galen as saying rocket is hot and dry in the third degree, hence "not fit or accustomed to be eaten alone." Its "vertues," in order, according to Gerard are as "a good sallet herbe, if it be eaten with Lettuce, Purslane, and such cold herbes" while with the others, it is good and wholesome and "causeth that such cold herbes do not over-coole the frame." Its use, especially the seed, "stirreth up bodily lust . . . provoketh urine and causeth good digestions." "The root and seed stamped and mixed with vinegaer and the gall of an Oxe, taketh away freckles . . . blacke and blew spots, and all such deformities of the face."

And he includes an interesting use reported by Pliny, that "whoever taketh the seed of Rocket before he be whipt, shall be so hardened, that he shall easily endure the paine."

Parkinson agrees with the salad composition and complexion remedy properties, but does not quote Pliny. Instead he introduces Matthiolus who "saith that the leaves boyled, and given with some sugar to little children cureth them of the cough." And he says it helps "spleneticke persons" and is used to kill "the wormes of the belly."

Culpeper repeats these uses and adds that it cures serpent bites, takes away "the ill scent of the arm-pits," and increases milk in nurses.

Evelyn says Rocket, *Erica Spanish* is hot and dry and must be "qualified" with Lettuce and "Purcelain and the rest."

R O C K E T ,   S W E E T   *Hesperis matronalis*
D A M E S   V I O L E T S

(Naturalized)

"Dames violets or Queenes Gillofloures," says Gerard, "have great large leaves of a darke greene colour, somewhat snipt about the edges, among which spring up stalkes of the height of two cubits, set with such like leaves; the floures come forth at the tops of the branches, of a faire purple colour, very like those of the stocke gillofloures, of a very sweet smell, after which come up long cods. . . ." Some, like the "Queenes white Gillofloures" are white, and some, "by the industrie of some of our Florists are double white. "They are sown in gardens for the beauty of their flowers" and are called in Latin "*Viola matronalis, Viola Hyemalis*, and *Viola Damascena*. And "is thought to be the *Hesperis* of Pliny," "for that it smells . . . more pleasantly in the evening or night than any other time." The French call them *Violettes des Dames*, and *Giroffles des dames* or *Matrones Violettes*. In English they are: Damaske Violets, winter Gillofloures, Rogues Gillofloures, and close Sciences. The leaves are hot and sharp in taste, like those of *Eruca*, or Rocket, of which this seems to be a kind. And the only "vertue" listed: "The distilled water of the floures hereof is counted to be a most effectuall thing to procure sweat."

Parkinson lists *Hesperis, sive Viola Matronalis*. "Dames Violets, or Queenes Gilloflowers" describes them as above and then somewhat surprisingly, even to himself, lists *"Lysimachia lutea*, the tree Primrose of Virginia" since he confesses he has no other place to put it. He says the English name "although it be too foolish . . . may pass for a time till a fitter be given." If you prefer to follow the Latin, call it "Virginia loose-strife."

ROCKET WINTER *Barbarea vulgaris*

(Naturalized)

"Yellow-rocket" or "winter cress" or "St. Barbara's wort."
Culpeper says it is good for provoking urine, to expel the stone, and also against scurvy. It cleanses inward wounds and heals outward ones by its drying quality.
See CRESSES.

ROSEMARY *Rosemarinus officinalis*

J.J.
"1 oz Rosemary seed at 8d per oz" –J.W.Jr.

Gerard describes it as "a woodie shrub" growing very tall when set by a wall, with "slender, brittle branches," "verie many long leaves, narrow, somewhat hard, of a quicke, spicy taste, whitish underneath, and of a full greene colour above," with "little floures of a whitish blew colour." The "vertues" are interesting to read, remembering Shakespeare's and Ophelia's "rosemary for remembrance". . . . "Rosemary is given against all fluxes of blood; it is also good, especially the floures thereof, for all infirmities of the head and braine, proceeding of a cold and moist cause; for they dry the brain, quicken the senses, and memories, and strengthen the sinewie parts." Serapio is quoted as saying that Rosemary is a remedy against "the stuffing of the head, that cometh through coldness of the braine, if a garland be placed about the head." Rosemary is good against jaundice, for provoking urine, and for opening stoppings of the liver. The distilled water drunk morning and evening "taketh away the stench of the mouth and the breath." Tragus is quoted as saying it is a spice in German kitchens "and other cold countries." The Arabians write that Rosemary "comforteth the brain and memorie, the inward senses, and restoreth speech unto them that are pos-

sessed with the dumbe palsie, especially the conserve made of the floures
and sugar, or any other way connected with sugar." The flowers made up
with sugar "comfort the heart and make it sorry, quicken the spirits, and
make them more lively." The oil "chymically drawne, comforteth the cold,
weake and feeble braine in most wonderful manner." Gerard says the people of Marchia put it into their drink, the sooner to make their clients
drunk, and also into chests and presses among clothes, to preserve them
from moths or other vermin. Gerard says, "In 'Eastern cold countries' it
is 'carefully and curiously kept in pots, set into the stoves and sellers against
the injuries of their cold Winters.'"

Parkinson says Rosemary is "in every womans garden" and "in Noble-
mens and great mens gardens against brick walls." "Rosemary," says Parkinson, "is almost of as great use as Bayes . . . both for inward and outward
remedies, and as well for civill as physical purposes. Inwardly for the head
and heart; outwardly for the sinewes and joints; for civill uses, as all doe
know, at weddings, funerals etc. to bestow among friends; and the physicall
are so many. . . . I will onely give you a taste of some. . . ." He then gives a
recipe for what he calls "an excellent oyle drawne from the flowers alone by
the heate of Sunne" which is a "soveraigne Balsame," good to restore sight,
and to take away spots, marks and scars from the skin. The method in-
volves more than the heat of the sun, however, as one must take a quantity of the flowers, put them into a strong glass "close stopped" and set
them in hot horse-dung "to digest" for fourteen days. Then, taken out of
the dung and unstopped, a fine linen cloth is tied over the mouth of the
glass and it is upended over the mouth of another "strong glasse" set in
the sun and an oil will distill down into the lower glass, "which preserve
for the uses before recited, and many more . . ." Another oil "chymically
drawne" serves the same purposes.

Culpeper says, "The Sun claims privilege in it and it is under the celestial
Ram," and adds, "It is a herb of as great use with us as any whatsoever . . ."
By its warming and comforting heat it helps all cold diseases "of the head,
stomach, liver and belly . . . and brain." He repeats all that has been quoted
above, and adds that "to burn the herb in houses and chambers correcteth
the air in them." The dried leaves, "smoked, help those that have a cough,
phthysic, or consumption. The leaves are used in bathings and made into
ointments and oiles. And he mentions the "chymical oil" and the one so
easily made in the sun, as described by Parkinson.

We have already quoted Josselyn as saying rosemary is "no Plant for this Country" and we assume that, like fennel, it was potted and kept in cellars for the winter.

Hammond says for "comforting the head and braine . . . take Rosemary, Sage and both sorts of both, with flowers of Rosemary if to be had, and Borage with ye flowers. Infuse in Muscadine or in good Canary 3 days, drink it often."

Evelyn says the flowers of rosemary are "always welcome in vinegar."

ROSES *Rosa*   J.J.   Bradford
                                                   J.W.Jr. Whipple

Gerard's entire list of roses does not exceed cuts of thirteen garden roses, counting as two one cut repeated for two of them, and five cuts of wild roses. Even with these, the list is made smaller by some of the sorts being counted twice, as single or double, so that the actual list becomes: the white rose, the red rose, the "Province or Damaske Rose," the rose without prickles, the "great Holland Rose, commonly called the great Province Rose," the "Muske" rose, the velvet rose, the yellow rose, the cinnamon rose, and, for wild roses, the eglantine, brier and pimpernel.

Gerard's list of "vertues" is longer than his descriptions of roses. Briefly, roses can be used in various forms: decoctions, conserves, dried, in waters, in ointments, syrup, honey. The leaves, petals, buds, "nails," are all used. Roses strengthen the heart, liver, kidneys. They comfort a "weak stomacke." Honey of roses is good for ulcers and old wounds. The bud-heads stop bleeding. The "nailes or white ends" of the petals are good for weak eyes. The syrup "carries down choleric humours. The conserve strengthens the heart. The distilled water cools and refreshes and flavors "junketting dishes, cakes, sauces and many other pleasant things. . . ."

Parkinson says, "The Rose is of exceeding great use . . . the Damaske Rose, besides the superexcellent sweete water . . . or the perfume of the leaves being dryed, serving to fill sweete bags, serveth to cause solublenesse . . . The red Rose hath many Physicall uses . . . many sorts of compositions . . . cordiall and cooling . . . binding and loosing. . . ."

Culpeper says red roses are under Jupiter, damask under Venus, and white under the moon. The white and red are cooling and drying; the white

3 *Rofa Holoferitea.*
The veluet Rofe.

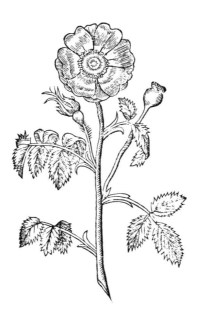

are seldome used medicinally. The decoction of red roses made with wine is good for the headache and pains in the eyes, ears, throat, gums, fundament and matrix. The same decoction with the rose "leaves" remaining in it, is good to apply over the inflamed heart. His recommendations and recipes are infinite yet all within the bounds set by Gerard and Parkinson.

Parkinson's list runs to twenty-four roses by his own count, beginning also with the "English white rose," the carnation rose, the "English red rose," called so because it is "more frequently used in England," the damask, having its "pale yellow threads" showing in the middle of its flowers which are of a "fine deep blush colour" and have "the most excellent sweet pleasant sent for surpassing all other Roses or Flowers." This is the rose most preferred by gentlewomen for their rose water. Then we have his "great double Damaske Province or Holland Rose," blush, red and white, then the "party coloured rose" called "Yorke and Lancaster," *
a smaller variety he calls the "chrystall gilloflower," another dwarf rose

*See Shakespeare, *Henry VI*, and Sonnet XCIX.

6 *Rosa Hollandica siue Bataua.*
The great Holland Rose, commonly called the great Prouince R

called the gilloflower for the same reason, the "Franckfort Rose" with purple shoots* the "Hungarian Rose" with green shoots. He then follows Gerard's listing with single and double sorts of the thornless rose, the cinnamon, the yellow, musk, eglantine and one new one at the end, the "ever green rose bush," and, finally, an eglantine whose leaves remain all winter.

John Josselyn, sampling his way about "this wildernesse" announces that New England wild roses, which so rejoiced the hearts of June arrivers, are "somewhat styptic." Governor Winthrop writes to his son John Winthrop, Junior, who is about to leave England with his stepmother and the rest of the family, to be well supplied with "conserve of redd roses" for the voyage.

William Bradford mentioned "the fragrant rose" in Plymouth gardens.

In 1634 Joseph Downing writes to John Winthrop, Junior, "If you write word that you have no roses there, I will send you over some damaske, red, white, and province rose plants, of all these 3 or 4 a peece or more. . . ." And we have numerous "receits" for roses, and a reference to rose water in well-equipped households.

*Which we now know as "Empress Josephine."

ROSE PENNYWORT *Saxifraga virginiensis*
ROSE PENNIWORT

Josselyn mentions this plant twice and says it is the "second kind of Navel Wort in Johnson upon Gerard." Josselyn also calls it the "New England Daysie or Primrose." The "vertues" of the plant pictured in Gerard are "of a certaine obscure binding qualitie, it cooleth, repelleth . . . scoureth and consummeth, or wasteth away. . . ." The juice . . . is a "singular remedie against all inflammations and hot tumours. . . ." Only the water pennywort is unwholesome, especially to cattle.

All navelworts are cold and moist, says Parkinson.

Culpeper says that Navelwort or Penny-wort is called *umbilicus veneris* and *herba coxendicum* and there are seven kinds, all but one found only on the Alps. That one grows on walls in England and is cool and binding.

Hydrocotyle americana grows in meadows and damp woods.

On the whole, Josselyn was hoping to have found an English plant and was actually no nearer than the two "robins" are to each other.

1 *Rosa alba.*
The white Rose.

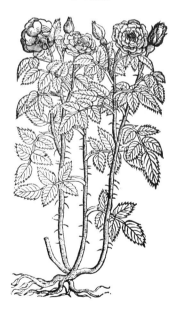

RHUBARB *Rheum rhaponticum*   "Cases" J.W.Jr.

Rhubarb, generally considered, would seem to have been a fairly movable feast.

Rheum rhaponticum is *"Rha,"* "the true Rubarbe of the Antients," of which Johnson places a correct picture in his edition of Gerard, who seems in this case to have used the wrong one from Matthiolus, together with a "Historie which was not much pertinent." The best rhubarb, says Johnson firmly, is that which is brought from China, fresh and new. The rhubarb or *Rha*, of the ancient physicians grows in Thrace and Romania and Hungary, and in "our gardens." Rhubarb was commended by Dioscorides and Galen for all "griefes" of the stomach, liver, for convulsions, and against bites. Purging was a use introduced by the Arabs, according to Gerard, who conceded the second-best rhubarb came from "Barbarie."

According to Parkinson, who deals first with Patience Rhubarb, which see, "the other Rubarbe or *Rhaponticum* . . . I have tried and found by experience to purge gently, without that astriction that is in the true Rubarbe . . . from the East Indies or China . . . A syrupe . . . made with the juice and sugar . . . effectuall in dejected appetites and hot fits of agues. . . ."

Culpeper loyally describes the "Rhubarb or Rhapontic" grown in England as good as any out of China and he describes how to dry the roots to be kept for a year. He then deals with "Garden Patience, or Monks Rhubarb" and then with "Great Round-leaved Dock or Bastard Rhubarb" which he prefers for all the uses mentioned by Dioscorides and Galen as quoted by Gerard, above: cleansing, against black and blue spots; jaundice, dropsy, fevers, melancholy; inward bruises; and, finally, purging the stomach.

RUE *Ruta graveolens*
HERBE GRACE   J.J.
                                                         How   C.M.

Rue is hot and dry, says Gerard, and he gives a long list of uses for rue, made up boiled or in an oil, juice, with honey, in clysters, and as a poultice.

It is a powerful anti-poison and good for provoking urine, expelling the afterbirth, opening the matrix, helping dim eyes, windy stomaches, shingles, earache, all pains, and fevers. The wild rue is much more violent,and to be avoided.

Parkinson sums it up: "The many good properties whereunto Rue serveth hath I think in former times caused the English name of Herbe Grace to be given unto it . . . without doubt it is a most wholesome herbe, although bitter and strong, and could our dainty stomackes brooke the use thereof, it would work admirable effects . . . weake eyes . . . glisters or drinks against the winde or the collicke . . . to procure urine . . . but beware of the too frequent or overmuch use thereof. . . ."

Culpeper says rue is an herb of the Sun and under Leo. It provokes urine. The seed, taken in wine, is an antidote against poisons. The leaves are called "Mithridates' counter-poison against the plague." Besides all the uses noted above it is good for the gout and takes away warts. Gerard quoted Dioscorides as saying the leaves drive away serpents. Culpeper quotes Mithridates as eating the leaves mashed with juniper berries every morning as an antidote to poisons or infections during the day.

Josselyn observed "rew" growing in gardens in New England.

RUPTUREWORT   *Herniaria vulgaris*   J.J. (Adventive)

Gerard considers it a "kind of Knot grasse" and says it "doth notably-drie, and thoroughly closeth up together and fasteneth." He says it is reported that, drunk, it cures ruptures, and the powder in wine washes away the stones in the kidneys.

Parkinson disregards it as a garden plant.

Obviously not a garden flower, Culpeper adds to the above virtues that it will kill worms in children and heal wounds and ulcers.

Asa Gray calls it Herniary, "an Old World genus of matted herbs . . ." named for "its reputed medicinal virtues."

SAFFLOWER   *Carthamus tinctorius*
FALSE SAFFRON
SAFFRON   *Crocus officinale*   C.M.
                                                                    How   Dr. Oliver
                                                                    "Cases" J.W.Jr.

The official, medicinal saffron of long standing comes from the *Crocus sativus*, or *Crocus officinale* grown extensively and commercially in Europe. The stigmas of 4000 are required to yield one ounce of dried saffron. Occasionally attempts are made to adulterate the product with the petals of *Calendula officinalis*. However, Culpeper speaks of the saffron crocus of Walden, grown commercially, and warns against overdosing.

There is another source for what passed for "saffron" in the seventeenth century. This is a plant which seems to have been very closely allied, in people's minds at least, to the so-called "blessed thistle," "Carduus benedictus," or *Cnicus benedictus*. Gerard calls it "Bastard Saffron" and says, rather cautiously, that it may well be reckoned among the thistles, and, equally cautiously, that its flowers are "of a deepe yellow shining colour, drawing neere to the colour of Saffron." Gerard says only the seed is used for purgations, and Parkinson agrees that "it serveth well to purge melancholicke humours." Gerard advises combining the seed with other ingredients, such as anise. He says the seed will curdle milk, which makes it "of great force to loose and open the belly." Parkinson does not feel that "Bastard or Spanish Saffron," cultivated in Europe since Theophrastus and Dioscorides, fully matures in cold climates like England. On the whole, perhaps the settlers relied upon the dried product being imported.

SAGE   *Salvia officinalis*   Hammond
See **CLARY** also   How

Gerard says that Agrippa and Aetius have called sage "the Holy-herbe, because women with childe if they be like to come before their time . . . do eat thereof to their great good." Sage, says Gerard, is "singular good for the head and braine; it quickeneth the senses and memory, strengtheneth the sinewes . . . cleanseth the bloud." It is good against the biting of serpents, spitting of blood, pains in the sides, makes a wholesome addition to ale and a good body-wash, and it helps those with the palsy.

Parkinson says sage is much used in the month of May fasting, with butter and parsley "to conduce to the health of mans body." It is also used "to bee tunned up with Ale . . . and also for teeming women, to helpe them the better forward in their childe bearing . . . also used to be boyled among other herbes to make gargles . . . bathings. . . ." "The Kitchen use is either to boyle it with a Calves head . . . minced, to be put with the braines, vinegar and pepper to serve as an ordinary sawse . . . among other fasting herbes to serve as a sawce for peeces of Veale. . . ."

Culpeper says Jupiter claims sage. It is good for the liver and to "breed good blood." He also commends it against snakebite and the plague, for palsy, gargles, washing, against the cough and stitches in the sides. He says it will turn hair black. His uses for women seem the opposite of those mentioned by Gerard and Parkinson.

Evelyn says the tops of the red sage particularly have "the noble Properties of the other hot Plantes especially of the Head, Memory, Eyes, and all Paralytical Affections. In short, it is a Plant endu'd with so many and wonderful Properties, as that the assiduous use of it is said to render Men Immortal; We cannot therefore but allow the tender . . . leaves . . . Flowers in our cold Sallet. . . ."

SAINT JOHNSWORT *Hypericum perforatum* Boorde
J.J.
(Naturalized)

Gerard includes several sorts of Saint Johnswort, one tall, one woolly and one creeping, all apparently with the same virtues, being good to stop bleedings, provoke urine, heal ulcers and cure burns. The leaves, flowers and seeds stamped and put in a glass of olive oil in the hot sun for weeks will yield an oil the color of blood which will heal any wound whatsoever. It is not for Parkinson's garden, but Culpeper hails it as under Leo and the sun. Gerard also recommends it against sciatica and the ague. Culpeper also recommends it as an excellent vulnerary, and good for all the uses mentioned above. He also uses it against worms, melancholy and madness.

Josselyn mentioned that cheeses put up in it would keep at sea. He listed it as one of the plants common also in England which he saw growing in New England.

SAINT PETERSWORT *Ascyrum stans*   J.J.
(Naturalized)

Gerard sees Saint Peterswort as a "square S. Johns Grasse" and put it directly after the chapter on Saint Johnswort. He says it is "endued with the same vertues as Saint Johnswort . . ." The seed in "Meade doth plentifully purge. . . ."
Culpeper says the same of its virtues and describes it as being like Saint Johnswort in blossoms, but browner and larger and without perforations. It is good for burns.

SANTOLINA See LAVENDER COTTON

SARSAPARILLA *Aralia nudicaulis* (wild sarsaparilla)   J.J.
W.W.   C.M.

Gerard speaks of having a great plenty of roots of this "Binde-weed of Peru, which we usually call Zarza or Sarsa Parilla," and yet no description of what the plant looks like. Only Monardus has said that the roots are deep, which Gerard scornfully asserts is like saying crows are black. Gerard says it grows in Peru, in Virginia and in both the Indies. He thinks it is a Smilax. The roots are a remedy against long-continuing pains of the joints and the head, and against cold diseases. They are good to provoke sweating. The European variety, of less worth, was recommended by Dioscorides against poisons.
Culpeper says there are three kinds of sarsaparilla, the first two from Spain and Italy, the second only from the West Indies. They are all plants of Mars, and of a healing quality. The leaves and berries are an anti-poison, provoke urine, heal sore eyes, promote sweating, expel wind, help all swell-

ings and pains. The powdered root has been proved useful in treating venereal diseases, especially in infants who have received the infection from
their nurses.

Josselyn said there was a lot of this "rough bind-weed" growing upon
the banks of ponds. The leaves pounded with hogs' grease and boiled to an
unguent proved excellent in the curing of wounds.

S A V O R Y ,   S U M M E R   *Satureia hortensis*   J.J.
                              "1/2 oz summer savory at 2d" – J.W.Jr.
S A V O R Y ,   W I N T E R   *Satureia montana*
                              1 oz winter savory at 6d" – J.W.Jr.

Gerard describes winter savory as looking like hyssop, but lower, with
many branches "compassed on every side with narrow and sharpe pointed
leaves," among which grow the flowers from the bottom to the top, "of
colour white tending to light purple." This is a perennial. The summer
savory he describes as growing a foot high, with the leaves set more thinly
upon the branches than winter savory, the flowers "of a light purple tending towards whiteness." It is an annual.

Gerard says winter savory is hot and dry in the third degree and so
"maketh thin, cutteth, it cleanses the passages; to be briefe, it is altogether
of like vertue with Time." Summer savory is not so hot and is therefore
"saith Dioscorides," fitter to be used in medicine. Both do "marvelously
prevail against the winde," and are consequently "with good successe
boiled and eaten with beanes, peason, and other windie pulses."

Parkinson says winter savory is one of the "farcing herbes" and is put
into puddings, sausages and "such like kindes of meates." Many, especially in other countries, use summer savory in sauces and garnish "as we do
Parsley," or boil it with peas to make pottage. He summarizes, "They are
both effectual to expell winde."

Culpeper says Mercury claims dominion, and lists the same cures as
Gerard except that he adds that the juice will clear the eyes and, with heated
oil of roses, help deafness, and "is much commended for women with child
to take inwardly, and to smell often to." The juice snuffed up into the
nostrils will quicken "dull spirits in the lethargy." Outwardly applied in a

poultice it will take away the pains of stings by bees, wasps and "any venomous reptile."

"Some doe use the pouder of the herbe dryed (as I sayd before of Tyme) to mixe with grated bread, to breade their meate . . . to give it the quicker relish."

Josselyn includes both Winter and Summer Savory as among those "Garden Herbs (Amongst Us) as do thrive there. . . ."

SCABIOUS *Knautia arvensis* J.J.

SCABIOUS, DEVILS BIT *Succisa australis* How
J.J.
(Naturalized)

Gerard describes seventeen varieties of scabious and then puts "Devils bit" in the next chapter. The first kind of scabious is the best known, having hoary, hairy, jagged leaves and at the top of the stalke "blew floures in thicke tufts or buttons." The devil's bit scabious looks like this with flowers of a deep purple but these being ripe "are carried away with the winde."

They both scour the chest and lungs and are good for all infirmities of the chest. They purge the bladder, cure scabs, are good against snakebite and pestilent fevers, drive out wind, and are especially recommended against the swelling of the "almond" and upper parts of the throat.

Parkinson says that they "yeeld not flowers of beauty or respect fit to be cherished in our Garden of delight," so he leaves them to the fields and woods, "there to abide." He has three garden-worthy varieties which he presents, one white and two red, but he does not know if they have any of the virtues of the wild kinds.

Culpeper says Mercury owns the plant and it is effectual in all diseases of the breast and lungs. It is good against pestilence and plague sores, wounds, swellings, freckles, pimples, leprosy, dandruff and the itch. The herb bruised and applied will draw out splinters.

Josselyn observed the women mistaking one of the New England plants for scabious.

How used it in his measles cure.

SCURVY GRASS *Cochlearia officinalis* J.J.
Cochlearie groenlandica

Scurvy-grass in all its varieties has been an anti-scorbutic since the beginning of voyages. It does not belong in Parkinson's gardens, but Evelyn admits into his salad "a few of the tender leaves," tasting like those of nasturtium.

SELF-HEAL *Prunella vulgaris* (Naturalized)

Gerard says that "Prunell or Brunell" grows in English fields and is called "Carpenters herb" because the decoction with wine or water will "joine together and make whole and sound all wounds both inward and outward even as Bugle doth." Bruised with oil of roses and vinegar and laid to "the forepart of the head," it assuages pain. In short, it and "Bugle" are the two best wound herbs in the world, "as has been often proved."

Bugle is *Ajuga reptans*, middle consound or middle comfrey, which see. Parkinson does not have self-heal in the garden.

Culpeper says self-heal is under Venus and is a special remedy for all inward and outward wounds. Compounded with bugle and other wound herbs, it makes an effectual wash for all wounds and sores anywhere.

SHEPHERDS PURSE *Capsella bursa-pastoris* J.J.
C.M.

Gerard says that "Shepheards purse stayeth bleeding in any part of the body . . . drunke . . . pultesse-wise, or in bath. . . ." It stays bleeding in any form, and is "marvelous good for inflammations new begun."

Obviously Parkinson does not entertain it in his garden.

Culpeper says it is under the dominion of Saturn and is cold, dry and binding. It helps all fluxes of the blood. Bound to the wrists or the soles of the feet, it helps jaundice. The juice dropped into the ears heals them. The ointment made of it is good especially for wounds in the head.

Josselyn lists it as having sprung up in New England since the English planted and kept cattle here.

Cotton Mather puts it in the ear for the toothache.

SKIRRET *Sium sisarum*   J.J.

"3 oz skerwort seed 3d per oz" –J.W.Jr.

Gerard thinks highly of "Skirrets." *Sisarum* he calls them in Latin. "In English, Skirret and Skirwort," "a medicinable herb" "which Tiberius the Emperor commanded to be conveyed unto him from Gelduba a castle about the river of Rhene, as Pliny reporteth."

"The roots of the Skirret be moderately hot and moist; they be easily concocted; they nourish meanly, and yield a reasonable good juice, but they are something windie, by reason whereof they also provoke lust." "They be eaten boiled, with vinegar, salt, and a little oile, after the manner of a sallad, and oftentimes they be fried in oile and butter, and also dressed after other fashions, according to the skil of the cooke, and the taste of the eater. . . . The women in Susula . . . prepare the roots for their husbands, and know full well wherefore and why. . . ." "The juice of the roots drunk with goats milke stoppeth the laske . . . with wine, putteth away windiness . . . and helpeth the hicket or yeoxing. They stir up appetite, and provoke urine."

Parkinson says that "after all the herbes . . . fit for sallets, there must follow such roots as are used to the same purpose; and first skirrets." "The rootes being boyled, peeled and pithed" . . . are then "stewed with butter, pepper and salt . . . or as others use them, to roule them in flower and fry them with butter" . . . or "as a sallat, cold with vinegar, oyle etc . . ." "any way that men please to use them they may finde their taste to be very pleasant, far beyond any Parsnep, as all agree that taste them.". "They doe help to provoke urine, and it is thought, to procure bodily lust, in that they are a little windy."

Culpeper says skirret is under Venus and is opening and cleansing, good against dropsy, jaundice and troubles of the liver. The young shoots are wholesome as well as the roots, easily digested and provoking urine.

Evelyn says that "Skirrets, *Sisarum*," are "hot and moist, good for the stomach, exceedingly nourishing, wholesome and delicate; of all the Root-kind, not subject to be Windy, and so valued by the Emperor Tiberius that he accepted them for Tribute. This excellent root is seldom eaten raw, but being boil'd, stew'd, roasted under the embers, baked in Pies, whole, sliced, or in pulp, is very acceptable to all Palates."

SMALLEDGE   *Apium graveolene*
SMALLAGE   J.J.

Gerard treats of "water Parsley or smallage," called "in shops, *Apium*, absolutely without an addition." It has "smooth and glittering leaves, cut into very many parcels, yet greater and broader than those of common parsley," and "little white floures" with seeds, "somewhat lesser than those of common parsley." It is called in English "smallage, Marsh Parsley, or water Parsley." The juice is good for many things, from all the cures effected by parsley to being a remedy for "long-lasting agues . . . ulcers of the mouth . . . bites of spiders . . . malignant sores . . . felons and whitlowes in the fingers. . . ." It is not as safe to use as garden parsley and "it hurteth those that are troubled with the falling sicknesse. . . ."

Parkinson says that although "Smallage" grows wild in moist grounds, it is also "much planted in gardens and although his evil taste and savour doth cause it not to be accepted in meates as Parsley, yet it is not without many special good properties, both for outward and inward diseases. . . ." He adds, "The juice cleanseth ulcers; and the leaves boyled with Hogs grease, healeth felons on the joynts of the fingers."

Culpeper says it is under Mercury, hotter than parsley and more medicinable. It opens obstructions and helps jaundice. The leaves eaten in the spring purify the blood.

Evelyn bridges the gap for us, since our celery is *Apium graveolens var. dulce*, and celeriac is *Apium graveolens var. rapaceum*. Evelyn's "Sellery . . . formerly a stranger with us," is "*Apium italicum* of the Petroseline Family . . . a more generous form of Macedonian Parsley, or Smallage. The tender leaves of the Blancht Stalk do well in our Sallet, as likewise the slices of the whiten'd Stems, which being crimp and short, first peeled and slit long wise, are eaten with Oyl, Vinegar, Salt and Peper, and for its high and grateful Taste is ever placed in the middle of the Grand Sallet . . . as the Grace of the whole Board."

Josselyn observed "Smalledge" as "of such garden herbs among us as do thrive there."

SNAKEROOT, BLACK *Cimicifuge racemosa*
Sanicula marilandica

J.W.

CANADIAN *Asarum canadense*
SENECA *Polygala seneca*
VIRGINIA *Aristolochia serpentaria*
WHITE *Eupatorium rugosum*

It was considered a wise precaution by such worthies as John Winthrop to have a bit of "snakeroot" in one's pocket when traveling about in the New England wilderness, whether as a preventive or a cure I do not know. I find no reference to growing it in gardens.

SNEEZEWORT *Achillea ptarmica*   (Introduced)
(Naturalized)

Sometimes called "false pellitory," according to Culpeper and "wild Pellitory," according to Gerard, this plant is "of great beautie" with its "single floures like unto double Fetherfew." "Its smell procureth sneezing."

SOLOMONS SEAL
Josselyn's "three kinds"
Polygonatum biflorum (native, very like the next)
Polygonatum multiflorum ("common in England")
Polygonatum virginianum (Virginia Solomons Seal)
Smilacina racemosa (Treacle Berries)
(False Solomons Seal)

Gerard describes several sorts of "Solomons Seale." The third variety in Gerard, Josselyn noted, was popular with the early settlers, as the berries were edible.

1 *Polygonatum.*
Solomons Seale.

Gerard says of the plant generally that Dioscorides wrote the "roots are excellent good for to seale or close up greene wounds, being stamped and laid thereon," which Gerard says is the reason for its name, not — as some say — because of marks upon the roots. The root, says Gerard, stamped while it is fresh and green and applied will take away in a night or two "any bruise, blacke or blew spots gotten by falls or womens wilfulnesse, in stumbling upon their hasty husbands fists." Although Galen warned against taking it inwardly, Gerard knows of many cases especially in Hampshire where it is given in ale to mend broken bones inwardly and has been used in this way with success upon cattle. Indeed, says Gerard, while what "might be written of this herbe as touching the knitting of bones, and that truely, would seeme to some incredible . . . common experience teacheth that in the world there is not to be found another herbe comparable to it for the purposes aforesaid." He adds that Italian women use water from the roots to take away sunburn.

I cannot find it in Parkinson's garden.

Culpeper says Saturn owns the plant and he runs through all the above "experience" of it in knitting bones, adding that it is good also for staying vomitings and bleedings and for ruptures. He also notes that it is the principal ingredient of "beauty wash."

SORREL   *Rumex acetosa*   J.J.
                                                           W.W.
"1 oz Sorrell seed at 2d" –J.W.Jr.

Gerard calls it *Oxalis, sive Acetosa* as does Parkinson, in referring to the "common Garden Sorrell." Gerard says, "in shops commonly Acetosa." In Gerard there are several kinds listed and pictured, but Parkinson deals with only one and describes it well, with "tender greene long leaves full of juice, broade, and bicorned as it were, next unto the stalke, like as Arrach, Spinach, and our English Mercurie have, of a sharpe, sower taste." Gerard says, "Sorrell doth undoubtedly coole and mightily dry, but because it is soure it likewise cutteth tough humours. . . . The leaves sodden and eaten in manner of a Spinach tart, or eaten as meate . . . doth attemper and cool the bloud. . . ." It is a "profitable sauce in many meats."

Parkinson says, "Sorrell is much used in sawces, both for the whole and the sicke, cooling hot livers and stomackes . . . procuring . . . an appetite unto meate. . . . It is also of a pleasant relish for the whole in quickening up a dull stomacke that is over-loaden with every daies plenty of dishes. It is divers waies dressed by Cookes, to please their Masters stomacks."

Culpeper says it is under Venus and concludes his list of virtues with ". . . the leaves eaten in a salled are excellent for the blood." Before this, he has run through most of the ills flesh is heir to, from the itch, agues, worms and fluxes, to scorpion bites, sores, and jaundice.

One imagines Winthrop intended it as a salad herb, which is where Meager puts it, in both a French and English variety.

Evelyn says of "Sorrel, *Acetosa*, of which there are divers kinds . . . Acid, sharpening Appetite asswages Heat, cools the Liver, strengthens the Heart, is so grateful a quickness to the rest, as supplies the want of Orange, Limon, and other Omphacia, and therefore never to be excluded. Vide Wood-Sorrel."

2 *Oxys lutea.*
Yellow wood Sorrell.

SORREL WOOD *Acetosella oxalis*   J.J.
                                                                     W.W.

Gerard illustrates two sorts of wood sorrel, one white, one yellow. He says "Wood Sorrell or Cuckoo Sorell" is called *Trifolium acetosum;* but that "Apothecaries and Herbarists call it *Alleluya*, and *Panis cuculi* or Cuckowes Meate, because either the Cuckow feedeth thereon, or by reason when it springeth forth and flourisheth the Cuckoo singeth most, at which time also Alleluya was wont to be sung in churches." "Stamped" and used for green sauce, it strengthens the stomach and "procureth appetite, and of all Sorrel sauces is the best, not only in vertue but also in the pleasantnesse of his taste."

Parkinson, having listed sorrel with the docks, does not treat of wood sorrel, but Culpeper considers it even more effectual than the true sorrel in "hindering the putrefaction of the blood," in pestilential fevers and as a gargle.

Evelyn lists "Wood-Sorrel, *Trifolium acetosum*, or *Alleluja*, of the nature of other Sorrels."

SOUTHERNWOOD *Aretmisia abrotanum*   J.J.
                                                                     (Introduced)

Gerard says of "Southernwood" or "Abrotanum" that there are several kinds. Dioscorides affirmed there were two, female and male, or the greater and the lesser, and Gerard says there is also one of a sweet smell and lesser than the others, and others "of a bastard kinde." The greater can grow as big as a shrub, with many branches and finely indented somewhat white leaves "of a certaine strong smell," with, instead of flowers, "little small clusters of buttons . . . of colour yellow." It is hot and dry in the third degree and has force to distribute and rarefy. The tops boiled and drunk help those short of breath, with a cramp or with sinews drawn together. It helps those with sciatica. It kills worms and, taken in wine, is a remedy against deadly poisons. It helps against the stingings of scorpions and spiders. Strewed around the bed or smoked upon embers, it drives away serpents. Put under the bed's head, they say, it "provoketh venerie." It is good to stop baldness, and against all stoppings of the spleen, kidneys and bladder. It digests all cold humors, and takes away the shivering cold that comes from ague fits. It is good also against cold swellings.

Parkinson deals only with the Lavender Cotton, which he considers the feminine of the above, and says it is used chiefly for bordering knots in gardens, although it is good in baths and ointments "used for cold causes," and the seed is used for worms.

Culpeper says it is a Mercurial plant and lists the same uses as Gerard's, adding that it will take out splinters and that it is "put among ointments that are used against the French disease." The Germans esteem it as a wound herb and call it stab-wort.

S P A R A G R A S   *Asparagus officinalis*   J.J.

Gerard treats of "Sperage, or Asparagus" and gives his first sort as "Garden Sperage, *Asparagus sativum*." We may quote his "vertues" in full.

"The first sprouts or naked tender shoots hereof be oftentimes sodden in flesh broth and eaten, or boyled in faire water and seasoned with oyle, vinegar, salt and pepper, then are served at mens tables for a salled, they are pleasant to the taste, easily concocted, and gently loose the belly. They somewhat provoke urine, are good for the kidnies and bladder, but they yeald unto the body little nourishment, and the same moist, yet not faultie; they are thought to increase seed and stir up lust."

Parkinson says that the first shoots are "a Sallet of as much esteeme with all sorts of persons as any other whatsoever . . . almost wholly spent for the pleasure of the pallate." He mentions also that it is good for opening "the reines or kidneys."

Culpeper says "Asparagus, Sparagus, or Sperage" is under Jupiter. To all the above uses, he adds that it makes a good bath or wash against all pains, and that a decoction of the roots is good to clear the sight and against toothache. He insists it is an aphrodisiac "whatsoever some have written to the contrary."

Evelyn says that "next to Flesh" nothing is more nourishing than "Sparagus." Seldom eaten raw, it is "with more delicacy . . . being so speedily boil'd as not to lose the verdue and agreeable tenderness; which is done by letting the Water boil before you put them in."

SPEARMINT *Mentha spicata*   J.J.
                                          (Introduced)

Gerard puts "Speare Mint" third in his mints and calls it *"Mentha Romana."* It is a "mint with a narrow leaf and in English is called Speare Mint, garden Mint, our Ladies Mint, browne Mint and Macrell Mint."

For the rest, see MINT.

SPEEDWELL *Veronica officinalis*   J.J.
SPEEDWELL, GERMANDER *Veronica chamaedrys*
MALE FLUELLIN
                                          (Naturalized)

Gerard gives us several sorts of "Fluellen or Speedwell": "Female," "Male," "sharpe pointed," "shrubby," "upright," "tree" and "leaning." He divides them into two chapters, apparently by sex, the first being the female and including the sharp-pointed "Elatine," said to be especially good for the eyes. Both these seemingly quite unlike plants are of "a singular astringent facultie." Gerard's second chapter treats of "Fluellen the male, or Paul's Betonie" and includes also the sorts of veronica we are used to seeing in gardens today. Johnson, obviously hoping to restore a sort of order to random collections of different plants, announces firmly that while "these plants are comprehended under this generall name *Veronica*, Dodonaeus, Turner and Gesner have all disagreed about which is truly the "Betonica of Paulus." But Johnson says, "We do call them in English, Paul's Betony or Speedwell; in Welch it is called Fluellin and the Welch people do attribute great virtues to the same. . . . These are of a meane temperature, between heate and drinesse. The decoction of *Veronica* drunke, sodereth and healeth all fresh and old wounds, clenseth the bloud from all corruption, and is good to be drunke for the kidnies, and against scurvinesse and foule spredding tetters, and consuming and fretting sores, the small pox and measels. The water of *Veronica* distilled with wine and re-distilled so often untill the liquor wax of a reddish colour, prevaileth against the old cough, the drinesse of the lungs, and all ulcers and inflammation of the same."

Parkinson seems to feel veronicas are not for his garden.

Culpeper, in the Physicians Library refers *Veronica* to *Betonica Pauli* "or male Lluellin to which add *Elatine* or female Lluellin." They are temperate, stop defluxions or humors that fall from the head into the eyes and are profitable in wounds. He tells a story which Gerard quoted, about a woman who had nearly lost her nose from the ravages of the Pox, when she was able to cure herself "by only taking her own Countrey Herb" inwardly and outwardly. Gerard made much more of this tale, having the surgeons holding ready knives to amputate the remains of the nose when an illiterate "Barbar" got them to halt until he found this herb, "which he knew well though not by name," and he cured her.

Culpeper's *English Physician* is brief. Governed by Venus, speedwell or *Veronica* is a vulnerary plant used inwardly and outwardly, also pectoral, and helpful against the stone, strangury, and pestilential fevers.

S P I N A C H   *Spinacia oleracea*

                                    "1 oz spynadg at 2d" – J.W.Jr.

Gerard says of "*Spinacia* . . . Spinach is a kind of Blite, after some notwithstanding I rather take it for a kind of Orach." "It is called in these daies Spinachia." "It is one of the pot-herbes whose substance is waterie, and almost without taste, and therefore quickly descendeth and looseneth the bellye. It is eaten boiled, but it yeeldeth little or no nourishment at all;

it is sometimes windie . . ." "It is used in sallades when young and tender. This herbe of all other pot-herbes and sallade herbes maketh the greatest diversitie of meates and sallades."

Spinach, or Spinage, says Parkinson, "is an herbe fit for sallets and for divers other purposes for the table only. . . . Many English that have learned it of the Dutch people, doe stew the herbe in a pot or pipkin without any other moisture then it owne, and after the moisture is a little pressed from it they put butter and a little spice unto it and make therewith a dish that many delight to eate of." "It is used likewise to be made into Tartes, and many other variety of dishes as Gentlewomen and their cookes can better tell then my selfe."

Culpeper is willing to let it go as chiefly a salad herb, "more for food than for medicine."

Evelyn does not hold with it raw in salads, in fact, "the oftener kept out the better," but boiled "to a Pult, and without other Water than its own moisture, [it] is a most excellent Condiment with Butter, Vinegar, or Limon, for almost all sorts of boil'd Flesh . . . Laxative and Emollient and therefore profitable for the Aged. . . ."

S T I T C H W O R T   *Stellaria graminea*   J.J.
                                                      (Naturalized)

Josselyn lists stitchwort as fourth in the plants he found in New England which were common also in England. He says it is commonly taken by ignorant people in New England for eyebright. "It blows in June."

Gerard says that its "vertues" are such that "they are wont to drink it in wine with the powder of acornes, against the paine in the side, stitches and such like"; and that it was reported by Dioscorides that the seed of stitchwort being drunk "causeth a woman to bring forth a man childe, if after the purgation of her sicknesse, before she conceive, she do drinke it fasting thrice in a day, half a dram at a time, in three ounces of water many dayes together."

Chickweed is *Stellaria media.*

Starwort is *Stellaria pubera.*

This would seem to be one of the plants everyone had, but considered beneath garden notice.

STOCKE GILLOFLOURES   *Matthiola incana*
"1/2 oz stockielliflower 3d" –J.W.Jr.

Gerard calls "Stocke Gillofloures" "*Leucoium,*" as does Parkinson. (*Leucojum* is now the name applied to the bulbous herbs known as Snowflakes.) Gerard has a pretty description: "The stalke of the great Stocke Gillofloure is two foot high . . . and parted into divers branches. The leaves are long, white, soft and having upon them as it were a downe like unto the leaves of willowe, but softer . . . the floures . . . white . . . purple . . . double floures of divers colours, greatly esteemed for the beautie of their floures and pleasant sweet smell." And he adds, "They are not used in Physicke, except amongst certaine Empericks and Quack-salvers, about love and lust matters, which for modestie I omit."

Parkinson lists a great many from many places, and says of its name, "Leucoium (which is in English the white Violet) is referred to divers plants; we call it in English generally Stocke gilloflower, or Stocke gillour, to put a difference between them and the Gilloflowers and Carnations, which are quite of another kindred as shall be shewne in place convenient."

Culpeper does not consider them useful.

Gerard precedes his "Stocke Gilloloures" with a chapter on "Wallfloures, or yellow Stocke-Gilloloures," which see under WALLFLOWER.

STRAWBERRY   *Fragaria virginiana*   Everyone
                  Fragaria vesca var. americana

Gerard says there are "divers sorts of Strawberries; one red, another white, a third sort greene, and likewise a wilde Straw-berrie, which is altogether barren of fruit." The leaves and roots cool and dry. The berries are cold and moist. The leaves boiled and in a poultice take away the heat in wounds. The decoction strengthens the gums, fastens the teeth and is good to hold in the mouth. It stays the bloody flux. The berries quench thirst. The distilled water with white wine is good "against the passion of the heart, reviving the spirits and making the heart merry. The distilled water is said to make the face fair and smooth. The leaves are good in washing waters. The ripe berries quench thirst and, if they be often used, they take away the heat of the face."

Parkinson describes various sorts, among them the "Virginia Strawberry which has the largest leaves of all. Parkinson says the leaves are always used among other herbs in cooling drinks and in lotions and gargles. He repeats Gerard's remedy for making the heart merry, the distilled water drunk alone or with wine. He says it cleanses the skin. Of the berries he says they "are often brought to the Table as a reare service whereunto claret wine, creame or milke is added with sugar . . . a good cooling and pleasant dish in the hot Summer season."

Culpeper says Venus owns the herb and repeats all the uses as given by Gerard, adding only that the juice is good for inflamed eyes.

We have seen how the early settlers welcomed the wealth of strawberries in New England.

SUCCORY See CHICORY and ENDIVE

SUMACH, STAGHORN *Rhus typhina*   J.J.
W.W.

The sumach which the settlers found and used for dyeing was not the same as the one Gerard describes as good for stopping all sorts of fluxes and against toothache. Gerard calls it *Rhus coriaria*, and Culpeper describes what we know as *Rhus cotinus*, which Pliny knew, but which is not the same as ours. So we omit their comments.

SUNFLOWER *Helianthus*

"flower of the sonne 1d" – J.W.Jr.

Gerard in "Of the floure of the Sun, or the Marigold of Peru," calls it *Flos Solis major* and *Flos Solis minor*, differing chiefly in height, and says they both come from Peru "and divers others provinces of America from whence the seeds have been brought into these parts of Europe." He says the name is taken from the fact it is "reported to turne with the Sun, the which" he "could never observe." He prefers to think it is because it looks like the sun. It is also called *"Sol Indianus"* or *"Chrysanthemum Peruvianum,* or the golden floure of Peru." They are "thought" to be "hot and dry of complexion." The "vertues" are not known. "There hath not anything been set downe either of the antient or later writers concerning the vertues of these plants, notwithstanding we have found by triall, that the buds before they be floured, boiled and eaten with butter, vineger and pepper, after the manner of Artichokes, are exceedingly pleasant meat, surpassing the Artichoke in provoking bodily lust. The same buds with the stalks near to the top (the hairiness being taken away) broiled upon a gridiron, and afterward eaten with oile, vineger and pepper, have a like property."

Parkinson begs leave to bring into his Garden of pleasant Flowers, this plant "in regard of his statelinesse." "*Chrysanthemum Peruvianum* or *Flos Solis*," "the golden flower of Peru, or the Flower of the Sunne," he says, has no use "in Physicke with us but that sometimes the heads of the Sunne Flower are dressed, and eaten as Hartichokes are, and are accounted of some to be good meate, but they are too strong for my taste."

SUNFLOWER, AMERICAN   *Helianthemum strumosus*

J.J.

(See Josselyn's sketch.)

TANSY   *Tanacetum vulgare*   Boorde
                                                                    Sewall
                                                    "tansy seeds 2d" – J.W.Jr.
                                                                    (Naturalized)

Gerard says, "Tansie groweth up with many stalkes, bearing on the tops of them certain clustered tufts with floures like the round buttons of yellow Romane, Cammomill or Feverfew (without any leaves paled about them) as yellow as gold. The leaves be long, made as it were of a great many set together upon one stalke, like those of Agrimony. . . . The whole plant is bitter in taste and of a strong smell, yet pleasant." He is brief about the "vertues." "In the Spring time are made with the leaves thereof newly sprung up, and with egs, cakes or tansies, which be pleasant in taste and good for the stomacke. For if any bad humours cleave thereunto, it doth perfectly concoct them and scowre them downward. The root preserved with hony or sugar, is an especial thing against the gout. . . . The seed of Tansie is a singular and approved medicine against Wormes. . . . The same pound and mixed with oile Olive, is very good against the paine and shrinking of the sinewes. . . . Also being drunke with wine it is good against the paine of the bladder. . . ."

Parkinson has tansy in his kitchen garden and mentions that the "gold yellow flowers like buttons . . . being gathered in their prime, will hold their colour fresh a long time." He says the leaves are "used while they are young, either shred small with other herbes, or else the juice of it and other herbes fit for the purpose, beaten with egges, and fryed into cakes (in Lent and the Spring of the yeare) which are usually called Tansies, and are often eaten, being taken to be very good for the stomack, to help to digest from thence bad humours that cleave thereunto: As also for weak raines and kidneys. . . . The seed is much commended against all sorts of wormes in children."

Culpeper's "Garden Tansey" is governed by Venus. Bruised and applied to the navel, it stays miscarriages. Drunk in a decoction boiled in beer, it does the like. Bruised and "often smelled to," it is profitable in the same cases. Culpeper includes the same advice as Gerard and Parkinson, describing "the herb fried with eggs which is called a tansey" and adds that the principal use of the herb is for worms in children, taken either in the seed or as juice.

Evelyn says, "Tansy, *Tanacetum*; hot and cleansing, but in regard of its domineering relish, sparingly mixt with our cold Sallet, and much fitter (tho' in very small quantity) for the Pan, being qualified with the Juices of other fresh Herbs, Spinach, Green Corn, Violet, Primrose-leaves, etc. at entrance of the Spring, and then fried brownish, is eaten hot with the Juice of Orange and Sugar, as one of the most agreeable of all the boil'd Herbaceous Dishes."

Josselyn reports "Tansie" as a garden herb growing well in New England.

Samuel Sewall has recorded observing the body of a friend long dead but well preserved in his coffin packed full of tansy.

Considering the New England countryside today, it is wonderful to think of a time when anyone would need to buy tansy seed for twopence to be sent to New England.

TARRAGON *Artemisia dracunculus*

This is a plant for which I find no direct references and yet I think it best to include it. Gerard treats of it only as "the sallade herbe" called *Dracunculus* in Latin and *Dragon* in French, hot and dry, not to be eaten alone but with lettuce, purslane, and the like, to temper their coldness "like as Rocket doth." He knows of no other use. Which is odd, as Evelyn calls it the herb of "Spanish extraction," recommends it "especially where there is much lettuce," and says, "Tis highly cordial and friendly to the Head, Heart, Liver, correcting the weakness of the Ventricle, etc."

Neither Parkinson for his garden nor Culpeper for his remedies are interested in it.

T E A S E L *Dipsacus fullonum* (Naturalized)
F U L L E R ' S H E R B

Teasel is another plant, like Bouncing Bet, for which I have no recorded sponsor and yet its usefulness and long life in this country argue that it must have had many.

Gerard describes teasels, the tame and the wild, and says the first is grown in gardens "to serve the use of Fullers and Clothworkers." There is small use of them in medicine, but "the heads are used to dresse wollen cloth with." Gerard scorns uses which involve hanging teasel heads full of worms about the neck to cure a fever, as ranking with wearing spiders in nutshells to the same end, the advice of "fantasticke peoples."

Culpeper sees many uses for this "herb of Venus" besides its being "used by the clothworkers." It clears the sight, reduces redness, removes warts, and kills worms in the ears.

T H I S T L E , B L E S S E D See C A R D U U S B E N E D I C T U S

T H R O A T W O R T *Trachelium caeruleum*

Called "Canturbury Bels" by Gerard and recommended as a gargle, it belongs to the Campanulaceae, has hairy, brittle stalks, leaves like nettles and "hollow floures hairy within and of a perfect blew colour, bell fashion." Parkinson calls this: "Great Canterbury Bels, *Trachelium maius*," and says it is used in salads "beyond the Seas" and "may safely bee used in gargles and lotions of the mouth, throat, or other partes."

Note: No documentary evidence to date of this plant in seventeenth-century New England, but I imagine someone else will supply it in time.

T H Y M E   *Thymus serpyllum*   J.J.
"1 oz Thyme seed 6d" – J.W.Jr.
(Naturalized)

Gerard has two chapters on "Time," the first wild and the second "Garden Time," and he gives us five pictures of the first kind and four of the second. However, his first illustration is of "Serpillum vulgre, Wilde Time" which, he says, comes with purple flowers and with white, and there is another "kinde of Serpillum which groweth in gardens" and is upright rather than creeping. Parkinson makes it easier for us to identify the sort of thyme probably brought by John Winthrop, Junior, by announcing that the "true Tyme of the ancient Writers" is not hardy in England. All the other sorts called garden thymes are, indeed, he says, "but kindes of wilde Tyme . . . in the defect of or want of the true Tyme they are used in the stead of it."

Gerard says "Wilde Time" is hot and dry, cutting and biting. It "bringeth downe the desired sicknesse, provoketh urine, applied in bathes and fomentations it procureth sweat; being boyled in wine it helpeth the ague, it easeth the strangurie, it stayeth the hicket, it breaketh the stones in the bladder, it helpeth the Lethargie, frensie and madnesse, and stayeth the vomiting of bloud . . . is good against the wambling and gripings of the bellie, ruptures, convulsions, and inflammations of the liver . . . helpeth against the bitings of any venomous beast . . . and . . . such as are grieved with the spleene. . . ."

Gerard's garden thyme has much the same properties except that he stresses its use against coughs, sciatica, worms and melancholy "or any other humour of the spleene."

Parkinson says the substitute thyme has the "vertues" of the true: to help melancholic and splenetic diseases, flatulent humours and the toothache.

Culpeper says "Thyme," so commonly known, is under the government of Venus, strengthens the lungs, is an unsurpassed remedy for the "chin cough" or whooping cough. It is good for those with sciatica, worms, warts, dull sight, shortness of breath, gout and hardness of the spleen. Culpeper says that "Wild Thyme, or Mother of Thyme" is equally well known, also under Venus, but, being under the sign of Aries, is "chiefly appropriated to the head," although he adds also all the uses it may be put to for the spleen, worms, gout, cough, frenzy and lethargy.

Evelyn does not mention it.

And John Josselyn, though he saw "time" growing in New England gardens, did not use it in his eel "receipt" where he did use all the other flavoring herbs.

Anne Bradstreet, in her extreme modesty, requested a wreath of parsley and of thyme.

TOADFLAX   *Linaria vulgaris*
FLAXWEED   J.J.
                                              (Naturalized)

Gerard gives "Tode-flax" a chapter to itself, describing it as a "kinde of Antyrrhinum," called by the "Herbarists of our time, Linaria or Flax-weed." The decoction takes away yellowness of the skin when used as a wash. "The same drunken" opens the liver and spleen, the kidneys and the bladder, and is good against jaundice.

Parkinson says that, while the true flax is not fit for a garden, there are several pretty sorts of "Tode flax" "whose pleasant and delightfull aspect doth entertaine the beholders eyes with good content," which are also good for the kidneys.

Culpeper says Mars owns the herb, in Sussex called gallwort and given to hens in their water to cure them of the gall. He approves of it also as a skin-cleanser, for the kidneys, dropsy, for inflamed eyes, foul ulcers, to expel poison and the dead child or afterbirth.

No wonder it covers our New England roadsides today.

TOBACCO   *Nicotiana Tabacum*   Endecott   J.W.
                                           J.J.   C.M.

Gerard treats of "Tabaco, or Henbane of Peru," and says it was brought from the "West Indies, in which is the countrey or province of Peru." He says the people of America call it Petun, Dodoens called it Henbane of Peru, and Monardes called it Tabacum. Gerard thinks it to be a kind of henbane as it has like qualities: "It bringeth drowsinesse, troubleth the sences and maketh a man as it were drunke by taking of the fume."

Gerard lists above twenty-five virtues of tobacco, six of which have to do with smoking it. It is used in salves and ointments and juice and oil against fits of the mother, old ulcers, worms, ulcers, scabs, dropsy, kibed heels, bites of venomous beasts, spots before the eyes, headaches, swooning, piles, sciatica, burns and scalds and deep wounds. One quarter of its uses are when the "drie leaves are used to be taken in a pipe set on fire and suckt into the stomacke, and thrust forth again at the nostrils." The whole purpose of this is "against the pains of the head, rheumes, aches . . . ," which the leaves "do palliate or ease for a time but never performe any cure absolutely . . . although they empty the body of humours. . . ." Gerard, indeed, does not think much of it. "Some use to drinke it (as it is tearmed) for

‡ 1 *Hyoſcyamus Peruvianus.*
   Tabaco or Henbane of Peru.

‡ 2 *Sana Sanĉta Indorum.*
   Tabaco of Trinidada.

wantonesse or rather custome, and cannot forbeare it, no not in the midst of their dinner . . . unwholesome and very dangerous. . . ." He commends the "syrup above this fume or smokie medicine."

Parkinson calls it "Indian Henbane, or Tabacco," and says that "the Herbe is out of question an excellent helpe and remedy for divers diseases, if it were rightly ordered and applied, but the continual abuse thereof in so many doth almost abolish all good use in any." He feels sure that if men could find out its real virtues, "many strange cures" would result. In the meantime he recommends a salve for ulcers and old wounds, and a syrup for a gentle vomit. The ashes also help cuts in the hands and small green wounds.

Culpeper says it is a martial plant and a good expectorant. Besides the same virtues as those quoted above, he adds that it will kill lice. When smoked, it should be taken fasting.

Roger Williams in *Key into the Language of the Indians of New England*, says, "They generally take tobacco and it is commonly the only plant which men labour in, the women managing all the rest. They say that men take tobacco for two causes: first against the rheum which causeth the tooth-ake, which they are impatient of, secondly to revive and refresh them, they drinking nothing but water."

Roger Williams says also, "They take their wuttammauog — that is, weak tobacco — which the men plant themselves, very frequently. Yet I never see any take so excessively as I have seen men in Europe, and yet excess were more tolerable in them, because they want the refreshing of beer and wine, which God had vouchsafed Europe."

What the settlers really thought of tobacco is a moot question, as hard to answer as it would be for people today. The only difference is that today we do not use it for dressing wounds. Endecott suggested it for inducing sneezing. Cotton Mather was rather ahead of his time in asserting that the "caustick Salt in the Smoke . . . may lay foundations for Diseases in Millions of unadvised People." John Josselyn saw it grown tenderly in Maine in special beds. John Winthrop asked for it dried. The settlers knew, by repute at least, of Monardes and his early accounts of its use among the South American Indians as a drug, inhaled from ashes in a bonfire through very long pipes, which they sniffed standing and were then knocked out for several hours, to awake most refreshed. Their priests used it to induce hallucinations in themselves which were regarded as most remarkable until the natives caught on and used it for the same end. It would be as hard to generalize about tobacco then as now.

TORMENTILE *Potentilla erecta*   J.J.
                                                (Adventive)

Gerard says "Of Setfoil, or Tormentill" "one of the Cinkfoiles," that the root "doth mightily dry and is binding. It is used against pestilence, "for it strongly resisteth putrefaction and procureth sweate." The leaves and roots boiled in wine provoke sweat and "by that means driveth out all venom from the heart . . . and preserveth the bodie in time of pestilence. . . ." The roots powdered and taken in the water from a smith's forge where hot steel has been quenched cure "the laske and bloudy flix" even when the patient has a fever. It stops "all issues of bloud" and the decoction is good for all wounds, heals the liver and lungs and is good for the jaundice. "The root beaten into a pouder . . . tempered with the white of an egge . . . staieth the desire to vomite, and is good against choler and melancholie."

Parkinson does not consider it a garden plant.

Culpeper says it is "an herb of the Sun," excellent to stay all fluxes. The juice resists all poisons, even the plague itself, the "French disease," measles and so on and on. Culpeper has a very high opinion of its drying qualities and recommends it for "any sharp rheum," sciatica, gout, ruptures, piles and running sores.

Geoffrey Grigson says that the roots were boiled in milk and given to calves and children to bind their loose insides.

Josselyn listed it as common to both countries.

TURNIP *Brassica rapa*   J.W.
                                                J.J.

Gerard begins the second section of his herbal with the title, "Containing the description, place, time, names, nature and vertues of all sorts of Herbes for meats, medicine, or sweet smelling use, etc." And the first entry is "Chap. I. Of Turneps." "The root may be eaten raw," he says, "as is done by the poor people in Wales, but it is usually boiled when it yields more nourishment. . . . They do increase milke in women's breasts, and naturall seed, and provoke urine." The decoction of turnips is good, hot or cold, is good for kibed heels. The young shoots make a good salad. The seed is a remedy against poisons. The oil pressed from the seed is given to children for worms and, "washed with water," it "doth allaie the fervent heat and ruggedness of the skin."

Parkinson and Culpeper disregard turnips.

We have seen John Winthrop writing to his wife to get the turnips in. John Josselyn saw them in gardens. And there is a court record in Salem of a woman who caused trouble while her husband was working in the fields, trying to get other men to come into her house. She was eventually shipped back to England as undesirable, but part of the evidence against her concerned turnips — she challenged any man to come in and kiss her and to say if she had or had not been eating turnips.

Evelyn commends them "in composition" and gives a recipe for turnip bread.

V A L E R I A N   *Valeriana officinalis*
G A R D E N   H E L I O T R O P E
(Introduced)
V A L E R I A N ,   B L U E   G R E E K   *Polemonium caeruleum*
J A C O B ' S   L A D D E R
(Introduced)

While these are dissimilar plants they were both considered valerians before Linnaeus sorted them out. Gerard and Parkinson treat of them together. Gerard describes "Valerian or Setwall" beginning with the "tame or garden valerian" whose root has "a pleasant, sweet smell when it is broken," Gerard says this sort, which grows tall and has a whitish flower, and the "Greekish Valerian," which is shorter and has a blue flower, are both grown in gardens. Gerard says it is called in English, besides valerian, "Capon's taile, and Setwall." He does not explain this last name which he says belongs properly to "*Zedoaris* which is not a Valerian." However, he quotes Dioscorides as saying the garden valerian is hot, "provoketh urine, bringeth down the desired sicknesse, helpeth the paine in the sides, and is put into counterpoisons . . . whereupon it hath been had (and is to this day among the poore people of our Northern part) in such veneration amongst them that no broths, pottage, or physicall meats are worth anything if Setwall were not at an end; whereupon some woman Poet or other hath made these verses:

> They that will have their heale,
> Must put Setwall in their keale.

Gerard says it is used for "sleight cuts, wounds, and small hurts." The extraction of the roots is given for the yellow jaundice, and the leaves for ulcers and sores of the mouth. Parkinson says that many valerians are fitter for a garden of simples than one of delightful flowers, but he selects the "Greeke Valerian" as one fit for a garden. He says that the mountain valerian is of most use in physic, the rest having little use "or none that I know" except for the "great garden kinde, or the Indian Nardus, in whose stead anciently it was used in oyles, ointments etc."

Culpeper deals only with the garden sort and quotes Dioscorides and Pliny throughout, after saying it is under the influence of Mercury. He adds to all the uses mentioned above that it will draw splinters.

3 *Valeriana minor.*
Small Valerian.

VERONICA   See   SPEEDWELL

VERVAIN   *Verbena officinalis*
VERVAIN, BLUE   *Verbena hastata*   J.J.
(Naturalized)

"Vervaine and his kindes" is one of the plants about which Gerard felt bound to report: "Many odd olde wives tales are written of Vervaine tending to witchcraft and sorcerie," which, Gerard says, "you may reade elsewhere. . . ." He says physicians have given vervain for the plague, but he feels truth in the report that the devil deceived them into it. Still, Gerard is able to recommend vervain for "paines of the mother," wounds, tumors, sores, fevers, falling hair, the stone, gripes, the bloody flux, headache, sore throat and toothache. He quotes Pliny as saying if the "dining room be sprinkled with water in which the herbe hath been steeped, the guests will be the merrier."

Parkinson does not have vervain in his garden.

Culpeper says vervain is "an herb of Venus," excellent to strengthen the womb, and "remedy all the cold griefs of it, as plantane doth the hot." He says that, bruised and hung about the neck, it helps the headache, so we we understand Gerard's doubts. There is practically nothing according to Culpeper, that vervain cannot cure, and he adds to a list of all the troubles of every organ in the body, that vervain is an antipoison, and will remove freckles.

The "vertues" of the English vervain as recorded above explain why Josselyn described what was probably our *Verbena hastata* as "Clownes all heal, of New England" which, he says, "some of our English practitioners take . . . for Vervene, and use it for the same, wherein they are grossly mistaken."

We can see, also, why the common name for vervain was "Simpler's Joy," because it grows everywhere and cures everything.

VINE   *Vitis novae-angliae*   Everyone

Of the twelve varieties of native American grapes noted by Asa Gray, the one he calls *Vitis novae-angliae* will have to stand for all the grapevines the early explorers and settlers rejoiced to see, and that John Josselyn recorded as growing well. What grapes were sent, with the hope that this country, being warmer in the summer than England, might offer an opportunity to make wine and raisins, we cannot know. But Gerard must have encouraged these hopes by devoting ten pages of his herbal to what he modestly calls "The Manured Vine." As this is immediately followed by a short page on "Hops," we can see why the settlers, disappointed in the quality of New England grapes, however well manured, and realizing that wines and raisins were beyond their powers, settled for hops and beer and bought dried blueberries from the Indians in lieu of raisins.

V I O L E T S   *Viola odorata*   "violett seeds 2d" – J.W.Jr.

Gerard remarks upon how many flowers are called violets, and yet he will address himself to the "blacke or purple violets, or March Violets of the Garden, which have a great prerogative above others, not only because the minde conceiveth a certaine pleasure and recreation by smelling and handling of those most odoriferous flours, but also for that very many by these Violets receive ornament and comely grace . . . Garlands, Nose-gaies, and poesies . . . yea Gardens themselves receive by these the greatest ornament of all, chiefest beauty and most gallant grace . . . for they stir up a man to that which is comely and honest. . . . For it would be an unseemely thing . . . for him that doth look upon and handle faire and beautifull things, and who frequenteth and is conversant in faire and beautiful places, to have his minde not faire. . . ."

The "vertues" are myriad. The flowers are good for all inflammations; an oil made of violets is cool and moist, and if the violets are fresh, and it is applied to the afflicted parts, will provoke sleep "which is hindered by a hot and dry distemper." "The later Physitians . . . mix dry Violets with medicines that are to comfort and strengthen the heart." The leaves are helpful used inwardly and outwardly. And there is a most useful syrup made by boiling sugar to a "meane thicknesse," into which one puts clean violet flowers and sets them "upon a gentle fire to simper." One must then, "strain it and put more violets, three or foure times . . . so have you it simply made

Early American Gardens

of a most perfect purple colour, and of the smell of the flowres themselves." Lemon juice added will enhance the beauty but not the virtue. This is good against the pleurisy and cough and agues in young children, especially with eight or nine drops of "Oyle of Vitriol" to an ounce of syrup.

Sugar Violets "comforteth the heart and other inward parts."

Parkinson includes under "*Viola* Violets" two violets — the Single March and the Double March — and then goes on to "Harts eases or Pansies," single and double, and "The great yellow Pansie," which has to be grown from slips as it does not seem to bear seed. He says the first are called "March violets," and hearts ease which he considers a "kind of Violet," are called *Viola* tricolor. The properties of violets, Parkinson says, are "sufficiently known to all, to cool and moisten . . ." and forbears to "recite the many virtues which may be set downe . . . opening . . . purging. . . ." Hearts ease is said to be beneficial "in the Frenche disease" . . . "which how true it is I know not."

Culpeper says of violets, "They are a fine pleasing plant of Venus, of a wild nature, no way harmful." Culpeper lists several ways of using violets and says, "The powder of the purple leaves of the flowers only, picked and dried, and drunk in water, is said to help the quinsey and the falling sicknesse in children. . . ."

VIPERS-GRASS *Scorzonera hispanica* (Black Salsify)
　　　　　　　　Tragopogon porrifolium (Salsify)
　　　　　　　　　　　　　　　　　　　　　(Introduced)
GOATS-BEARD *Tragopogon pratensis*　　　　(Naturalized)

These dissimilar plants are put together here because of Evelyn who deals with "Viper-grass, Tragopogon, Scorzonera, Salsifex, etc." and seems to regard them as much the same plant, "excellent against the Palpitations of the Heart, Faintings, Obstruction of the Bowels, etc. and are besides a very sweet and pleasant Sallet; being laid to soak out the bitterness, then peeled, may be eaten raw, or Condited, but best of all stew'd with Marrow, Spice, Wine, etc. as Artichoak, Skirrets etc. sliced or whole. They likewise may bake, fry or boil them; a more excellent Root there is hardly growing."

It would seem that the early settlers used goats-beard as their "excellent root." Gerard compares "Scorzonera," the Spanish remedy against vipers and plague, with goats-beard, of which there are a purple and a yellow-flowered sort, and considers them similar. Parkinson has both viper-grass and goats-beard in his garden and prefers the yellow goats-beard for the kitchen garden and the purple for medicine.

In all fairness, Gerard says of "Goats Beard or Go to Bed at Noone," the roots boiled in wine "asswageth the paine and pricking stitches of the sides. The same boyled in water untill they be tender, and buttered as parseneps and carrots, are a most pleasant and wholesome meate, in delicate taste far surpassing either . . . which meate procures appetite, warmeth the stomacke, prevaileth greatly in consumptions, and strengtheneth those that have been sicke of a long lingring disease."

V I P E R S   B U G L O S S   See   **B U G L O S S**

W A L L F L O W E R S   *Cheiranthus cheiri*
"1/2 oz Wallflower seed at 4d per oz" – J.W.Jr.

Gerard calls them *Leucojum*, but says there is confusion with another *Leucojum*, thought to be that of Theophrastus, which is called *Leucojum bulbosum* (as see our snowdrops). Gerard says the single yellow wallflower grows "upon bricke and stone walls, in the corners of churches every-where. . . . ," the double . . . "in most gardens of England." Dioscorides said the "yellow Wall-flowre is most used in physicke." The juice in a lini-ment, the herb boiled in wine, the leaves "stamped" with a little bay salt, and a decoction of the flowers will cure ulcers, cankers, the ague, tumors, and gout. A strong decoction "expelleth the dead child."

Parkinson calls it *"Keiri sive Leucojum luteum"* or *"Cheiri"* by which name it is chiefly known in our Apothecaries shops, because there is an oyle made there called *Cheirinum.*" "The sweetnesse of the flowers causeth them to be generally used in Nosegays and to deck up houses, but physically they are used in divers manners; as a Conserve made of the flowers is used for a remedy both for the Appoplexie and Palsie. . . . The oyle made of the flow-ers is heating and resolving, good to ease paines of strained and pained sinews."

Culpeper says the Moon rules them, lists all the above remedies and par-ticularly mentions the conserve against apoplexy and palsy.

WOAD-WAX   *Genista tinctoria*
DYERS-GREENWEED   Endecott
                                                 (Naturalized)

Gerard calls it "Base Broome or greening weed," and has the plant Endecott is reputed to have brought over as "Genistella tinctoria." He says these
brooms are supposed to have the same properties as "common Broome . . .
notwithstanding their use is not as well known, and therefore not used
where the other may be had; we shall not need to speak of that use that
Diers make thereof, being a matter impertinent to our Historie."

When the plant imported for dyeing "and colours" escaped to grow wild
on Gallows Hill near Salem, it was called "witches' blood." It has become a
strong and very pretty weed today in all that part of Essex County.

WOOD SORREL   See SORREL

WORMWOOD   *Artemisia absinthum*
           Artemisia abrotanum (Southernwood)
           Artemisia dracunculus (Tarragon)
           Artemisia pontica (Roman wormwood)
           Artemisia vulgaris (Mugwort)

Wormwood, says Gerard generally, beginning with the "common" and
the "Ponticke," is hot and dry, bitter and cleansing, and has power to bind
and strengthen. It is good for a weak stomach and to be taken either before
or after a "surfeit," or "large eating and drinking." It is profitable for those
troubled with inward bleeding, worms, and a "stinking breath." It withstands all putrefactions, "keepeth garments also from the Mothes; it
driveth away gnats, the bodie being annointed with the oile thereof." It
helps those who are "strangled with eating of Mushrooms or Toadstools if
it be drunk with vineger." It is a good anti-poison, and helps dim eyes and
sore ears.

Gerard goes on in other chapters to list lesser wormwoods with the same
qualities as the common, variously called: sea, creeping sea, "Holie" (which
is a foreign plant and the one that yields wormseed) white, "of Aegypt,"

unsavory, "Small Lavander Cotton," mugwort, and then a group of "Sothernwood." Of these, we have dealt with many separately. It is enough here to see what a lot of plants were supposed to share the special property of expelling worms.

In spite of the use of *Artemisia abrotanum* for keeping borders "in fashion," Parkinson includes no wormwoods in his garden.

Culpeper entertains three wormwoods: sea, common and Roman. Wormwood, he says, is "an herb of Mars," hot and dry, cleansing the body of choler. Besides all the virtues mentioned by Gerard, Culpeper adds that wormwood added to ink will keep mice from destroying paper on which it has been used. It is an antidote to the stingings of all "martial creatures," like wasps, hornets and scorpions. It is a cure for the colic and for black-and-blue spots.

YARROW *Achillea millefolium*   J.J.
                                                          W.W.   How
                                                                 Sewall

Gerard says *Millefolium* is the Latin Herbarists' name for it, although Pliny said it was *Achillea*, as Chiron's pupil Achilles discovered it, and Dioscorides agreed with this. It is also called *Militaria* in Latin. The French call it *Millefeuille* . . . in English it is Yarrow, Milfoile and Nose-bleed. In any case, Gerard grows the red-flowered variety which is less common than the white.

The leaves, Gerard says, close up wounds and staunch blood in any part of the body. It is put in baths for women. "Being drunke it helpeth the bloody flixe. . . . The leaves being put into the nose do cause it to bleed and ease the pain of the megrim."

Parkinson does not admit it as a garden plant.

Culpeper says it is under Venus, and that its virtues in wound healing and against inflammations and toothache, and for stopping the bloody flux, show it to be dry and binding. The decoction, used to bathe the head, "stayeth the shedding of hair." "There is an ancient charm," he says, "for curing tertian agues with yarrow. . . . A leaf of it is to be pulled off with the left hand, pronouncing at the same time the sick man's name; and this leaf is to be taken." So much for that, apparently, but he goes on to say that

† 2 *Achillea, siue Millefolium nobile.*
Achilles Yarrow.

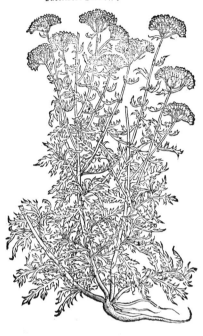

the same thing has been said of feverfew. "In old times," he says, "the names of plants, as well as now, were too much confounded." His conclusion is surprising. The charm stands unquestioned, but "feverfew seems best for the purpose."

We do not know why Samuel Sewall gathered it in March. Abraham How's recipe combined it with brandy and gunpowder and either comfrey or borage for a pain in the back.

Magnus Deus in minimis.

Frontispiece for *The Newe and Admirable Arte of Setting of Corne*, 1600.

Bibliography

For the convenience of those who, like myself, turn to the Bibliography before reading the book, I have made a record of where each volume that I used was found, and which edition it was. The key to the abbreviations of the sources follows.

A.	Author's Library
A.A.S.	American Antiquarian Society
B.A.	Boston Athenaeum
B.N.	Bibliothèque Nationale
B.M.L.	British Museum Library
H.	Harvard, Houghton Library
M.H.S.	Massachusetts Historical Society
Mass. Hort. Soc.	Massachusetts Horticultural Society
S.V.	Sturbridge Village
Y.	Yale, Beinecke Library

Bibliography

Allan, Mea. *The Tradescants* (London: Michael Joseph, 1964).　　　　　　　　　　　　　　　　　　　　A.

Andrews, Charles M. *The Colonial Period of American History* (New Haven: Yale University Press, 1934).　　A.

Arber, Agnes. *Herbals, Their Origin and Evolution* (Cambridge [England]: Cambridge University Press, 1953).　　A.

Ary, S. and M. Gregory. *The Oxford Book of Wild Flowers* (Oxford: Oxford University Press, 1960).　　A.

Bailey, L. H. *The Standard Cyclopedia of Horticulture* (New York: MacMillan Company, 1950).　　A.

Baker, George. *The New Jewell of Health* (London: Henrie Denham, 1576).　　B.M.S.

Bay Psalm Book. (Cambridge [Mass.]: Stephen Daye, 1640). Facsimile reprint by New England Society, New York: 1903.　　A.

Beall, Otho T. and Richard H. Shryock. *Cotton Mather, First Significant Figure in American Medicine* (Baltimore: Johns Hopkins Press, 1954).　　A.

Bennett, Joan. *Sir Thomas Browne* (Cambridge [England]: Cambridge University Press, 1962).　　A.

Bigelow, Jacob, M.D. *Florula Bostoniensis*, Second Edition (Boston: Cummings, Hilliard and Company, 1824).　　A.

Black, Robert C. III. *The Younger John Winthrop* (New York: Columbia University Press, 1966).　　A.

Blunt, Wilfrid. *The Art of Botanical Illustration* (London: Collins, 1950).　　A.

Boorde, Andrew. *The Breviarie of Health* (London: Thomas East, 1546).　　A.

Boorde, Andrew. *The Second Book of the Breviary of Health, named The Extravagantes* (London: Thomas East, 1587).   A.

Bowen, Catherine Drinker. *Francis Bacon* (Boston: Atlantic–Little, Brown, 1963).   A.

Boyceau, Jacques. *Traite Du Jardinage* (Paris: 1638).   B.N.

Bradford, William. *Of Plymouth Plantation*, ed. by Samuel Eliot Morison (New York: Alfred A. Knopf, 1952).   A.

———. *Letter Book* (Boston: Massachusetts Historical Society, n.d.).   M.H.S.

Bradstreet, Anne. *The Works of Anne Bradstreet*, ed. by John Harvard Ellis (Charlestown [Mass.]: Abram E. Cutter, 1867).   A.

Brewster, Charles W. *Rambles About Portsmouth* (Portsmouth [N.H.]: Lewis H. Brewster, 1869).   A.

Brooklyn Botanic Garden. *Handbook on Herbs*, "Plants and Gardens," Vol. 14, No. 2.   A.

———. *Dye Plants and Dyeing*, "Plants and Gardens," Vol. 20, No. 3.   A.

Browne, C. A. "Scientific Notes From The Books and Letters of John Winthrop, Jr. 1606–1676" (Bruges: *Isis*, December, 1928).   B.M.L.

———. *A Colonial Manuscript Volume Relating to Alchemy*.   B.M.L.

Burrage, Henry S. *Early English and French Voyages* (New York: Charles Scribner's Sons, 1906).   B.A.

Burton, Robert ("Democritus Junior"). *The Anatomy of Melancholy*, Twelfth Edition (London: Longmans Green & Company, 1821).   A.

Chute, Marchette. *Shakespeare of London* (New York: E. P. Dutton, 1949). Paperback.   A.

Clayton, John. *Flora Virginica*, ed. by Gronovius (Cambridge: [Mass.], 1946). Reprint for Arnold Arboretum.   A.

Clifford, Derek. *A History of Garden Design* (London: Faber and Faber, 1962).   A.

Coats, Alice M. *Flowers And Their Histories* (New York: Pitman Publishing Corporation, 1956).   A.

Coles, William. *The Art of Simpling* (London: N. Brook, 1657). First Reprint: Milford, Clarkson, 1938).   Mass. Hort. Soc.

Cornut, Jacques Philippe. *Canadensium Plantarum* (New York: Johnson, 1966). Reprinted from Paris, 1635.   A.

Covey, Cyclone. *The Gentle Radical, Roger Williams* (New York: MacMillan Company, 1966).   A.

Creevey, Caroline. *Flowers of Field, Hill and Swamp* (New York: Harper and Brothers, 1897). A.

Culpeper, Nicholas. *The English Physician, or an Astrologo-Physical Discourse of the Vulgar Herbs of This Nation* (London: Nathaniel Brook, 1652). H.

————. *Galens Art of Physic* (London: Peter Cole, 1652). H.

————. *Pharmacopoeia Londinensis, or the London Dispensatory* (London: Hannah Sawbridge, 1683). A.

————. *English Physician Enlarged* (London: McQueen, Laikington, Mathew, 1788). A.

————. *Culpeper's English Physician and Complete Herbal*, Illustrated, ed. by E. Sibley, M.D. (London: British Directory Office, 1798). A.

————. *Culpeper's Complete Herbal* (London: Foulsham and Co., Ltd. Modern. n.d.). A.

Cummings, Abbott Lowell. *Rural Household Inventories 1675–1775* (Boston: Society for the Preservation of New England Antiquities, 1964). A.

Dana, Mrs. William Starr. *How to Know the Wildflowers* (New York: Charles Scribner's Sons, 1893). A.

Davenport, Elsie G. *Yarn Dyeing* (London: Sylvan Press Ltd., 1955). A.

Deganawidah. *The Ranger's Guide to Useful Plants of the Eastern Wilds* (Boston: Christopher Publishing House, 1964). A.

Dexter, Franklin B. "Early Private Libraries in New England," American Antiquarian Society *Proceedings*, New Series, Vol. XVIII (Worcester: 1907). B.A.

Dow, George Francis. *Two Centuries of Travel in Essex County* (Topsfield [Mass.]: Topsfield Historical Society, 1921).

Dunton, John. *Letters From England, 1686* (Boston: Prince Society, 1867). First Published London, 1705. B.A.

Earle, Alice Morse. *Margaret Winthrop* (New York: Charles Scribner's Sons, 1896). A.

Emmart, Dr. Emily Walcott. *The Badianus Manuscript* (Baltimore: Johns Hopkins Press, 1940). A.

Evelyn, John. *Acetaria, A Discourse of Sallets* (London: D. Tooke, 1699). Reprinted Brooklyn Botanic Garden, 1937. S.V.

Felter, Harvey Wicker. *Genesis of American Materia Medica* (Cincinnati: 1927). B.A.

Bibliography

Fiennes, Celia. *Journeys 1685-1703*, ed. by Christopher
Morris (London: Cresset Press, 1959).   A.

Finch, Jeremiah S. *Sir Thomas Browne* (New York: Henry
Schuman, 1950).   A.

Fleischauer, Warren L. "On Nature and Art in the Pleasures
of the Imagination" in *Addison and Steele, Selected Essays
From the Tatler and the Spectator* (New York: Gateway
Editions, 1956).   A.

Flint, Martha Bockee. *A Garden of Simples*. (New York:
Charles Scribner's Sons, 1900).   A.

Flower, Barbara and Elisabeth Rosenbaum. *The Roman
Cookery Book of Apicius* (London: Peter Nevill, 1957).   A.

Ford, W. C. *The Boston Book Market 1679-1700* (Boston:
Club of Odd Volumes, 1917).   B.A.

Freeman, Margaret B. *Herbs* (New York: Metropolitan
Museum, 1943).   A.

Georgia, Ada. *Manual of Weeds* (New York: MacMillan
Company, 1914).   A.

Gerard, John. *The Herball or General Historie of Plantes,
Gathered by John Gerarde of London Master in Chirur-
gerie, Very Much Enlarged and Amended by Thomas
Johnson Citizen and Apothecarye of London* (London:
Adam, Issip Joice Norton and Richard Whitakers, 1633).   A.

Gordon, Dr. Maurice Bear. *Aesculapius Comes to the Colonies*
(Ventnor [N.J.]: Ventnor Publishers, 1949).   A.

Gorges, Sir Ferdinando, and Ferdinando Gorges. *America
Painted to the Life* (London: Nathaniel Brook, 1659). Said
to have been written by the senior and "publisht" by the
grandson. Actually almost all of it is an unacknowledged
reprint of *Sions Saviour in New England*, by Edward
Johnson.   H.

Gray, Asa. *Manual of Botany*. Eighth Edition, expanded by
Merritt Lundon Fernald (New York: American Book
Company, 1950).   A.

Grieve, Mrs. M. *A Modern Herbal* (New York: Hafner
Publishing Company, 1967).   A.

Grigson, Geoffrey. *The Englishman's Flora* (London: Phoe-
nix House, Ltd., 1955).   A.

———. *A Herbal of All Sorts* (London: Phoenix House,
Ltd., 1959).   A.

Gunther, Robert T. *The Greek Herbal of Dioscorides* (New
York: Hafner Publishing Company, 1959). Englished by
John Goodyer, 1655. Edited and first printed A.D. 1933.   A.
Hadfield, Miles. *Gardening in Britain* (London: Hutchin-
son, 1960).   A.
Hakluyt, Richard. *Divers Voyages Touching the Discovery of
America and the Islands Adjacent* (London: Hakluyt
Society, 1850).   H.
Hammond, Captain Lawrence. *Journal.*   M.H.S.
Hare, Caspari and Rusby. *National Standard Dispensatory*
(Philadelphia and New York: Lea Brothers, 1905).   A.
Hare, Lloyd C. M. *Thomas Mayhew, Patriarch to the Indians*
(New York: D. Appleton and Company, 1932).   A.
Hariot, Thomas. *A Briefe and True Report of the Newfound-
land of Virginia, Diligentlye Collected and Drawne by John
White* (Frankfort: Theodore de Bry, 1590). Reprint.   A.
Hausman, Ethel Hinckley. *Beginner's Guide to Wild Flowers*
(New York: G. P. Putnam's Sons, 1948).   A.
Herbert, George. *The Temple, Sacred Poems and Private
Ejaculations* (London: Pickering, 1835). *The Temple* first
published 1633.   A.

————.

Higginson, Francis. *A True Relation of the Last Voyage to
New England, Written From New England, July 24, 1629.*   A.
Hole, Christina. *The English Housewife in the Seventeenth Cen-
tury* (London: Chatto and Windus, 1953).   A.
Hollingsworth, Buckner. *Flower Chronicles* (New Brunswick
[N.J.]: Rutgers University Press, 1958).   A.
Holmes, Oliver Wendell. *Medical Essays* (Boston: Hough-
ton Mifflin Company, 1861).   A.
Holmyard, E. J. *Alchemy* (New York: Penguin Books,
1957).   T. H. Cain
Johnson, Edward. *Wonder Working Providence of Sions
Saviour in New England* (London: 1654).
Reprint, Andover [Mass.]: Warren F. Draper, 1867.   A.
Josselyn, John. *New-Englands Rarities* (London: G. Wid-
dowes, 1672). Reprint, Boston: William Veazie, 1865.   A.
————. *Account of Two Voyages to New England* (London:
1673. Second Edition, London: G. Widdowes, 1675).
Reprint, Boston: William Veazie, 1865.   A.

Keble, Martin W. *The Concise British Flora in Colour* (London: John Rainbird, Ltd., 1965). A.

Lawson, William. *A New Orchard and Garden*, with *The Country Housewifes Garden for Herbs of Common Use* (London: W. Wilson, 1648). A.

Lechford, Thomas. *Plain Dealing, or News From New England* (London: Nathaniel Butter, 1642). Reprint, edited by J. Hammond Trumbull, Boston: Wiggin and Lunt, 1867. A.

Leyel, Mrs. C. F. *The Truth About Herbs* (London: Andrew Dakers, Ltd., 1943). Mass. Hort. Soc.

————. *The Magic of Herbs* (London: Jonathan Cape, 1926). Mass. Hort. Soc.

————. *Culpeper's English Physician and Complete Herbal* (London: Culpeper House, 1947). Arranged by Mrs. C. F. Leyel. A.

Lockwood, Mrs. Alice B., for Garden Club of America. *Gardens of Colony and State* (New York: Charles Scribner's Sons, 1931). B.A.

Macaulay, Rose. *Milton* (London: Gerald Duckworth and Company, Ltd., 1934). A.

MacDonell, Anne. *The Closet of Sir Kenelm Digby, Knight, Opened* (London: Philip Lee Warner, 1910). Newly edited by Anne MacDonell. M.M.S.

Mather, Cotton. *Magnalia Christi Americana, or the Ecclesiastical History of New England* (Hartford [Conn.]: Silus Andrus, 1820). First American edition from the London edition of 1702. A.

————. *Diary of Cotton Mather* (New York: Ungar Publishing Company, 1911). Ed. by Worthington Chauncey Ford, Boston, 1911. A.

————. "The Angel of Bethesda" (Worcester [Mass.]: American Antiquarian Society). Typescript. A.A.S.

————. Manuscripts and Papers (Boston: Massachusetts Historical Society, n.d.). M.H.S.

Meager, Leonard. *The English Gardener* (London: T. Pierrepoint, 1682). A.

Meyer, Joseph E. *The Herbalist* (Hammond [Indiana]: Hammond Book Company, 1934). A.

Miller, Perry. *Orthodoxy in Massachusetts* (Boston: Beacon Paperback, 1933). A.

_____. *The New England Mind, From Colony to Province* (Cambridge [Mass.]: Harvard University Press, 1953). A.

_____. *The American Puritans* (New York: Anchor Books 1965). A.

Mollet, André. *Le Jardin de Plaisir* (Paris: 1651). B.N.

Mollet, Claude. *Le Théâtre des Plans et Jardinages* (Paris: 1618). B.N.

Monardes, Nicholas. *Joyfull Newes Out of the New-Found World*, Englished by John Frampton Marchant (London: Bonham Norton, 1596). Reprinted New York: Alfred A. Knopf, 1925. Mass. Hort. Soc.

Moreton, C. Oscar. *Old Carnations and Pinks* (London: George Rainbird, 1955). A.

Morgan, Edmund S. *The Puritan Dilemma, The Story of John Winthrop* (Boston: Little, Brown and Company, 1958). A.

Morison, Samuel Eliot. *The Puritan Pronaos* (New York: New York University Press, 1936). A.

_____. *Builders of the Bay Colony* (Boston: Houghton Mifflin Company, 1930). A.

Morton, George. *Mourt's Relation, or Journal of the Plantation of Plymouth* (London: John Bellamie, 1622). Reprint ed. by Henry Martyn Dexter, Boston: John Kimball Wiggin, 1865. B.A.

Morton, Thomas. *The New English Canaan* (Boston: Prince Society, 1883). B.A.

Murdock, Kenneth B. *Cotton Mather, Selections* (New York: Hafner Publishing Company). Reprint, American Author Series. A.

"M. W." *The Queens Closet Opened. Incomparable Secrets in Physick, Chirurgery, Preserving, Candying and Cookery.* (London: Nathaniel Brook, 1656). H.

Napheys, George H. *Prevention and Cure of Disease* (Chicago: W. J. Holland and Company, 1871). A.

Nicoll, Allardyce. "Sir Kenelm Digby," *Johns Hopkins Alumni Magazine*, June 1933, pp. 330–350. B.M.S.

Notestein, Wallace. *The English People on the Eve of Colonization* (New York: Harper, Torchbook Edition, 1962). A.

Parkinson, John. *Paradisi in Sole Paradisus Terrestris* (London: Methuen and Company, 1904). Faithfully reprinted from the edition of 1629. A.

Parkinson, John. *Theatrum Botanicum, The Theater of Plants Or An Herbal of a Large Extent* (London: Thomas Cotes, 1640). — Mass. Hort. Soc.

Pechey, John. *The English Herbal of Physical Plants* (London: Medical Publications Ltd., 1951). First published 1664. — A.

Plat, Sir Hugh. *Delightes for Ladies* (London: Lownes, 1609). Ed. by G. E. Fussell and K. R. Fussell, and reprinted London: Crosby Lockwood and Son, Ltd., 1948. — A.

Pliny, the Younger. Letter VI, Book V in *Letters*. Melmoth's Literal Translation (New York: Translation Publishing Company, 1925). — A.

Powell, Sumner Chilton. *Puritan Village* (New York: Doubleday Anchor Books, 1965). — A.

Purchas, Samuel. *Purchas His Pilgrimes* or *Hakluytus Posthumus* (London: H. Fetherstone, 1625). — H.

De la Quinitanye, Jean. *Jardins Fruitiers et Potagers* (Amsterdam: Henri Desbordes, 1597). — A.

Raffald, Elizabeth. *The Experienced English Housekeeper* (London: Eighth Edition, R. Baldwin, 1782). — A.

Ransom, Florence. *British Herbs* (London: Penguin, 1954). — A.

Rea, John. *Flora, or a Complete Florilege* (London: George Marriott, 1676). Second Impression. — A.

Rice, Charles. *Posological Table, all the officinal and the most frequently employed unofficinal preparations* (New York: William Wood and Company, 1879). — A.

Rohde, Eleanour Sinclair. *The Old English Gardening Books* (London: Martin Hopkinson and Company, 1924). — A.

Rosier, James. *A True Relation of the Most Prosperous Voyage 1605, by Captain George Waymouth* (Portland [Maine]: Gorges Society Publications, 1887). — B.A.

Sewall, Samuel. *Diary* (Boston: Collections of the Massachusetts Historical Society, Fifth Series, Vols. V, VI, VII). — A.

———. *Letter Book* (Boston: Collections of the Massachusetts Historical Society, Sixth Series, Vols. I, II). — M.H.S.

Shurcliff, Arthur A. "The Gardens of the Governor's Palace, Williamsburg, Virginia," *Landscape Architecture Quarterly Magazine*, January 1937. — A.

Smith, A. William. *A Gardener's Book of Plant Names* (New York: Harper and Row, 1963). — A.

Sonnedecker, Glenn and Sami Khalaf Hararneh. *A Pharma-ceutical view of Abulcasis Al-Zahrawi in Moorish Spain* (Leiden: E. J. Brill, 1963).   A.

Stearn, William T. *Botanical Latin* (London: Thomas Nelson, Ltd., 1966).   A.

Stevens, Charles and John Liebault. *Maison Rustique* or *The Country Farme*. Translated from French into English by Richard Surflet (London: Bollifant, 1600).   M.H.H.

Taylor, Edward. *Poems*, ed. by Donald E. Stanford (New Haven: Yale University Press, 1960).   A.

———. "Dispensatory." (Manuscript).   Y.

Taylor, Gladys. *Old London Gardens* (London: B. T. Batsford, Ltd., 1953).   A.

Taylor, Kathryn S. and Stephen F. Hamblin. *A Handbook of Wildflower Cultivation* (New York: MacMillan Company, 1963).

Thomas, Graham Stuart. *The Old Shrub Roses* (London: Phoenix House, 1961).   A.

Thomson, Gladys Scott. *Life in a Noble Household 1641–1700* (Ann Arbor [Mich.]: University of Michigan Press, 1959). Paperback.   A.

Turner, William. *Libellus De Re Herbaria, 1538* and *The Names of Herbes, 1548* (London: The Ray Society, 1965). Facsimile.   A.

Tusser, Thomas. *Five Hundred Points of Good Husbandry* together with *A Book of Huswifery*, ed. by William Mayor (London: Lackington, Allen and Company, 1812).   A.

Van de Pass, Crispin. *Hortus Floridus* (Utrecht: de Roy, 1615). Reprinted and edited by Eleanor Sinclair Rohde, London: Cresset Press Ltd., 1928.   A.

Viets, Dr. Henry R. *A Brief History of Medicine in Massachusetts* (Boston: Houghton Mifflin Company, 1930).   A.

———. "Some Features of the History of Medicine in Massachusetts During the Colonial Period," *Isis*, No. 66 (September, 1935).   A.

Ward, Nathaniel. *Simple Cobler of Aggawam in America* (London: Stephen Bowrell, 1647). Reprint edited by David Pulsifer, Boston: James Munroe and Company, 1843.   A.

Waters, Thomas Franklin. *John Winthrop the Younger* (Cambridge [Mass.]: Ipswich Historical Society, 1900).   A.

Wendell, Barrett. *Cotton Mather, Puritan Priest* (New York: Dodd, Mead and Company, 1891).   A.

Wheatland, Henry M.D., Cyrus M. Tracy, William E. Graves, and Henry M. Batchelder. *Standard History of Essex County, Massachusetts* (Boston: C. F. Jewett and Company, 1878).   B. H. Carpenter

Wheelwright, Edith Grey. *The Physick Garden* (Boston: Houghton Mifflin Company, 1935).   A.

Wherry, Edgar T. *Wild Flower Guide* (New York: Doubleday and Company, 1948).   A.

Wilkinson, Ronald Sterne. *The Alchemical Library of John Winthrop, Jr. 1606–1676.*

Williams, Roger. *Key Into the Language of the Indians of New England, 1643* (Boston: Collections of Massachusetts Historical Society for 1794, Vol. III, 1810).   B.A.

Winslow, Ola Elizabeth. *Samuel Sewall of Boston* (New York: MacMillan Company, 1964).   A.

―――. *John Bunyan.* (New York: MacMillan Company, 1961).   A.

Winthrop Papers. Five Volumes (Boston: Massachusetts Historical Society, 1929).

Index to Part I

Index